Morphological Control in
Multiphase Polymer Mixtures

MATERIALS RESEARCH SOCIETY
SYMPOSIUM PROCEEDINGS VOLUME 461

Morphological Control in Multiphase Polymer Mixtures

Symposium held December 2–5, 1996, Boston, Massachusetts, U.S.A.

EDITORS:

Robert M. Briber
University of Maryland
College Park, Maryland, U.S.A.

Charles C. Han
National Institute of Standards and Technology
Gaithersburg, Maryland, U.S.A.

Dennis G. Peiffer
Exxon Research and Engineering Company
Annandale, New Jersey, U.S.A.

PITTSBURGH, PENNSYLVANIA

Published by:

Materials Research Society
9800 McKnight Road
Pittsburgh, Pennsylvania   15237
Telephone (412) 367-3003
Fax (412) 367-4373
Website: http://www.mrs.org/

Library of Congress Cataloging in Publication Data

Morphological control in multiphase polymer mixtures : symposium held
     December 2–5, 1996, Boston, Massachusetts, U.S.A. / editors,
     Robert M. Briber, Charles C. Han, Dennis G. Peiffer
          p. cm—(Materials Research Society symposium proceedings ; v. 461)
          Includes bibliographical references and index.
          ISBN 1-55899-365-7
          1. Polymers—Structure—Congresses. 2. Polymerization—Congresses.
     I. Briber, Robert M. II. Han, Charles C. III. Peiffer, Dennis G.
     IV. Series: Materials Research Society symposium proceedings ; v. 461.
QD381.9.S87M67 1997   97-6800
620.1′92—dc21   CIP

Manufactured in the United States of America

CONTENTS

*Invited Paper

*Invited Paper

*Invited Paper

*Invited Paper

PREFACE

Recent advances in the properties of multiphase polymeric materials can be largely attributed to the ability of scientists and engineers to control the size, arrangement, morphology and crystallization of the components in the mixture. Many of the advances in this area have come from the modification of the interfacial properties of the mixtures, either directly or indirectly, through the addition of different types of copolymer architectures and/or the formation of copolymers *in situ* via reactive processing. In addition, control of the morphology in multiphase polymer mixtures can occur through the use of controlled-temperature histories, crosslinking and chemical modification (among others). This symposium brought together researchers from academia, government and industry to present recent results on morphological control in multiphase polymers. Topics covered during the four days of talks include: morphology of both compatible and incompatible polymer blends, crystallization of multicomponent mixtures, blends of copolymers and homopolymers, reactive processing, interfacial modification of polymer blends and structure-property relations in polymer mixtures. The symposium was a resounding success, with all participants expressing satisfaction with the high quality of the papers presented and submitted for publication in this proceedings.

Robert M. Briber
Charles C. Han
Dennis G. Peiffer

January 1997

MATERIALS RESEARCH SOCIETY SYMPOSIUM PROCEEDINGS

Prior Materials Research Society Symposium Proceedings available by contacting Materials Research Society

Morphological Control in Multiphase Polymer Mixtures

STRUCTURE PINNING DURING PHASE-SEPARATION
OF POLYAMIDE/IONOMER BLENDS

R. A. WEISS*, Y. FENG*, R. TUCKER*, R. XIE*, C. C. HAN** AND A. KARIM**
*Dept. of Chemical Engineering and Polymer Science Program, University of Connecticut,
Storrs, CT 06269-3136
**Polymer Section, National Institute of Standards and Technology, Gaithersburg, MD 20899

ABSTRACT

Blends of lightly sulfonated polystyrene and poly(N,N'-dimethylethylene sebacamide) (Li-SPS/mPA) are miscible as a result of strong ion-amide complexation. The blends exhibit LCST phase behavior and an increase of the sulfonation level from 4 to 9.5 mol% raises the critical temperature by 150°C. Phase separation may be thermally induced and is thermodynamically reversible. The phase separation kinetics that occur following a temperature-jump deep into the spinodal region of the phase diagram deviate from conventional Cahn-Hilliard theory and the phase separation process stalls after a couple of hours, essentially *pinning* the structure at that point.. The extent of phase-separation that occurs before pinning is temperature-dependent.

INTRODUCTION

The properties of polymer blends are intimately tied to their morphology, which depends on the miscibility of the components and the mechanism and kinetics of phase-separation. For some applications, phase-separated morphologies are required for improving mechanical properties such as impact toughness, while in other instances, a miscible blend is desired, e.g., for improving processability or extending an expensive material with a cheaper one. In either case, the development of the optimum morphology and properties requires knowledge of the phase diagram and the characteristics of the phase-separation process.

Compatibilization of polymer blends by promoting intermolecular complexation has recently attracted considerable interest. Intermolecular complexation, such as hydrogen bonding, proton-transfer, transition metal coordination or ion-dipole interaction, can result in miscible or partially miscible blends of otherwise immiscible polymers. Several laboratories have been particularly concerned with the compatibilization of polystyrene/polyamide blends by incorporating a small amount of metal sulfonate groups into the polystyrene [1-12]. The choice of the metal counterion significantly affects the phase behavior, due to differences in the relative strength of metal-amide interactions [3-5]. The strongest enhancement of miscibility occurs when a transition metal cation such as Mn^{2+} or Zn^{2+} is used. Among alkali-metal cations, lithium provides substantial improvement of miscibility, while sodium does not [3-11].

Previous studies of polymer blends exhibiting strong intermolecular interactions have been primarily concerned with the morphology and thermal-mechanical properties of either single-phase materials or phase-separated blends following an arbitrary and usually poorly characterized thermal history. As a result, knowledge of the effect of strong intermolecular interactions on the phase separation process is limited. Our research seeks to answer the following questions: (1) does complexation perturb the kinetics of phase-separation and (2) are there novel mechanisms that may provide opportunities for developing unique blend morphologies and application. This paper

Mat. Res. Soc. Symp. Proc. Vol. 461 © 1997 Materials Research Society

describes the effect of ion-dipole complexation on the miscibility and phase separation kinetics of polystyrene-polyamide blends, and in particular, on the kinetics of phase-separation within the spinodal region of the phase diagram.

The only report of the effect of specific intermolecular associations on the phase separation kinetics of a polymer blend was by He et al. [13], who studied blends of poly(butyl methacrylate) with a polystyrene modified with 1.5 mol% of a hydroxy-containing comonomer. That system exhibited lower critical solution temperature (LCST) behavior, and miscibility occurred because of hydrogen bonding between the ester and hydroxyl groups. For small excursions into the spinodal region, multiple structures developed in the blend which suggested that multiple mechanisms may be involved in the phase separation process. For most cases, however, the kinetics of phase separation followed Cahn-Hilliard theory [14,15] in the early stage of spinodal decomposition and a self-similar mechanism in the later stages, similar to non-associating polymer blends such as PS/PVME.

The failure to observe an effect of a specific intermolecular interaction on SD kinetics in ref. [13] may have been a consequence of the low concentration of hydroxyl groups on the polystyrene, ca. five per chain, and the weakening of the hydrogen bond at the elevated temperatures used to study phase separation. At elevated temperatures, the association-dissociation equilibrium shifts towards non-associated hydroxyl and ester groups such that the crosslinking effect by hydrogen bonding may not be significant. An objective of the present study was to utilize a polymer blend having a relatively higher degree of intermolecular association at the phase separation temperature and to investigate how physical crosslinks affect the phase separation kinetics accompanying spinodal decomposition. The polymers used in this study were the lithium salt of a lightly sulfonated polystyrene ionomer (Li-SPS) and an N-alkylated polyamide, poly(N,N'-dimethylethylene sebacamide). The latter is methylated nylon 2,10 and is hereafter referred to as mPA. The substitution of the methyl group for the amide proton prevents self-hydrogen bonding, which yields a relatively low melting temperature, ca. about 75 °C and very slow crystallization kinetics [16]. The low melting point also provided the distinct advantage of accessing the melt state of the polyamide at reasonably low experimental temperatures so that degradation of the polymers was not a concern. The slow crystallization kinetics made it relatively easy to produce completely amorphous samples. Characterization of the Li^+-amide complex in Li-SPS/mPA blends is reported in ref. [5].

EXPERIMENT

<u>Materials</u>

Poly(N, N'-dimethylethylene sebacamide) (mPA) with M_n = 25 kg/mole and M_w = 65 kg/mole was prepared by a nucleophilic acyl substitution reaction following the procedure of Hang and Kozakiewicz [16]. Free acid derivatives of lightly sulfonated polystyrene (H-SPS) were prepared by sulfonating a polystyrene homopolymer (M_w = 288 kg/mole; M_w/M_n = 2.7) in 1,2-dichloroethane at 50 °C using acetyl sulfate as the sulfonating reagent. The sulfonation reaction adds sulfonic acid groups randomly to the para-position of the styrene rings. Three samples with sulfonation levels of 4.0, 5.2, and 9.5 mol% were prepared, and the corresponding lithium salts were prepared by neutralizing the H-SPS in a 90% toluene/10% methanol solution with a 20% excess of lithium acetate. The Li-SPS ionomers were isolated from solution by steam stripping, filtered, washed, and dried under vacuum. Other salts were prepared by the same procedure using the appropriate metal hydroxide or acetate.

Samples were prepared by mixing the Li-SPS/mPA blends in a mixed solvent of methanol/dichloroethane (10/90) and casting films onto glass microscope slides for cloud point and optical microscopy studies or onto quartz plates for time-resolved light scattering (LS) measurements. The films were dried in vacuum for two days at temperatures below the cloud point and were then covered by another quartz plate or a microscope cover slide with a metal spacer between the two plates to maintain the film thickness. The thickness of the specimens was 0.1 mm for LS and 0.05 mm for cloud point measurements and optical microscopy.

Cloud Point Measurements

Cloud point (CP) measurements were made using a Mettler FP80 microscope hot stage to control the sample temperature at a constant heating rate and a photo diode was to monitor the light intensity at a 45° scattering angle. The CP was taken as the temperature at which the scattered light intensity deviated from the baseline as the temperature was increased. CP data were obtained at two heating rates, 0.2 and 2°C/min, and the cloud point temperature (T_{cp}) was calculated by a linear extrapolation of the two experimental CPs to a temperature corresponding to zero heating rate.

Time Resolved-Light Scattering Measurements

The kinetics of phase separation following a temperature jump were studied by time-resolved light scattering (LS). The light source was a 5 MW He-Ne laser and the detector was a one-dimensional photodiode array detector. The data were collected with an optical multichannel analyzer.

The blend specimen was first annealed below the cloud point in a preheating block for several hours, and then phase separation was induced by quickly transferring the sample from the preheating block to a second heating block set at the phase-separation temperature. Both blocks were controlled by PID controllers to within ±0.02°C, and the experimental temperature equilibrated within about 1 minute.

Optical Microscopy Measurements

Optical micrographs were obtained with a Zeiss transmission microscope with a Sony XC-77 CCD camera. The microscopic images were digitized using a frame grabber and software from Data Translation. A Mettler FP80 microscope hot stage was used to control the sample temperature during isothermal phase separation.

RESULTS AND DISCUSSION

Liquid-Liquid Phase Diagrams

Cloud point curves determined for the Li-SPS/mPA blends as a function of sulfonation level and composition are given in Fig. 1. PS and mPA are immiscible, but the introduction of the lithium sulfonate groups produces miscibility as a result of ion-dipole complexation [5]. Complete miscibility is achieved at room temperature with a sulfonation level as low as 4 mol%. The Li-SPS/mPA blends exhibit LCST phase behavior, and the critical point increased by ca. 150°C when the sulfonation level was increased from 4 to 9.5 mol%.

Similarly, the phase diagrams in shown in Fig.1 for the Li-SPS/mPA blends are also consistent with de Gennes's prediction [17] that crosslinking improves miscibility. In this case, the strong intermolecular ion-dipole complex may be viewed as a virtual, i.e., physical, crosslink. Unlike a covalent crosslink, the ion-dipole complex is in equilibrium with the non-associated metal sulfonate and amide groups and the relative populations are expected to be temperature-sensitive. At relatively low temperatures, e.g., at room temperature, the equilibrium favors the complex, so that the level of "crosslinking" is relatively high and the blend is miscible. As the temperature increases, however, the equilibrium of the Li^+-amide interaction shifts towards the non-associated species, and the number of physical crosslinks falls below a critical value required to maintain miscibility and phase separation occurs. Increasing the sulfonation level is equivalent to increasing the absolute number of physical crosslinks at all temperatures, so that the critical crosslink concentration is maintained to higher temperature. As a result, the LCST increases with increasing sulfonation of the ionomer.

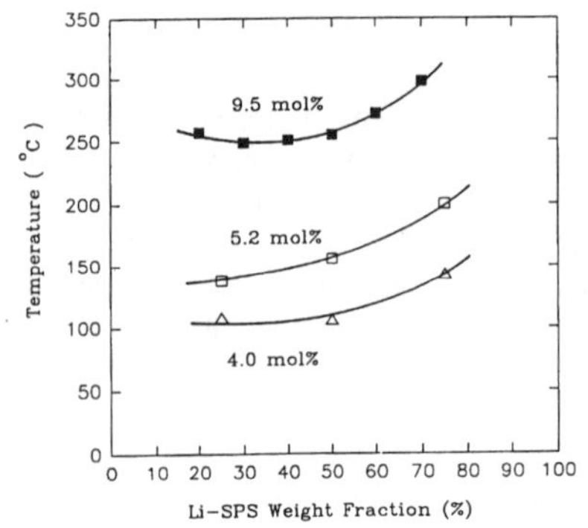

Fig. 1. Cloud point curves for Li-SPS/mPA blends. Numbers correspond to sulfonate content of the Li-SPS

Weiss and Lu [4] reported phase diagrams similar to Fig. 1 for blends of SPS ionomers and poly(ϵ-caprolactam) (PA6)blends. In that case, the LCST depended on the cation used, and for a fixed sulfonation level and composition, the LCST increased as the cation was changed in the order of $Mg^{2+} << Li^+ < Zn^{2+} < Mn^{2+}$. Blends of the Na^+-salt of SPS and PA6 were immiscible.

For the M-SPS/mPA blends, the ion-dipole complex involves the metal cation, the carbonyl group and the amide nitrogen [5]. Coordination of the metal cation with the carbonyl oxygen induces charge transfer at the carbonyl group that lowers the order of the C=O bond and induces a $2P_z$ electron redistribution from the nitrogen to the carbonyl group. The extent of nitrogen $2P_z$ electron migration toward the carbonyl group depends mainly on the extent of charge transfer at the carbonyl group, which is determined by the electron-withdrawing power of the metal cation. Fig. 2 shows the FTIR spectra of the amide I region (the absence of the amide II bands is due to the N-methylation of the polyamide) for blends of mPA with various salts of toluene sulfonic acid, which is a model compound for the ionomer. The new absorption to the right of the amide I absorption at 1645 cm^{-1} is due to the ion-dipole complex and the lower the frequency of that band, the stronger is the complex. In this case, the strength of the ion-dipole complex in M-SPS/mPA blends increases in the order of $Li^+ < Cd^+ \leq Mn^{2+} < Cu^{2+} \leq Zn^{2+}$. Similar results were observed for the ionomer/mPA blends, so one would expect that the phase behavior of

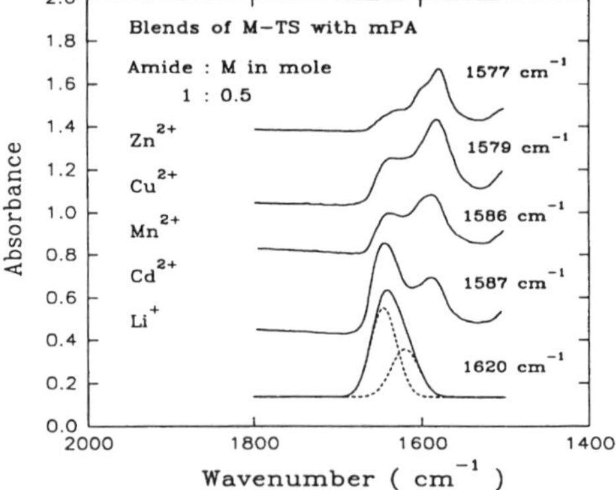

Fig. 2. FTIR spectra of blends of mPA and salts of toulene sulfonic acid (2 moles ion/mole amide)

the ionomer/mPA blends may be controlled by the sulfonation level and the choice of the cation.

<u>Phase Separation Kinetics</u>

Time-reolved LS experiments following temperature jumps from 150°C to 175°C and 195°C were done for a 50/50 blend of mPA and the Li-SPS (5.2 mol% sulfonation). The data for the T-jump to 175°C are shown in Fig. 3 ($q = 4\pi\sin\theta/\lambda$, where λ is the wavelength of the incident light and θ is one-half the scattering angle). The lines through the data in Fig. 3 have no significance; they are merely to guide the eye. For this blend, T_{cp}=156°C, and it was expected that isothermal time-resolved light scattering measurements were carried out at two temperatures, 175°C and 195°C, that were chosen to be sufficiently greater than the cloud point so as to probably be within the spinodal region of the phase diagram.

There are two important observations related to the evolution of the scattering profile in Fig. 3: 1) the scattering intensities increased with time over the entire q-range, which indicates that concentration fluctuations of various size scales were formed simultaneously during the process, and 2) the scattering intensities and the profile did not change significantly beyond the longest time shown in the figure (122 min.), which indicates that the growth of the phase separated structure was retarded at long times and appeared to stall.

The scattering profile of a polymer blend undergoing spinodal decomposition (SD) usually exhibits a distinctive peak at a constant q, which corresponds to a dominant concentration fluctuation, during the early stages of phase separation, and the peak moves towards lower q at later stages of SD as the phase separated structure ripens. The absence of a peak in Fig. 3 indicates that no dominant domain size occurred, which is confirmed by the optical micrograph of the blend obtained after phase separation stalled, Fig. 4. The phase separated structure in Fig. 4 shows spatial steric regularity, but also a fairly broad distribution of domain sizes. The larger domain sizes are on the order of only 1 μm which is much smaller than the typical domain size of a macrophase separated binary polymer blend. The data for the T-jump to 195°C were qualitatively similar to the results shown in Fig. 3, except that scattering intensities at 195°C were higher and the structure factor had a steeper slope. In that case, phase separation stalled after 153 min.

The stalling of the phase separation process is unusual. The structure factor became nearly constant after 2 - 2.5 hours and scattering intensities at high q remained comparable to those at low q, which indicates the persistence of a significant amount of small domains. In normal SD, however, the phase-separated domains coalesce into larger ones at late stages of phase separation so as to reduce the interfacial energy. It is clear from Figs. 3 and 4 that this is not the case in the Li-SPS/mPA blends, and some new mechanism acts to "pin" the morphology that was initially formed.

This *pinning* effect is better illustrated in plots of the scattering intensity vs. time at constant q in Fig. 5 for the phase separation at 175°C. The scattered light intensity at any q approaches an asymptotic value. Even at early times, the growth of the scattered light intensity does not follow

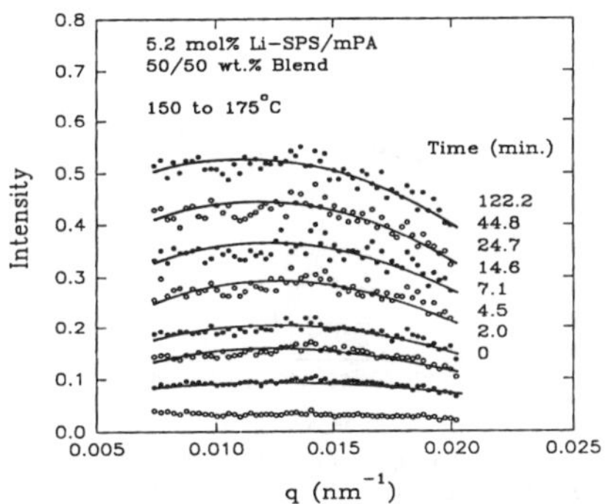

Fig. 3. Light scattering curves following a T-jump from 150°C to 175°C for a (50/50) 5.2Li-SPS/mPA blend.

conventional Cahn-Hilliard theory [14,15], i.e., eqn. (1),

$$I(q,t) = I(q,0)e^{2R(q)t} \tag{1}$$

where q is the scattering vector, t is time, $I(q,t)$ is the scattering intensity, $I(q,0)$ is the initial scattering intensity, and $R(q)$ is the growth rate of the fluctuation with a wavevector q. Instead, the experimental data follow a stretched exponential relationship,

$$I(q,t) = I_\infty(q)[1 - e^{-(t/t_o)^\alpha}] \tag{2}$$

with $\alpha = 0.65 - 0.75$.

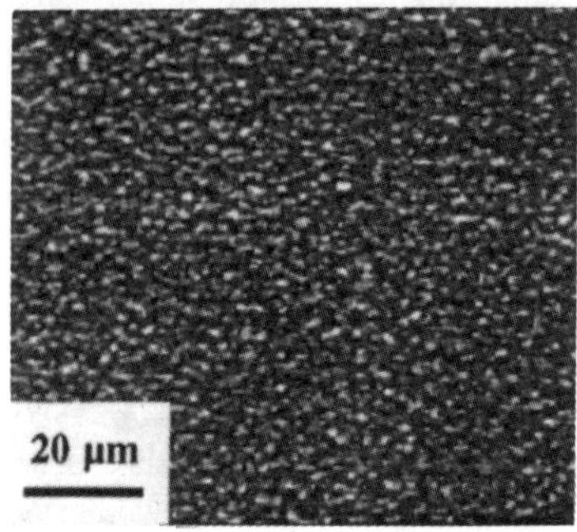

Fig. 4. Optical micrograph of morphology of sample in Fig. 3 two hrs. after T-jump.

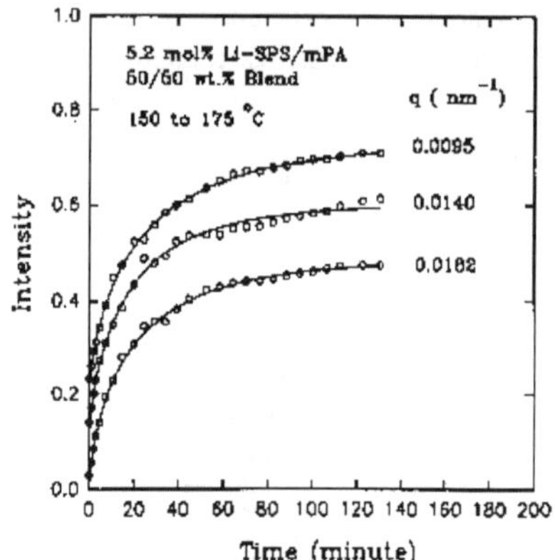

Fig. 5. LS intensity vs. time for several values of q cross-plotted from Fig. 3.

Glotzer et al. [19] suggested that a random, strong interaction within one of the two resulting phases may produce structure pinning during phase separation. For the phase separation of gelatin, where phase separation and gelation occur concurrently, the formation of hydrogen-bond crosslinks arrests phase separation.

In contrast to the gelatin system, the physical network in the Li-SPS/mPA blends due to the ion-dipole complex exists prior to phase separation, and presumably it is the shift of the complex equilibrium towards non-associated species at elevated temperatures that initiates phase-separation. Phase separation produces an ionomer-rich phase and a polyamide-rich phase, which suggests two possible origins of pinning in this system. First, the stoichiometry of Li$^+$-ions to amide groups increases within the ionomer-rich phase compared with the original single-phase melt which may produce sufficient physical crosslinks between the ionomer and polyamide to effectively gel that phase. Alternatively, if the ion-amide interaction is inhibited at the phase-separation temperature, the increased concentration of lithium sulfonate groups in the ionomer-rich phase may result in microphase separation of the salt groups, which is common in bulk ionomers. Microphase-separated ionic aggregates also act as physical crosslinks, which may gel the ionomer-rich phase. In either case, the lack of molecular mobility due to local gelation of the ionomer-rich phase may stall phase separation and pin the morphology. Of course, there is no real crosslink in the Li-SPS/mPA system. Both the ion-amide complexation and the ionic aggregation are in dynamic equilibrium with non-associated and non-aggregated species. As a result, the "strong bonds" are continually breaking and reforming and the relative time spent in each state is dictated by the equilibrium constant. The dynamic nature of the strong bond may still permit a slow continuation of phase separation. However, as the local concentration of strong bonds increase, the rate of the phase separation will become very slow and limited by the dynamics of the strong molecular associations.

Within a given Li-SPS rich domain, the local Li-SPS concentration may not be homogenous throughout that domain, since the functional groups in Li-SPS chains are randomly distributed and the minimum Li-SPS concentration needed to achieve the local gelation depends on the sulfonation level within that particular domain. In other words, the extent of phase separation depends on the local sulfonation level, and neither a distinctive interphase nor a monodisperse ionic concentration in all the Li-SPS rich microdomains is expected. This may explain the relatively flat scattering profiles in Fig. 3 and the absence of a characteristic scattering peak.

The phase separation data for the Li-SPS/mPA blend do not exhibit a linear growth region at early times as predicted by Cahn-Hilliard theory, though that result is consistent with the results of Glotzer et al. [19]. This may be due to the fact that the mixture was well inside the unstable region when phase-separation was initiated after a temperature jump. In that case, the higher order terms of the Ginzburg-Landau equation may no longer be neglected and, therefore, the linearized Cahn-Hilliard model is not valid.

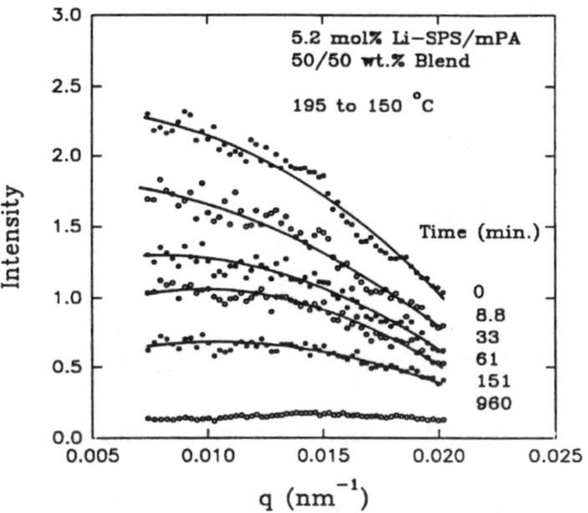

Fig. 6. Light scattering curves following a T-jump from 195°C to 150°C for a (50/50) 5.2Li-SPS/mPA blend.

Thermal-Reversibility of Phase Separation

The phase separation represented by Fig. 3 is reversible as was demonstrated by the data in Fig. 6, which show the evolution of the scattering profile after first allowing a 50/50 blend of Li-SPS/mPA to phase separate at 195°C and then quenching the temperature to 150°C which is in the single-phase region of the phase diagram in Fig. 1. The scattering intensities over the entire q-range decreased with time, which indicate that remixing of the blend occurred at 150°C. The remixing process was much slower than phase-separation, c.f., Figs 3 and 6, which is a consequence of the lower mobility of the polymers at the lower temperature.

CONCLUSIONS

Li-SPS/mPA blends are miscible as a result of strong ion-amide complexation. Phase separation blends may be thermally induced and is thermodynamically reversible. The phase separation kinetics that occur following a temperature-jump deep into the spinodal region of the phase diagram deviated from conventional Cahn-Hilliard theory and a structure pinning phenomenon similar to that reported for the spinodal decomposition of gelatin [18] was observed. The pinning, i.e., stalling of the phase-separation process, is attributed to either concentrating ion-amide complexes in the ionomer-rich phase or the formation of microphase-separated ionic aggregates within the ionomer during phase-separation. Either mechanism is expected to gel the ionomer-rich phase, and the lack of molecular mobility arrests the phase separation process. The extent of phase-separation that occurred before pinning was temperature-dependent. The structure pinning mechanism may be useful for controlling the development of a two-phase morphology, and it may have particular application in processing of polymer blends so as to achieve well-defined microstructures..

ACKNOWLEDGMENT

This research was supported by the Polymer Program of the National Science Foundation (Grant DMR 9400862).

REFERENCES

1. X. Lu and R. A. Weiss, Macromolecules, 24, 4381(1991).

2. X. Lu and R. A. Weiss, Macromolecules, 25, 6185(1992).

3. X. Lu and R. A. Weiss, Mater. Res. Soc., Proc. 215, 29 (1991).

4. R. A. Weiss and X. Lu, Polymer, 35, 1963(1994).

5. Y. Feng, A. Schmidt and R. A. Weiss, Macromolecules, 29, 3909 (1996).

6. Y. Feng, R. A. Weiss, A. Karim, C. C. Han, J. F. Anker and D. G. Peiffer, Macromolecules, 29, 3918 (1996).

7. Y. Feng, R. A. Weiss and C. C. Han, Macromolecules, 29, 3925 (1996).

8. A. Molnar and A. Eisenberg, Macromolecules, 25, 5774 (1992).

9. A. Molnar and A. Eisenberg, Polymer, 32, 370 (1991).

10. Z. Gao, A. Molnar, F. G. Morin and A. Eisenberg, Macromolecules, 25, 6460 (1992).

11. P. Rajagopolan, J.-S. Kim, H. P. Brack, X. Lu, A. Eisenberg, R. A. Weiss and W. M. Risen, J. Polym. Sci., Polym. Phys Ed., 33, 495 (1995).

12. C. W. A. Ng, M. A. Bellinger and W. J. MacKnight, Macromolecules, 27, 6942 (1994).

13. M. He, Y. Liu, Y. Feng, Y., M. Jiang and C. C. Han, Macromolecules, 24, 464 (1991).

14. J. W. Cahn and J. E. Hilliard, J. Chem. Phys., 28, 258 (1958).

15. J. W. Cahn and J. E. Hilliard, J. Chem. Phys., 31, 688 (1959).

16. S. J. Huang and J. Kozakiewicz, J. Macromol. Sci.-Chem., A15, 821 (1981).

17. P.G. de Gennes, J. Phys., Let. (Paris), <u>40</u>, 69. (1979).

18. R. Bansil, J. Lal and B. L. Carvalho, Polymer, <u>33</u>, 2961 (1994).

19. S. C. Glotzer, M. F. Gyure, F. Sciortino, A. Coniglio and H. E. Stanley, Phys. Rev. E , <u>49</u>, 247 (1994).

20. T. Hashimoto, Phase Transition, <u>12</u>, 47 (1988).

DEVELOPMENT OF MISCIBLE BLENDS FROM INTRAPOLYMER REPULSIVE INTERACTIONS: POLYCARBONATE/POLYSTYRENE IONOMERS

R. XIE, R. A. WEISS
Polymer Science Program and Department of Chemical Engineering, University of Connecticut
Storrs, CT 06269-3136

ABSTRACT

The phase behavior of blends of zinc sulfonated polystyrene (ZnSPS) and bisphenol-A polycarbonate (PC) was studied as a function of the sulfonation level and the molecular weight of the ZnSPS ionomer. The system exhibits upper critical solution temperature (UCST) behavior. The cloud point temperatures increased with increasing ZnSPS molecular weight and decreased with increasing sulfonation level. No strong interactions between ZnSPS and PC were detected by FTIR. The composition dependence of the glass transition temperatures of the miscible blends exhibited negative deviation from linear additivity, which is consistent with no or weak interactions between ZnSPS and PC. Miscibility is believed to arise from strong repulsive interactions between the charged and uncharged species on the ionomer chain.

INTRODUCTION

Many studies have shown that the introduction of attractive intermolecular interactions in a polymer blend, such as hydrogen bonding,[1] ion-dipole interactions,[2] acid-base interactions,[3] or transition metal complexation[4], is effective at enhancing the miscibility of otherwise immiscible polymers. Although it has been shown that the miscibility of random copolymers and homopolymers may occur as a result of *intra*molecular repulsive interactions within the copolymer[5-8], relatively few reports of such systems are available in the literature. None of those discuss blends of homopolymers with ionomers, for which one would expect exceptionally strong repulsive interactions between the non-ionic and ionic repeat units.

Ionomers appear to be exceptional candidates for developing miscible blends with other polymers or for compatibilizing two-phase polymer blends. On the one hand, the ionic group provides a rich chemistry for attaining specific interactions with polymers that contain complementary polar groups, such as amine or amide groups. On the other hand, the strong repulsive interactions that occur between the ionic and non-ionic species of the ionomer, which is essentially a random copolymer, may be an especially effective mechanism for attaining miscibility between polymers where no specific intermolecular interactions occur. The contribution of the repulsive interactions to the free energy of mixing is expected to depend on the concentration of the ionic groups and the strength of the dipole, which is dependent on the choice of the anion and cation pair.

In this paper, we report the phase behavior of blends of bisphenol-A polycarbonate and the zinc salt of lightly sulfonated polystyrene determined by optical microscopy and light scattering. The absence of specific interpolymer interactions was confirmed using temperature-resolved Fourier transform infrared spectroscopy (FTIR).

EXPERIMENT

Materials. Bisphenol-A polycarbonate (PC) with M_n = 48,000 g/mole was obtained from

13

Mat. Res. Soc. Symp. Proc. Vol. 461 © 1997 Materials Research Society

General Electric Co. and was used as received. Lightly sulfonated polystyrene (SPS) was prepared according to the procedure of Makowski et al.[9] The reaction substitutes a sulfonic acid group at the para-position of the phenyl ring and sulfonation is random along the chain. Three different molecular weight polystyrenes (PS) were used as the starting polymer. The sulfonation level was determined by titration of the sulfonic acid derivative of the SPS product in a mixed solvent of toluene/methanol (90/10 v/v) with methanolic sodium hydroxide. The zinc salt (ZnSPS) was prepared by neutralizing the acid derivative in toluene/methanol (90/10) with a methanol solution of zinc acetate dihydrate. The nomenclature used for the ionomers is x.yZn-SPS, where x.y denotes the sulfonation level in mol% of sulfonated styrene. The ionomer samples used in this study are summarized in Table I.

Table I. Characteristics of the Zinc Sulfonated Polystyrene

Sample	Sulfonation of the Ionomer (mol%)	M_w of starting PS (g/mole)	M_w/M_n
8.7ZnSPS	8.7	4000	1.06
10.3ZnSPS	10.3	4000	1.06
13.7ZnSPS	13.7	4000	1.06
5.88ZnSPS	5.88	135 000	1.13
5.89ZnSPS	5.89	280 000	1.09

Sample Preparation. Blends were prepared by solution mixing the two polymers in a common solvent. PC was first dissolved in tetrahydrofuran (THF) to form a 1% (g/100ml) solution, after which a small amount of methanol was added to achieve a ratio of THF and methanol of 10/1 v/v. The PC solution was then added dropwise to a stirred 1% solution of ZnSPS in a mixture of THF and methanol from 10/1 to 5/1 (v/v) depending on the molecular weight of the ZnSPS. Blend samples for DSC measurements were solution cast at 30°C. The solutions were allowed to evaporate over 3 days, and were further dried in a vacuum oven at 80°C for another 3 days. Thin films of the blends for FTIR spectroscopy were cast from solution onto KBr windows, and those for optical microscopy and light scattering measurements were cast from solution onto glass plates. All the films were then further dried in vacuum at 80°C for at least 3 days.

Measurements The morphologies and the phase transitions of the blends were observed with a Nikon microscope equipped with a hot stage whose temperature was controlled within ±0.5°C. Cloud points were also determined by light scattering using a custom built device[10] that measures the backscattered light from a blend contained within a sealed glass vial. The scattering intensity of the sample was recorded as the temperature was raised at a rate of 1°C/min.

Glass transition temperatures, T_g, were measured with a Perkin Elmer DSC-7 using a dry nitrogen atmosphere and a heating rate of 20°C/min. The glass transition temperature was defined as the midpoint of the change in the heat capacity at the transition. Miscibility of the blends at various temperatures was assessed by annealing the blend at the desired temperature under nitrogen for 30-60 minutes, quenching the sample rapidly to 0°C and then running a DSC heating experiment. The cooling rate was assumed to be fast enough so that the subsequent heating thermogram represented the state of the blend at the annealing temperature. The blend was considered to be miscible at the annealing temperature if the subsequent heating thermogram exhibited a single T_g and phase-separated if the thermogram showed two T_gs.

FTIR spectroscopy was used to assess any interactions involving either the carbonate or the sulfonate groups. FTIR spectra were measured with a Mattson Cygnus 100 FTIR spectrometer using a resolution of 1 cm^{-1}. A total of 320 scans was signal-averaged for each spectrum. A hot stage with temperature control of ±1°C over the range of 25°C -300°C was used for measuring spectra at elevated temperatures.

RESULTS

All the as-cast films exhibited a two-phase morphology at room temperature. Fig.1 shows optical micrographs for a (20/80) 13.7ZnSPS/PC blend for three different temperature histories. Fig. 1(a) shows a two-phase morphology for a film cast from solution at room temperature and then heated to 200°C. When the temperature was raised to 225°C, only a single, homogeneous phase was observed, Fig. 1(b). That result indicates that the two polymers were miscible at the higher temperature. When the temperature was then slowly lowered from 225°C to 200°C and annealed at 200°C for 20 minutes, phase separation was again observed, Fig. 1(c). The result clearly shows the reversibility of the phase behavior of ZnSPS/PC blends and demonstrates upper critical solution temperature (UCST) type phase behavior, which is unusual for high molecular weight binary polymer blends.

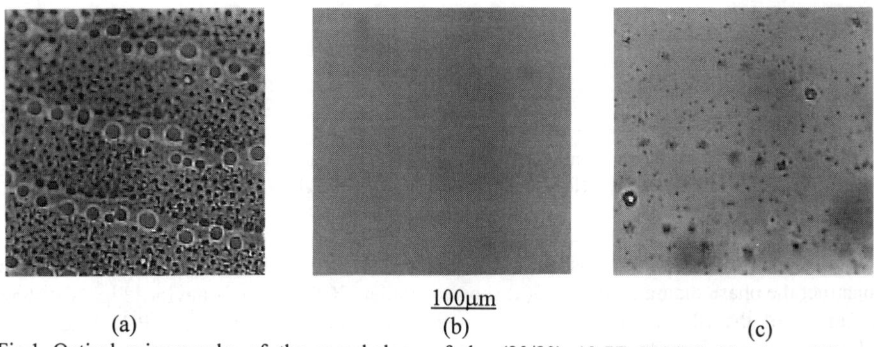

100μm

| (a) | (b) | (c) |

Fig.1 Optical micrographs of the morphology of the (20/80) 13.7ZnSPS/PC blend at different temperatures: (a) 200°C; (b) 225°C; (c) cooled from 225°C to 200°C and annealed for 20 minutes.

Further evidence of the UCST behavior was obtained from light scattering, as shown in Fig.2. Below the cloud point, the film was cloudy and the scattering intensity was nearly constant as the temperature was increased. Above the cloud point temperature, the film became transparent and the backscattered light intensity decreased sharply. The cloud point temperature, T_{cp}, was calculated from the intercept of the baseline below the cloud point and the linear section of the transient in the scattered intensity above the cloud point. The T_{cp} data are plotted in Fig 3. In

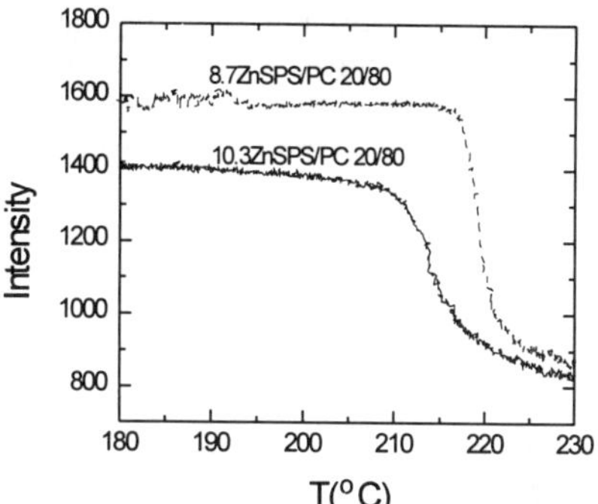

Fig. 2. Backscatterd light scattering intensity of ZnSPS/PC blends as a function of temperature.

Fig. 3 (a), the molecular weight of the ionomer used in the three blends was held constant, but the sulfonation level was varied from 8.7 to 13.7 mol%. Although the SPS molecular weight used for the blends in Fig. 3(a) was relatively low, M_w = 4000 g/mole, blends of PC and the starting PS were completely immiscible over a comparable temperature range. All the blends exhibited UCST-type phase behavior, and the cloud point temperature decreased with increasing sulfonation. This observation is in stark contrast with the effect of strong intermolecular interactions on the phase behavior of blends, such as SPS/polyamide blends, which exhibit LCST-type phase behavior and a cloud point that increases with increasing sulfonation level.[11,12] The critical point for the blends in Fig. 3(a) is skewed towards the higher molecular weight component, i.e., PC, which is what is usually expected. The critical composition, ca. 60 wt% PC, was relatively insensitive to the sulfonation level.

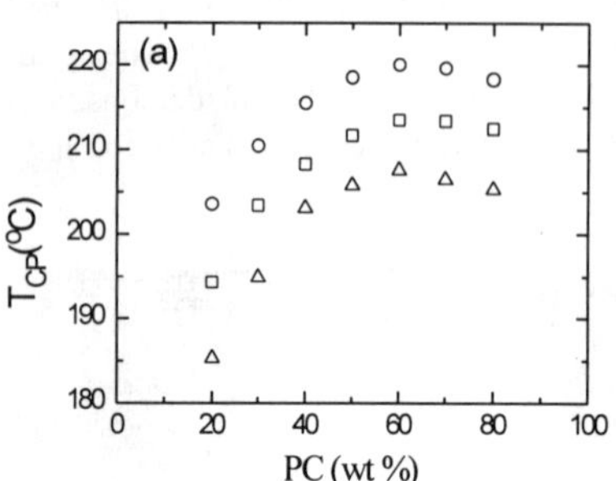
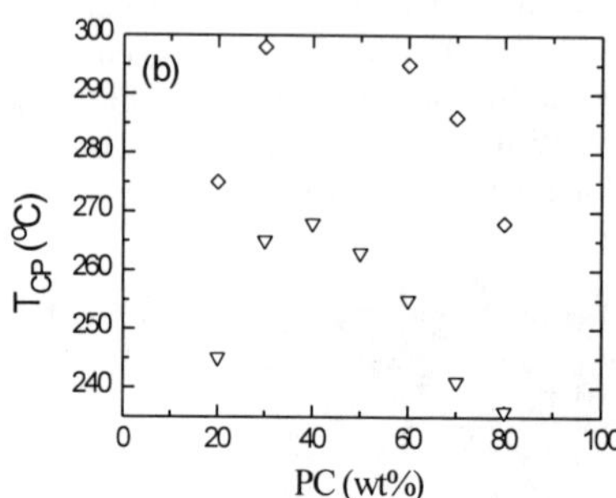

Fig.3 T_{cp} vs. composition for blends of ZnSPS and PC. (a) Effect of sulfonation: (o) 8.7ZnSPS, (□) 10.3ZnSPS, (Δ) 13.7ZnSPS. (b) effect of ionomer molecular weight: (◊)M_w =280 000, (∇)M_w = 135 000.

Blends of ZnSPS/PC based on a higher molecular weight ionomer than was used to construct the phase diagrams in Fig. 3(a) showed similar UCST-phase behavior. Fig. 3(b) shows the effect of the SPS molecular weight on the phase behavior of ZnSPS/PC blends. The sulfonation level of the PS was held constant at ca. 5.9 mol%, while the molecular weight of the ionomer was varied from M_w = 135,000 g/mole to 280,000 g/mole. As is normally expected for polymer blends with UCST-type phase behavior, increasing the molecular weight of either component reduces the miscibility, and T_{cp} increased when the molecular weight of the SPS was doubled. A comparison of Figs. 3(a) and (b) indicates that the critical composition shifted from PC-rich to ionomer-rich, i.e., from 60% PC to 40% PC, when the ionomer molecular weight was increased from M_w = 4000 to 135,000 g/mole, which again is the expected result for a polymer blend when the higher molecular weight component switches from one polymer to the other.

The T_gs obtained after annealing at a temperature within the single phase region (see Fig. 3) are plotted against composition in Fig. 4 for the blends based on the 10.3 mol% ionomer (M_w = 4000 g/mole and 5.88 mol% ionomer (M_w = 135,000). The T_g vs. composition curve for these blends, as well as for the other miscible blends investigated in this study, showed negative deviation from a simple, weighted average of the two component polymer T_gs. The Gordon-Taylor equation[13], which is indicative of very weak or no specific intermolecular interactions between the component polymers, i.e., $\chi \sim 0$, fit the data in Fig. 4 very well. This indicates that any attractive interaction between ZnSPS and PC is weak at best, and certainly much weaker than the intermolecular interactions that occur in ZnSPS/polyamide blends that exhibit positive deviation of T_g from a linear weighted average of the component T_gs.

Xie et al.[14] observed that although weak interactions between PS and poly(vinyl methyl ether) occurred in a miscible blend, no interaction was detected when the blend was phase-separated. To ascertain whether a similar result might be obtained for the Zn-SPS/PC blends, we measured the FTIR spectra of a (50/50)10.3ZnSPS/PC blend at 200°C, 235°C and 250°C. The lower temperature was below T_{cp} and the two higher temperatures were above T_{cp}. Fig.5 shows the FTIR spectra at those temperatures for the carbonyl and sulfonate stretching regions. The spectra were recorded after the film was annealed at the indicated temperatures for at least 30 minutes, which

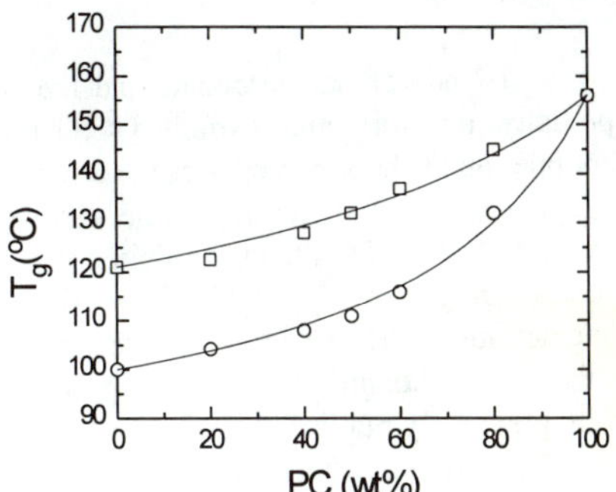

Fig.4 T_g vs. composition for (O) PC/10.3ZnSPS (M_w = 4000 g/mole) and (□) 5.88 Zn-SPS/PC (M_w = 135,000) . The solid lines are fits of the Gordon-Taylor equation.

was long enough to reach its equilibrium morphology according to the DSC data (After a two-T_g sample annealed at one-phase region for 30 minutes, it just showed one T_g). Although there is some broadening of the absorbance peaks at the higher temperatures, the maximum in the carbonate carbonyl absorption remains constant over the temperature range studied, Fig. 5(a). Similarly, the absorption peak for the symmetric S-O stretching vibration of the sulfonate anion at 1044 cm^{-1} remains constant with increasing temperature, even though there appears to be some changes to a lower frequency neighboring vibration at 1034 cm^{-1}, which is due to the S-O stretch of an isolated SO_3^-.However, this change, together with the intensity changes of the band at 1130 cm^{-1} due to the in-plane skeleton vibration of the disubstituted benzene ring, and those of the weak absorbances between 850 - 880 cm^{-1}, was found to have nothing to do with the addition of PC.[15] The constancy of the frequency of the carbonyl and S-O stretching vibration from room temperature to 250°C indicates that little change in the environment of both groups occurred over that temperature interval, which represented two-phase and single-phase blends. That result further supports the conclusion that no specific interaction occurred in the ZnSPS/PC system.

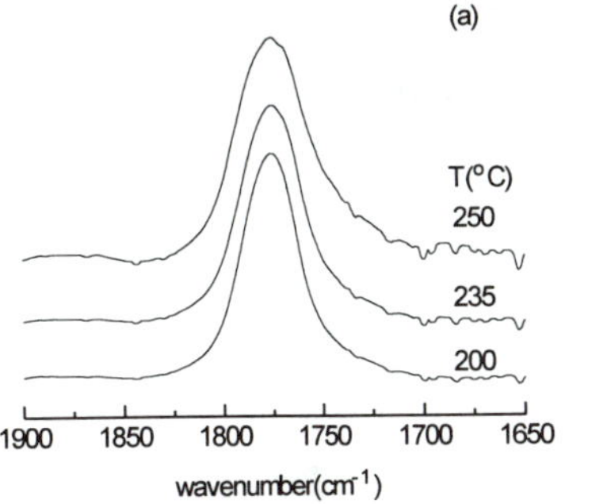

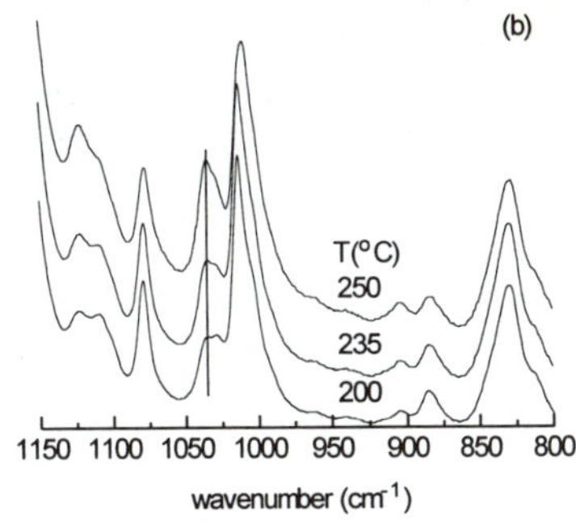

Fig. 5 FT-IR spectra for a (50/50) 10.3ZnSPS/PC blend in (a) the carbonyl stretching region and (b) the sulfonate stretching region.

CONCLUSION

Blends of polycarbonate and the zinc salt of light sulfonated polystyrene ionomers are partially miscible and exhibit UCST-type phase behavior. T_{cp} decreased with increasing sulfonation of the ionomer over the range of 8.7~13.7mol%, and for a constant sulfonation level, T_{cp} increased with increasing molecular weight of the ionomer. The T_g-composition behavior of miscible ZnSPS/PC blends exhibited negative deviation from a linear weighted average of the component polymer T_gs, which is consistent with a miscible blend with weak or no interpolymer interactions. FTIR analyses also demonstrated that no specific interaction occurred involving either the carbonate carbonyl group or the metal sulfonate group. These results are consistent with previously published results for blends of PC with SPS ionomers[6], and they strongly suggest that the miscibility between these two polymers arises from intrapolymer repulsive interactions between the ionic and non-ionic species within the ionomer.

ACKNOWLEDGEMENT

We gratefully acknowledge support of this research by the Polymer Program of the National Science Foundation (Grant DMR 9400862).

REFERENCES

1   E. M. Pearce, T. K. Kwei, T.K.; B. Y. Min, *J. Macromol. Sci. Chem.* 1984, A21, 1181.
2   A. Eisenberg, M. Hara, *Polym. Eng. Sci.* 1984, 24, 1306
3   Z. L. Zhow, A. Eisenberg, *J. Polym. Sci. Polym. Phys. Ed.* 1983, 21, 223.
4. X. Y. Lu, R. A. Weiss, *Macromolecules* 1992, 25, 6185
5   X. Y. Lu, R. A. Weiss, *Macromolecules* 1996, 29, 1216
6   R. P. Kambour, J. T. Bendler, R. C. Bopp, *Macromolecules* 1983, 16, 753.
7. D. R. Paul, J. W. Barlow, *Polymer* 1984, 25, 487.
8. G. Ten Brinke, G.; F. E. Karasz, W. J. Macknight, *Macromolecules* 1987,28,957
9. H. S. Makowski, R. D. Lunderg, G. H. Singhal,. *U.S. Patent* 3,870,841, 1975
10. D. J. Moonay, R. J. Wu, M. T. Shaw, *Proc. Soc. Plastics. Eng.*, ANTEC, 1994, 2038.
11. R. A. Weiss, X. Y. Lu, *Polymer* 1994, 35, 1963.
12 Y. Feng, R. A. Weiss; *Macromolecules*, 1996, 29, 3925.
13. M. Gordon, J. S. Taylor, *J. Appl. Chem.* 1952, 2, 495.
14. R. Xie, B. X Yang and B. Z. Jiang, B.Z. *J. Polym. Sci. Polym. Phys.* 1995, 33, 25.
15. R. Xie, R. A. Weiss, to be published results.

POLYMER BLENDS MADE FROM IONOMER AND IONOMER PRECURSOR POLYMER

M.A. BELLINGER, X. MA, L. TSOU, J.A. SAUER, M. HARA
Department of Chemical and Biochemical Engineering
Rutgers University, Piscataway, NJ 08855-0909

ABSTRACT

Rigid-rigid blends made of an ionomer and an ionomer precursor polymer, based on either polystyrene (PS) or poly(methyl methacrylate)(PMMA), were studied. A positive deviation from the rule of mixtures in both strength and toughness was observed over the entire composition range; and an especially significant enhancement was observed at low ionomer composition (e.g., 10 wt.%). This is attributed to a fine dispersion of the rigid ionomer phase in the matrix polymer together with good interfacial adhesion between the phases. The adhesion arises from entanglements due to athermal interactions between the same monomeric units of both the ionomer and the ionomer precursor. In addition, all the blend samples are found to be transparent due to similar refractive indices of the two components.

INTRODUCTION

In polymer blends, it is known that miscible, homogeneous blends generally show a negative deviation from the rule of mixtures for toughness. This is because molecular interactions responsible for miscibility are considered to reduce chain mobility, which leads to higher modulus and yield stress, but lower elongation, craze resistance, and toughness of the blends [1]. At best, additive mechanical properties are expected for homogeneous miscible blends that have athermal interactions [1]. However, blending of slightly immiscible, rigid, thermoplastic polymers (e.g., a creation of *rigid-rigid blends* [2,3]) leads to phase separation and this can, in some cases, provide a means for achieving better mechanical properties, especially better toughness [4].

We are investigating simple rigid-rigid blend systems made of *ionomers* (ion-containing polymers having a small number of ionic groups distributed along nonionic hydrocarbon chains [5-7]) and their precursor polymer [8-10]. In such a combination of (amorphous) ionomer/ionomer-precursor, the degrees of polymerization of both component polymers can be very close and no crystalline phase exits. Because of a difference in polarity between the two component polymers in the blend, even though the ionomer component may have only several mol% of ionic groups, there is a tendency to phase separation, leading to the creation of rigid-rigid blends. At low ionomer composition, the ionomer second phase particles are dispersed in the matrix polymer, and ionic aggregates are formed in the ionomer phase. In addition, athermal interactions between the identical repeat units; e.g., styrene units on the PS ionomer and on PS, may insure compatibility at the interface through interpenetration of polymer chains, and development of some molecular entanglements. It is well known that entanglements are a prerequisite for strength in polymer systems, and also for good interfacial adhesion [11,12].

Recently, ionomers have been used for various polymer blend studies: the major approach of these efforts has been to enhance miscibility by introducing ionic interactions, such as ion-ion, ion-dipole, and complex formation, between two otherwise immiscible polymers [13-16]. Our blend system results from a different approach, in that the attractive interactions are at

Mat. Res. Soc. Symp. Proc. Vol. 461 © 1997 Materials Research Society

best athermal interactions between the same repeat units (e.g., styrene), and the ionic groups, introduced into one component of the polymer blend, slightly decrease miscibility due to the difference in polarity between the ionomer and the ionomer precursor. This leads to phase separation, which is considered to be an essential condition for obtaining better mechanical properties [4].

In this study, rigid-rigid blends of ionomer/precursor polymer based on either PS or PMMA, were examined. It is found that synergistic enhancement in mechanical properties is produced by the addition of a rigid ionomer phase to the precursor polymer. The influence of composition and ion content on the mechanical properties of these blends was investigated and a strong correlation exists between the morphology and micromechanisms of deformation and fracture.

EXPERIMENT
Materials

PS ionomer, lightly sulfonated polystyrene (SPS), was prepared by sulfonation of starting PS according to the procedure described by Makowski et al. [17]. The ion content was controlled by adjusting the amount of sulfonating agent: the ion contents of the ionomers were 2.65, 5.26 and 7.45 mol%. PMMA ionomers (6.0 and 12.4 mol%) were made by neutralization of random copolymers of methyl methacrylate and methacrylic acid [18]. The counterion of all the ionomers discussed in this paper is Na.

Solution blending was used to make ionomer/ionomer precursor blend samples; a proper amount of each component polymer was dissolved in a solvent mixture, 1,2-dichloroethane (DCE)/methanol (90/10 v/v) for the PS ionomer blends and benzene/methanol (80/20 v/v) for the PMMA ionomer blends, and mixed under stirring for several days. Then, the blend material was recovered by steam stripping in boiling water for the PS ionomer blends and by freeze-drying for the PMMA ionomer blends. After vacuum drying at elevated temperature, the blend samples were compression molded to make bulk specimens for testing.

Testing

Tensile specimens were directly machined from compression molded rectangular bars into either hour-glass-shaped round specimens or dog-bone shaped specimens. Tensile testing was done either on an Instron tensile testing machine or Minimat Material Tester (Polymer Laboratories) at room temperature. The average values of at least three specimens were used for analysis. Fracture toughness was measured with the Instron by means of a 4-point bending test for notched specimens that were machined from compression molded bars. Fracture surfaces were examined by a scanning electron microscope (SEM) (ETEC or Amray) after coating them with a thin layer of gold.

RESULTS

Blends made of PS Ionomer and PS

The tensile strength and tensile toughness (or energy to fracture) are plotted as a function of ionomer content of the blend in Figure 1. A significant increase in tensile properties is apparent in the low ionomer composition range; the addition of 10 wt% of ionomer to PS has resulted in a 23% increase in tensile strength and a 65% increase in toughness as compared with unmodified PS. There is also a positive deviation from the rule of mixtures. Other blends

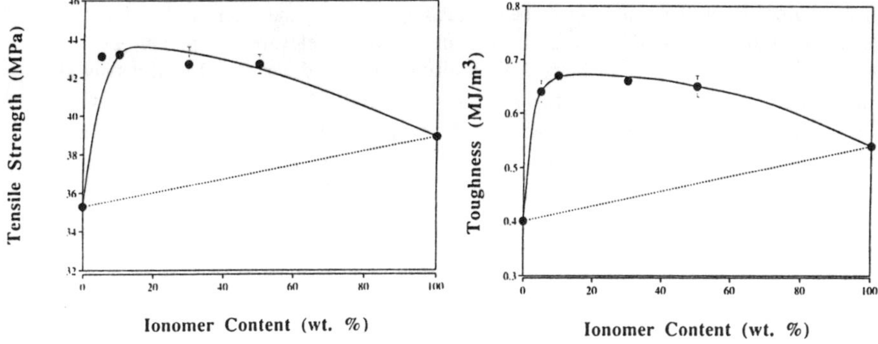

Figure 1. Composition dependence of (a) tensile strength and (b) tensile toughness of SPS ionomer (2.65 mol%; Na salt)/PS blends. The dotted lines represent the rule of mixtures.

of different ion contents (5.26 abd 7.48 mol%) show similar behavior.

The tensile fracture surfaces of the blends show two distinct regions: a smooth initiation region and an extended flat region [10]. The microstructures within the initiation region of the blends (10/90) are shown in Figure 2. In general, particle/matrix interactions are observed in all the blend materials, regardless of ion content or blend composition. However, the manner in which this interaction occurs differs according to the ion content. In the blends containing low ion content ionomer (2.65 mol%), the size of the second phase particles is ca. 1-2 μm (see Figure 2a), while the blends containing high ion content ionomer (7.45 mol%) have larger particles, ca. 5 μm (see Figure 2b). Figure 2 clearly show signs of adhesion between the ionomer particles and the PS matrix, especially for low ion content blends, in which the second-phase particles are completely embedded in the matrix polymer. Also, TEM pictures of strained thin films of SPS ionomer/PS blends show no voids at the poles of the second phase particles, reflecting the good interfacial adhesion [8,9].

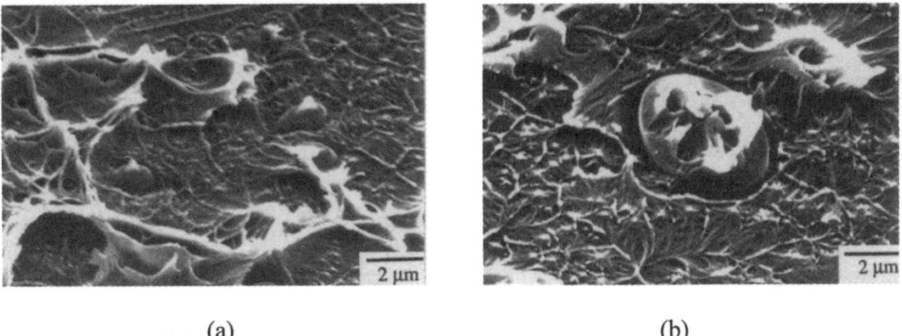

(a)   (b)

Figure 2. High magnification view showing particle/matrix interactions in SPS ionomer/PS (30/70) blends: (a) 2.65 mol%; (b) 7.45 mol%.

The relative transparency of the SPS ionomer (5.26 mol%)/PS blends, along with that of the homopolymers, is shown in Figure 3. All compression-molded samples are optically clear as evidenced by the visible grids below each specimen. Other blends of different ion contents (2.65 and 7.48 mol%) are also transparent, even at elevated temperatures. While optical clarity is an indication of blend miscibility on a scale down to 0.1 μm, it may also result if the refractive indices of the two polymers are similar ($\Delta n < 0.01$), or if the size of the second phase is too small compared with the wavelength of light (order of 0.1 μm) to scatter light [19]. Since the size of the ionomer phase is of the order of 0.1-1 μm, the former reason is applicable for our blend samples.

Blends made of PMMA Ionomer and PMMA

The behavior observed for the PMMA blend system is similar to that for the PS blend system, indicating the general applicability of the principle of ionomer/ionomer precursor blends.

First, a synergistic enhancement in mechanical properties is noted. Figure 4 indicates how the fracture toughness, obtained by a 4-point bending test of the PMMA ionomer (6.0 mol%)/PMMA blends varies with ionomer content. The curve clearly shows a positive deviation from the rule of mixtures. About a 30% increase in fracture toughness from that of PMMA, which is tougher than PS, is noted at a 30 wt.% ionomer composition. In these blend samples, a positive deviation from the rule of mixtures is also observed for tensile strength and tensile toughness.

Second, TEM pictures indicate that a two-phase morphology exists for the PMMA ionomer blends: for example, the ionomer second phase of the order of 0.1-1 μm is formed in the matrix of PMMA for the PMMA ionomer (6.0 mol%)/PMMA blends. It is also observed that the size of the second phase increases with increasing ionomer composition or ion content.

Finally, optical clarity of the PMMA is retained for the blend samples. Samples of PMMA, PMMA ionomer, and the blend, are all clear and transparent, in contrast to samples

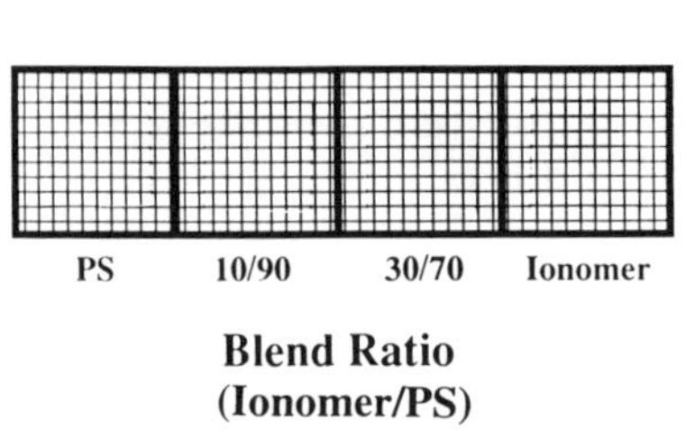

**Blend Ratio
(Ionomer/PS)**

Figure 3. Optical clarity of SPS ionomer (5.26 mol%; Na salt)/PS blends.

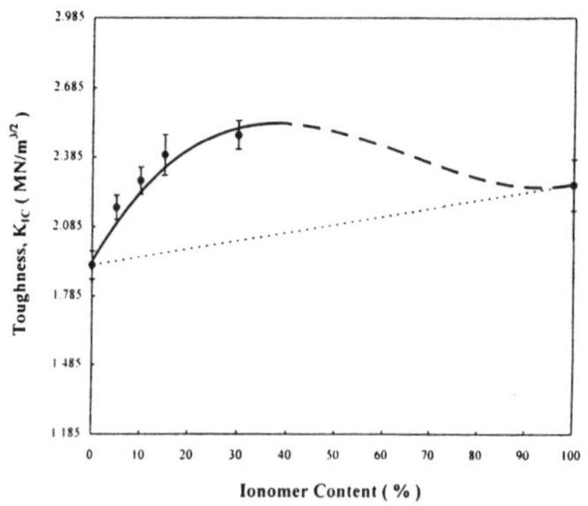

Figure 4. Composition dependence of fracture toughness, K_{IC}, of PMMA ionomer (6.0 mol%; Na salt)/PMMA blends.

of immiscible PMMA/PS blends (taken as a reference), which are opaque.

DISCUSSION

The ionomer/ionomer precursor polymer blends show significant enhancement in mechanical properties compared with the original polymer. The addition of a rigid ionomer to the precursor polymer causes a positive deviation from the rule of mixtures in strength and toughness. The behavior can be attributed to several factors: *enhanced fracture resistance of ionomer particles, good interfacial adhesion,* and *a stress concentration effect.*

The most important factor is the fracture resistance of the ionomer second phase particles. The ionomer particles have a greater strand density than unmodified polymers due to an ionic crosslinking effect [20]. Such an increase in strand density of the ionomer phase produces an increase in its craze stress, and when crazing occurs within the second phase particles, more cross-tie fibrils will be present. This should lead to a greater resistance to the development of damaging cracks in the craze and inhibit early fracture.

The good interfacial adhesion between ionomer particles and the matrix phase is also an important factor for enhanced toughness. Both TEM and SEM observations suggest that a sufficient level of adhesion is present between the ionomer particles and the precursor polymer matrix. This is due to favorable (athermal) interactions between the same monomeric units of both the ionomer and the ionomer precursor, which leads to the entanglement of the polymer chains of the blend components at the interface. On the contrary, in two-phase polymer blends with poor adhesion, smooth second phase particles are observed, reflecting fracture occurring though a path along the particle/matrix interface [21]. The formation of entanglements between two phases is a necessary condition for good interfacial adhesion in rigid-rigid blends, as shown in the work on the reinforcing effect of block copolymers in two-phase polymer blends, where at least one average "entanglement" is needed between the block copolymer and the homopolymers to have good stress transfer across the interface [12].

Another factor affecting the toughness of the blends is a stress concentration effect of the second phase particles: additional matrix deformation due to a mild stress concentration can contribute to increased toughness. Although the ionomer and the precursor polymer have rather comparable moduli, the stress concentrating effect of the ionomer second phase particles is not completely eliminated. TEM studies of thin films of the SPS ionomer/PS blend deformed under simple tension have shown that small crazes have been generated by the presence of the ionomer particles [8,9]. Similarly, Pearson and Yee have reported that rigid poly(phenylene oxide)(PPO) particles in an epoxy matrix can initiate shear bands in the matrix despite the small modulus difference between the particles and the matrix polymer [22]. In addition, ionomer chains in the glassy state can undergo crazing and deform significantly, unlike particulate fillers. Thus, in contrast to rigid particulate fillers, a second phase comprised of glassy ionomers can function as a more effective rigid second phase.

CONCLUSIONS

The addition of an ionomer to an ionomer precursor polymer results in creation of rigid-rigid blends with significant improvements in strength and toughness. The enhancement in mechanical properties is a result of a phase-separated blend morphology arising from the difference in polarity of the two component polymers; the ionomer second phase particles have more fracture resistance and act as stress concentrators, generating some matrix crazing and allowing some plastic energy to be absorbed before failure occurs. The elongation and fibrillation of the ionomer particles, along with the absence of voids at the poles, suggest that a sufficient level of interfacial adhesion exists to transfer the applied stress across the interface. This interfacial adhesion is a result of the favorable (athermal) interactions between the same monomeric units (e.g., styrene) of both the ionomer and the ionomer precursor, which allows molecular entanglements to occur at the interface. The rigid-rigid blends are transparent at all compositions and temperatures studied, because of the very close refractive indices of the two component polymers. Since these blend materials retain the transparency of the original polymer, and the total amount of repeat units having ionic groups included in the material is less than 1 %, synthesis of rigid-rigid blends of ionomer/ionomer-precursor seems an effective approach for improving mechanical properties of optically clear, glassy polymers.

ACKNOWLEDGMENT

Financial support from the U.S. Army Research Office and ACS-PRF is gratefully acknowledged.

REFERENCES

1. R. Kambour, in *Polymer Blends and Mixtures*, edited by D.J. Walsh, J.S. Higgins, and A. MaConnachie, (NATO ASI Series, Martinus Nijhoff Pub., Boston, 1985), p. 331.
2. M.E.J. Dekkers, S.Y. Hobbs, and V.H. Watkins, J. Mater. Sci. **23**, 1225 (1988).
3. H. Sue and A.F. Yee, J. Mater. Sci. **24**, 1447 (1989).
4. R. Fayt, R. Jerome, and Ph. Teyssie, in *Multiphase Polymers: Blends and Ionomers*, edited by L.A. Utracki and R.A. Weiss (ACS Symposium Series 395, American Chemical Society, Washington, DC, 1989), Chap. 2.
5. A. Eisenberg and M. King, *Ion Containing Polymers*, Academic Press, New York, 1977.
6. *Ionic Polymers*, edited by L. Holliday, Applied Science, London, 1975.
7. W.J. MacKnight and T.R. Earnest, J. Polym. Sci., Macrmol. Rev. **16**, 41 (1981).
8. M. Hara, M. Bellinger, and J.A. Sauer, Polymer Intnl. **26**, 137 (1991).
9. M. Hara, M. Bellinger, and J.A. Sauer, J. Colloid Polym. Sci. **270**, 652 (1992).
10. M. Bellinger, J.A Sauer, and M. Hara, Macromolecules **27**, 6147 (1994).
11. H.R. Brown, K. Char, V.R. Deline, and P.F. Green, Macromolecules **26**, 4155 (1993).
12. C. Creton, E.J. Kramer, and G. Hadziioannou, Macromolecules **24**, 1846 (1991).
13. P.Smith, M. Hara, and A. Eisenberg, in *Current Topics in Polymer Chemistry*, edited by R.H Ottenbrite, L.A. Utracki, and S. Inoue (Hauser Publ., Munich, 1987), Chap. 6.3.
14. X. Lu and R.A. Weiss, Macromolecules **24**, 4381 (1991).
15. A. Molnar and A. Eisenberg, Macromolecules **25**, 5774 (1992).
16. K. Sakurai, E.P. Douglas, and W.J. MacKnight, Macromolecules, **25**, 4506 (1992).
17. H.S. Makowski, R.D. Lundberg, and G.H. Singhal, U.S. Patent 3,870,841 (assigned to Exxon Research & Engineering Co.), 1975.
18. X. Ma, J.A. Sauer, and M. Hara. Macromolecules **28**, 3953 (1995).
19. *Polymer Blends*, edited by D.R. Paul and S. Newman (Academic Press, New York, 1978).
20. M. Hara and J.A. Sauer, J. Macromol. Sci., Rev. Macromol. Chem. Phys. **C34**, 325 (1994).
21. M. Lambla, R.X. Yu, and S. Lorek, in *Multiphase Polymers: Blends and Ionomers*, edited by L.A. Utracki and R.A. Weiss (ACS Symposium Series 395, American Chemical Society, Washington, DC, 1989), Chap.3.
22. R.A. Pearson and A.F. Yee, Polymer **34**, 3658 (1993).

MORPHOLOGY OF A BLEND OF ZINC NEUTRALIZED SULFONATED POLY(PHENYLENE OXIDE) OR POLYSTYRENE AND AN AMINO SILICONE

Alan A.Jones, Paul T. Inglefield, Changlai Yang and Pamela Bergquist, Clark University, Chemistry Department, 950 Main St.,Worcester, MA 01610
Jiefeng Shi, W.L.Gore & Associates, Inc., Polymer Products, 1320 Appleton Road, Elkton, MD 21922
Roger P.Kambour, General Electric Research and Development, Polymer Physics and Engineering, Schenectady, NY, 21301
George D.Wignall, Oak Ridge National Laboratory, Bethel Valley Road, Oak Ridge, TN, 37831
Henri-Noel Migeon and Jean-Paul Martin, Centre Univeristaire, Laboratoire D'Analyse Des Materiaux, 162a Avenue de la Faiencerie, Luxembourg City L-1511, Luxembourg

Abstract

A polymer blend of an ionomer based on zinc neutralized sulfonated poly(phenylene oxide) or polystyrene with a silicone copolymer containing 6.45% propylamine groups in place of one of the methyl groups on a backbone silicon was prepared. Carbon-13 magic angle spinning spectra show coordination of the amine by the zinc ions. Morphological characterization was made by NMR spectroscopy based on proton spin diffusion, by small angle x-ray scattering and by energy filtered transmission electron microscopy. All experiments show domains in the range of 1 to 1000 nm and domain size can be controlled by the extent of coordination of the amine groups by the zinc ions and by thermal history. The different morphological experiments lead to an apparent hierarchy of domain sizes. The NMR experiment yields the smallest domains, 2 to 10 nm, where contrast is produced by differences in chain mobility. Small angle x-ray scattering indicates domains of 16 nm while electron microscopy indicates domains in the range of 100 to 1000 nm. The variation of the domain size between experiments may reflect the different sources of contrast in each case.

Introduction

The motivation for preparing ionomers based on polystyrene and poly(phenylene oxide) is to blend these high glass transition polymers with an amino silicone copolymer which has a low glass transition temperature. These polymer blends would be incompatible except for the specific interactions between the zinc ions of the ionomers and the amine groups of the copolymer.

The extent of coordination of the amine group can be conveniently followed by carbon-13 magic angle spinning (MAS) spectra of the blend and blends are produced which have every amine group coordinated. Since the amine content is about one group every 16 repeat units, the coordination chemistry forces the two incompatible polymers to be in close proximity. The chain dynamics are radically altered; and to attempt to understand these dynamics, it is important to try to develop a morphological description of the blend.

Several different experiments were performed to attempt to provide an overall morphological description. First, the monitoring of the extent of coordination cross links by carbon-13 MAS is itself an important piece of information in developing a morphological description [1]. In addition Goldman-Shen [2] proton spin diffusion NMR experiments were performed to characterize the size of the mobile domains. Since the glass transition temperatures are so different for the two polymeric constituents, there are regions of the blend which have mobility typical of a rubber while other regions have mobility typical of a glass at a given temperature. The second morphological class of experiments is based on small angle x-ray scattering. This is a classic technique for characterizing the morphology of polymer blends as well as the morphology of ionomers. Lastly, transmission electron microscopy was performed on these blends. This technique has been extremely useful in characterizing the morphology of multicomponent polymer systems such as block copolymers though it has not been as useful in the field of ionomers.

Mat. Res. Soc. Symp. Proc. Vol. 461 © 1997 Materials Research Society

Experimental

Two amino silicone copolymer (a-silicone) from Genesee Polymers Corporation with 6.45 (GP4) and 4.0 (GP6) mole per cent amino propyl groups in place of one of the methyl groups on the backbone silicon is used as one component of the blend. A zinc neutralized sulfonated poly(2,6-dimethyl-1,4-phenylene oxide) ionomer (ZnSPXE) or a zinc neutralized sulfonated polystyrene (ZnSPS) was used as the other component of the blend. The sulfonation level is at 10 mole percent for the ZnSPXE and 4.4% for the ZnSPS. The ZnSPXE and the a-silicone were dissolved in dry dimethyl sulfoxide and tetrahydrofuran respectively at the level of a few weight per cent. The relative stoichiometry was set at one zinc ion per amine. A second synthesis was also carried out where an excess of the amine was employed. The two homogeneous solutions were mixed, at which point they became turbid. After a period of days a gel was formed which precipitated out of solution. The supernatant liquid was then poured off leaving the gel as a coherent film which was washed with dry acetone and vacuum dried for several days.

All NMR experiments were performed on a Bruker MSL 300 including magic angle carbon-13 spectra, proton wide line spectra, and proton Goldman-Shen [2] domain size experiments.

SAXS experiments were conducted at the Oak Ridge National Laboratory in Tennessee on the 10-m spectrometer [3] operating at 40kV and 100mA. The instrument was operated with a sample-detector distance of 1.119m. Corrections were made for instrumental background and detector efficiency on a cell be cell basis prior to radial averaging to yield a q range of 0.005 to 0.1 Angstroms^{-1}. The scattering profiles were corrected for sample absorption and incident X-ray fluctuations. The net intensities were converted to an absolute differential cross section per unit sample volume by comparison with precalibrated secondary standards.

TEM experiments were performed at the Laboratoire D'Analyse Des Materiaux in Centre Universitaire in Luxembourg on a Zeiss EM 912 Omega. Both standard TEM and energy filtered TEM experiments were performed.

Results

Carbon-13 magic angle spinning spectra of two of the starting materials (ZnSPXE and the amino silicone) as well as the polymer blend are shown in Figure 1 along with the structures of these starting materials. Note that the extent of reaction can be monitored by the resonance at 45.0 ppm from the amino silicone which corresponds to the terminal carbon of the propyl group bearing the amine. In the blend this resonance is absent and a new line appears at 42.3 ppm which is interpreted as indicating nearly complete coordination of the amine by the zinc ion. The blend spectrum shown in Figure 1 was prepared from a starting stoichiometry of 1 zinc ion per 1 amine group. In the excess amino silicone blend, a resonance corresponding to uncoordinated amine at 45.0 ppm is present as well as the resonance indicating coordinated amine at 42.3 ppm. This indicates that the blend can be prepared with different ratios of the two starting materials and the level of cross-linking produced by coordination of the zinc ion can be controlled and monitored. Thus the magic angle spectra provide direct evidence of the reaction producing this ionomer blend and a convenient monitor of the extent of the reaction.

Representative proton line shapes as a function of temperature for the starting materials and the blends are presented in Figure 2. The amino silicone displays typical proton line narrowing for an amorphous glass with a transition temperature around -115°C. Below that temperature, the line is a broad Gaussian and above that temperature it becomes a narrow Lorentzian. Figure 3 is a plot of line width versus temperature. The ZnSPXE line shape remains a broad Gaussian over the accessible temperature range (up to 100°C) and its line width is also given as a function of temperature in Figure 3. Similarly, the ZnSPS does not undergo significant line narrowing below 100°C. The 1 to 1 blend line shapes begin narrowing at -115°C but thereafter the line shape is neither a Gaussian nor a Lorentzian. It can however be fit as a sum of a broad Gaussian and a narrow Lorentzian with the fractional population of Gaussian component decreasing with temperature.

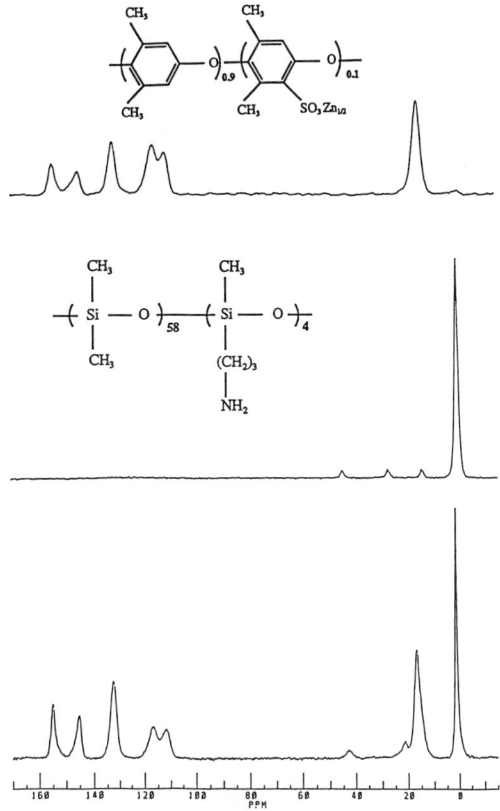

Figure 1 Structures and Carbon-13 Magic Angle Spinning Spectra of Two of the Starting Materials and the Corresponding Blend

In bimodal line shapes of the type seen in this blend, proton spin diffusion measurements can be performed to determine the size of the domain corresponding to the narrow or Lorentzian component. In the simple Goldman-Shen [2] (GS) experiment, discrimination times of 80 to 100 μs were found to separate the narrow component from the broad component. Free induction decays (FIDs) were taken with different mix times after the discrimination period ranging from 4 ms to 500 ms. At short mix times, an exponential decay corresponding to the Lorentzian component is observed. As the mix time is increased, a rapidly decaying component corresponding to the broad Gaussian shape reappears indicating spin diffusion from the more mobile, narrow component to the less mobile broad component.

Each FID in the GS experiment can be fit to an equation of the form

$$G(t) = A\,e^{-(t/T_{2l})^2} + (1-A)\,e^{-(t/T_{2m})} \tag{1}$$

The spin-spin relaxation time of the less mobile component in the GS experiment is T_{2l} and a Gaussian form for the time dependance is assumed. The spin-spin relaxation of the more mobile component in the GS experiment is T_{2m} and a time dependance corresponding to a Lorentzian is assumed. The quantity A is a function of mix time and is used to monitor the repopulation of the broad Gaussian component following the selection of the narrow

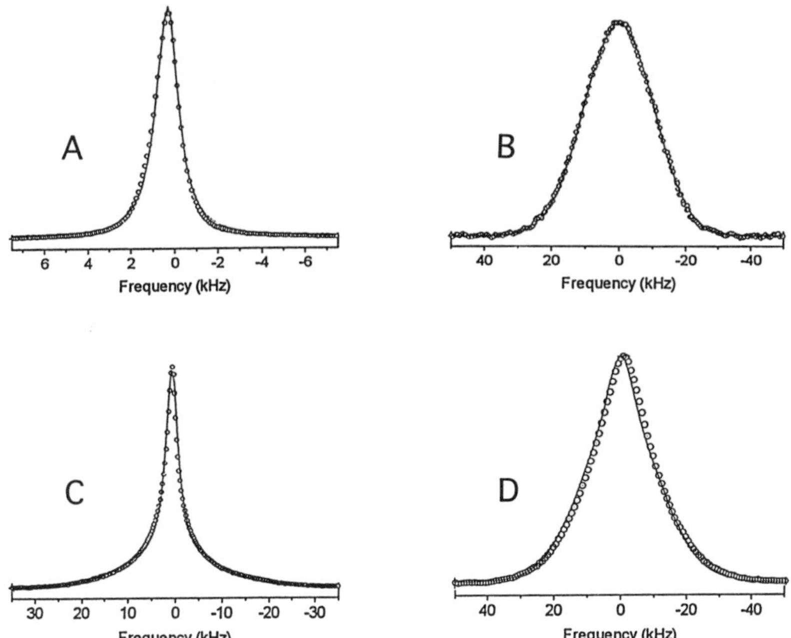

Figure 2 Proton Line Shapes at a Temperature of -93°C: (A) amino silicone starting material (B) ZnSPXE starting material (C) excess amino silicone blended withAnSPXE (D) one to one blend

component in the first part of the GS experiment. The recovery of the less mobile component in the GS experiment is monitored by the quantity $R(t_{mix})$ where

$$R(t_{mix}) = A(t_{mix})/A(\infty) \tag{2}$$

where $A(\infty)$ is the population of the less mobile Gaussian component at time infinity in the absence of any spin-lattice relaxation.

Expression for the recovery factor as a function of mix time based on spin diffusion have been developed by Cheung and Gerstein [4]. The key parameters are the spin diffusion constant, D, and the domain size b. The Cheung and Gerstein expression neglects spin-lattice relaxation and assumes a random distribution of more mobile domains in the sample. The three dimensional form of the recovery equation best suited for the sample under consideration is

$$R(t) = 1 - \phi_x(t)\ \phi_y(t)\ \phi_z(t) \tag{3}$$

where each $\phi(t)$ has the form

$$\phi(t) = \exp(Dt/b^2)\, \mathrm{erfc}\,(Dt/b^2)^{1/2} \tag{4}$$

Figure 3 shows a typical recovery curve fit to equations 3 and 4. To obtain a value for b as a measure of domain size, D will be estimated from the equation presented by Cheung and

Gerstein [4].

$$D = 0.13\ a^2/T_{2I} \tag{5}$$

where a is the average distance between adjacent protons in the less mobile domain.

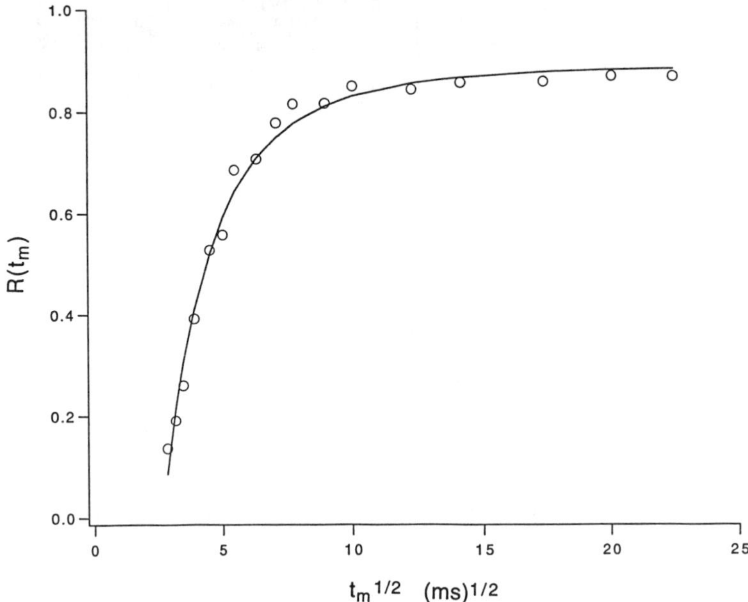

Figure 3 Typical Goldman-Shen Spin Diffusion Recovery Curve

Somewhat arbitrarily, a is set a 1.8 Angstroms corresponding to the intramolecular separation of methyl and protons. For the one to one blend of GP4 with ZnSPXE, a mobile domain size of 20 Angstroms was found and for the blend of the same two polymers with an excess of GP4, a domain size of 80 Angstroms was found. For the blend of GP6 with ZnSPS, a domain size of 35 Angstroms was found.

Two qualitatively different types of results were observed in the small angle X-ray experiments on these blends. The X-ray intensity as a function of q for the blend of ZnSPXE and an excess amino silicone shows no maxima in the scattering curve but the data do produce a linear Debye-Beuche plot. The characteristic length obtained is 165 Angstroms and the absolute intensity at q=0 can be accounted for by the scattering lengths for ZnSPXE and amino silicone with an assumption of complete phase separation. For comparison the proton spin diffusion experiment leads to a domain size of 80 Angstroms on the same sample. A qualitatively different result is observed on the blend of ZnSPS and the 4% amino silicone copolymer as shown in Figure 4. Here a maximum in scattering intensity is observed which can be best fit with a shell model [5] where the scattering length differences are 3×10^{11} cm-1 for the inner shell and 7×10^{8} cm-1 for the outer shell relative to the background. An inner radius of 35 Angstroms is used while the outer radius is set at 250 Angstroms. In this blend, proton spin diffusion gives a domain size of 35 Angstroms.

Figure 5 contains standard TEM results at two magnifications. Large domains of the order of 1000 Angstroms are seen in the lower magnification but at the higher magnification structure appears to exist within these larger domains. The structure within domains has a length scale of hundreds of Angstroms. Energy filtered TEM indicates the dark regions observed at lower the lower magnification are higher in ion content and silicon content. In simple ionomer systems, the identification of the ion cluster is generally not possible and ionomer systems are subject to a number of artifacts in TEM experiments [6].

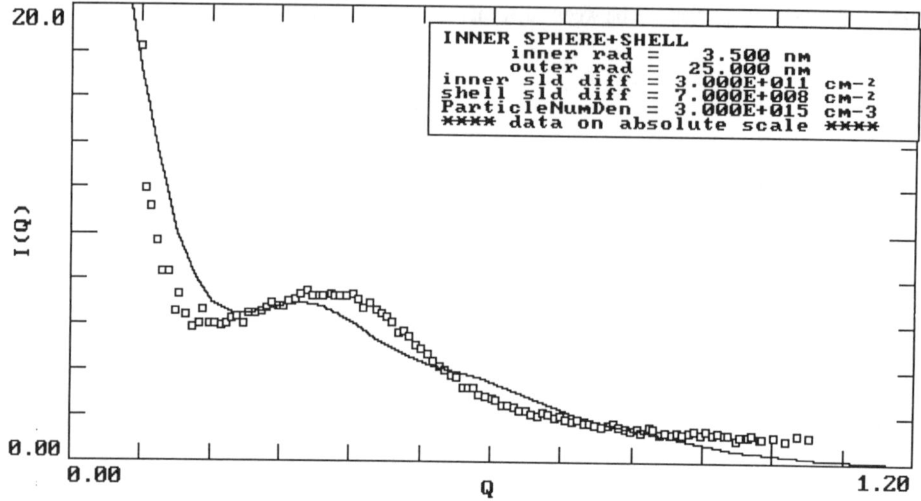

Figure 4 SAXS Pattern for the Blend of ZNSPS and Amino Silicone GP6

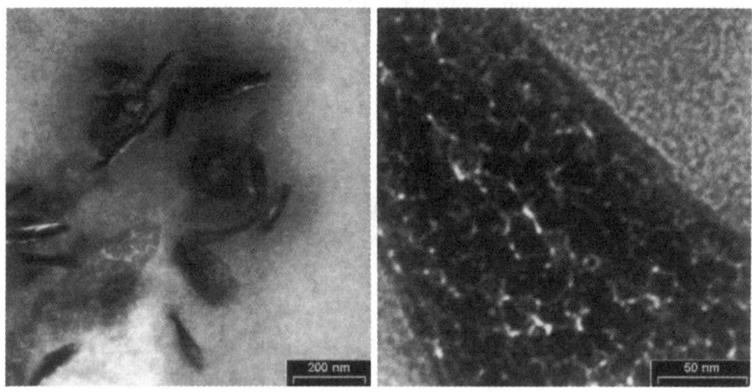

Figure 5 Bright Field Image of the ZnSPXE and Amino Silicone GP4

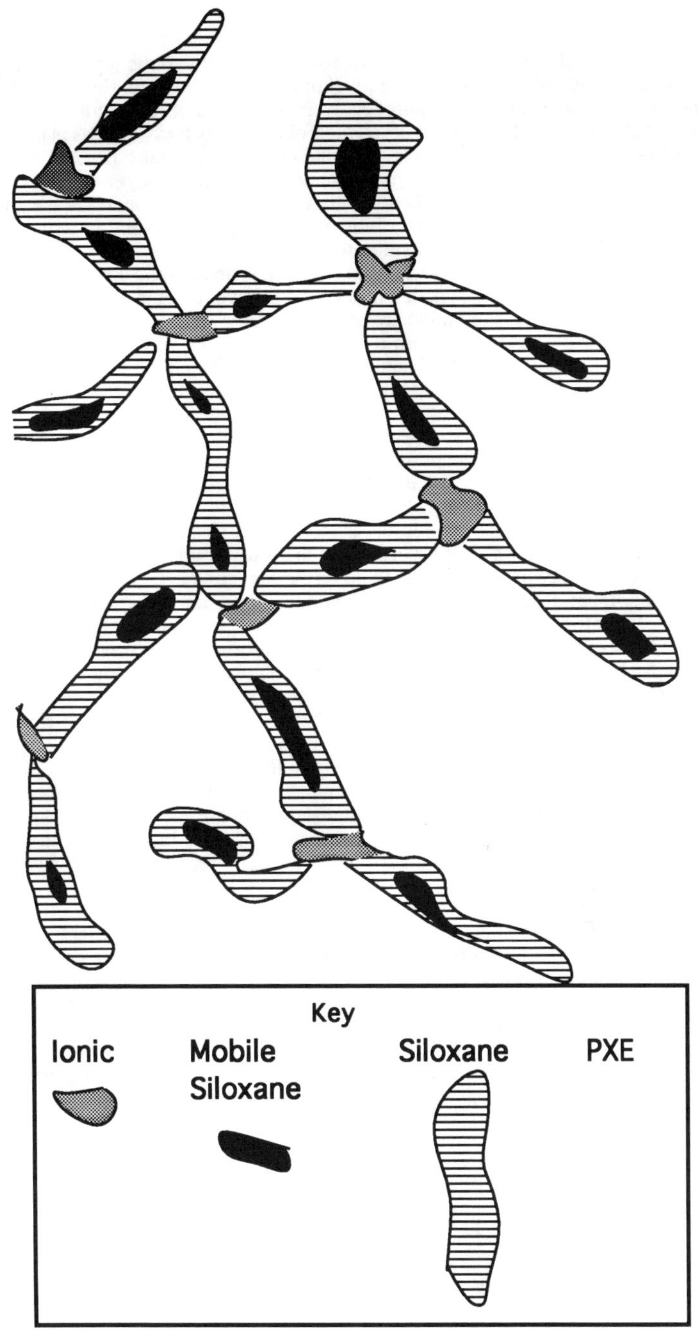

Figure 6 Ionomer Blend Morphological Model

Discussion

To try to develop an overall model to account for the various types of morphological information, it is convenient to start at the level of chemical structure. From the standpoint of the amino silicone, there is an amine nitrogen every 16 to 25 units on average which can cross link to the ZnSPXE. Thus if the extent of reaction of amines is fairly high as is observed in the carbon-13 MAS spectra, the level of mixing of the two polymers must be near the scale of the size of a polymer chain or 100 Angstroms. If domains were much larger than this, it would be hard to imagine a structure which would allow for nearly complete reaction of the amines.

The proton line shapes indicate a bimodal distribution of mobility which is plausible given the large difference in glass transition temperatures of the starting materials. The more mobile regions are associated with the silicone component and mobile domain sizes range form 20 to 90 Angstroms with larger domains observed for lower extents of reaction of the amine groups or with lower numbers of amine groups in the amino silicone.

The same size domains are seen by SAXS. For the blend of the ZnSPS and the 4% amino silicone copolymer, both SAXS and protons spin diffusion give 35 Angstrom domains. While it would be convenient to associate the same domain with the results of the two experiments, it is unlikely to be the same domain given the differing sources of the contrast in the two experiments. The SAXS experiment detects contrast in electron density with the highest density associated with the presence of ionic groups in the ionomer. The ionic groups are only present at a level of 4 to 10% so high electron density can only exist as a minority phase. Thus in the shell model the high electron density is best considered to reside in the smaller inner shell of 35 Angstroms. In the proton spin diffusion experiment, the contrast is based on mobility. The ionic regions would have the lowest mobility so that the 35 Angstrom domain seen in this experiments is not likely to be the same as the 35 Angstrom domain seen in the SAXS experiment. Both the SAXS and the proton spin diffusion experiment are interpreted in terms of concentric sphere models so an overall interpretation must imagine two sets of spheres, one for each experiment, which do not have a common origin. In a sense, each experiment, SAXS and NMR, take their own source of contrast as the center of the universe in a form of parochial bias to begin an interpretation.

Figure 6 shows a cartoon model which attempts to combine the results of both experiments thereby including both an ionic domain and a separate mobile domain of comparable size.In this model and most probably in the blends under consideration spherical domains are only approximate descriptions and the cartoon model attempts to avoid this simplification. The model leads to two co-continuous phases as the level of cross links is raised. Thus the description in the case of lower levels of cross links would be more isolated domains with a more spherical like character and would go over to a co-continuous description as the level of cross links increases.

The results of NMR and SAXS are relatively consistent with each other but neither senses the large domains seen in TEM. Either these large domains are artifacts [6] or there are larger scale morphology consisting of silicone rich domains and silicone deficient domains. Within the larger scale contrast seem in TEM, there are smaller domains of about the size of those observed by SAXS and NMR.

Acknowledgement

This research was carried out with the financial support of the National Science Foundation Grant DMR 9303193.

References

1. J.-F. Shi, P. Bergquist, J. Zhao, A. A. Jones, P. T. Inglefield, J. R. Campbell and R. P. Kambour, Polym. Prepr.(Am. Chem. Soc., Div. Polym. Chem.) 37(1), 371 (1996).
2. M. Goldman and L. Shen, Phys. Rev. 144, 321 (1966).
3. G. D. Wignall, J. S. Lin, and S. J. Spooner, J. Appl. Crystallogr. 23, 241 (1990).
4. T. T. P. Chenug and B. C. Gerstein, J. Appl. Phys. 52, 5517, (1981).
5. W. J. MacKnight, W. P. Taggert, R. S. Stein, J. Polym. Sci. Polym. Phys. Ed. 18, 1497 (1980).
6. D. L. Handlin, W. J. MacKnight, and E. L. Thomas, Macromolecules 14, 795 (1980).

MICROSTRUCTURE AND PHASE SEPARATION OF PEEK, PEKK AND THEIR BLENDS

Wu. Wang *[#], Jerold M. Schultz * and Benjamin S. Hsiao **
* Materials Science Program, University of Delaware, DE 19716.
** DuPont CR&D, Experimental Station, Wilmington, DE 19880-0302.
[#] Current address: Chemistry Department, Queen's University, Kingston, ON K7L 3N6, Canada.

ABSTRACT

Microstructures of PEEK|PEKK blends at both structural and lamellar levels were studied using simultaneously time-resolved small-angle and wide-angle x-ray scattering (SAXS and WAXS) with synchrotron radiations. Lamellar dimensions and unit cell parameters were determined, and phase separation within the blends were observed. A suitable two-phase growth model is proposed. It is concluded that when crystallized from the melt, PEEK and PEKK lamellae formed in separate lamellar bundles within a common spherulite.

INTRODUCTION

As a well-known high-performance engineering thermoplastic material, poly(aryl ether ether ketone) (PEEK) offers excellent chemical resistance and physical and mechanical properties at elevated temperatures, and has been widely used in thermoplastic composites and other applications [1-3]. Poly(aryl ether ketone ketone) (PEKKs) constitute a new set of materials and also possess high melting temperatures, high glass temperatures and good high temperature properties [4,5]. PEEK has received more attention in the past decade. Its crystal structure, spherulite and lamellar morphologies, crystallization kinetics and physicochemical properties have been well characterized [6-9]. Studies of the crystal structures, morphologies and crystallization behaviors of PEKKs have also been reported [10-12]. It is thus our objective to investigate the crystallization and microstructures of the PEEK|PEKK mixtures.

In the present work, studies of crystal structure and lamellar structures of the PEEK|PEKK blends were carried out using simultaneously time-resolved SAXS and WAXS techniques with synchrotron x-rays under various crystallization conditions. A two-phase growth model is proposed to explain the behavior of phase separation.

EXPERIMENTAL

Materials and Sample Preparation. Commercially available PEEK from ICI was used in our experiments. It has a glass transition temperature T_g of 145°C and an approximate equilibrium melting temperature $T_m°$ of 395°C. The weight-average molecular weight M_w and number-average molecular weight M_n are 38,000 and 14,000 respectively. Two types of PEKKs with different T/I ratios (30/70 and 50/50) were used in this study. The T_g of the PEKKs is about 153°C, and their melting temperatures are 305°C and 325°C respectively. The M_w and M_n of the PEKKs were 10,000 and 30,000 respectively. Void-free samples with thickness of 1.1 mm were prepared for the SAXS and WAXS experiments All the materials were first vacuum dried at 100°C for 24 hours prior to the sample preparation and measurements.

Mat. Res. Soc. Symp. Proc. Vol. 461 © 1997 Materials Research Society

Time-Resolved SAXS and WAXS measurements. SAXS and WAXS experiments were performed at the X3A2 beamline (wavelength 1.54 Å) at the National Synchrotron Light Source, Brookhaven National Laboratory. The collected scattering angular range 2θ was 0.21-2.81° for SAXS and 15-35° for WAXS. Typical data collecting time was selected between 10 and 30 seconds, depending on the crystallization rates of the materials. Lamellar dimensions (the long period of the lamellar structure L, the crystal lamellar thickness l_c and the thickness of interlamellar amorphous region l_a) were determined by the analysis of the correlation functions [13].

RESULTS

Lamellar-level structures and lamellar dimensions

The results of lamellar structures of neat PEEK and PEKKs have been reported by the authors elsewhere [12]. It was found that the lamellar dimension of PEEK depend on mainly crystallization temperature, and those of PEKKs vary with crystallization temperature and their T/I ratios. For comparison, Figure 1 shows the lamellar dimensions of PEEK and PEKKs crystallized at T_c=260°C.

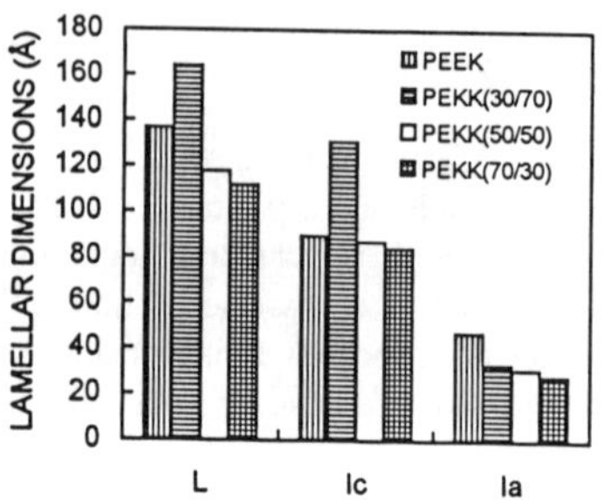

Figure 1. Lamellar dimensions of PEEK and PEKKs (T_c=260°C).

<u>PEEK|PEKK(30/70) blend</u>

Figure 2a shows the changes in dimensions of the lamellar stacks of PEEK|PEKK(30/70) blend during isothermal crystallization (T_c=260°C). The invariant Q (scattering power, the area underneath the Lorentz-corrected curve) is also plotted. Since the thickness of amorphous layers (l_a) stays almost unchanged throughout the crystallization process, it is not included in the figures. The curves in Figure 2a are divided into two regions. At the initial stage of crystallization (region I), the long period (L) and lamellar thickness (l_c) decrease with time; in region II, both of them increase as temperature increases. Corresponding to the changes in L and l_c, the invariant Q also experiences a two-step evolution. Within region I, Q increases at a higher rate; in region II its

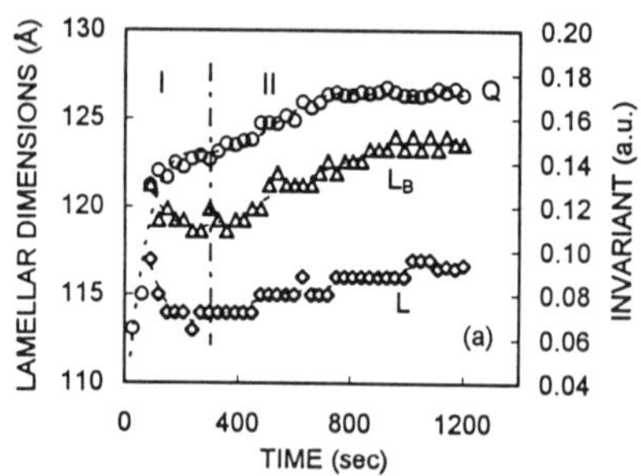
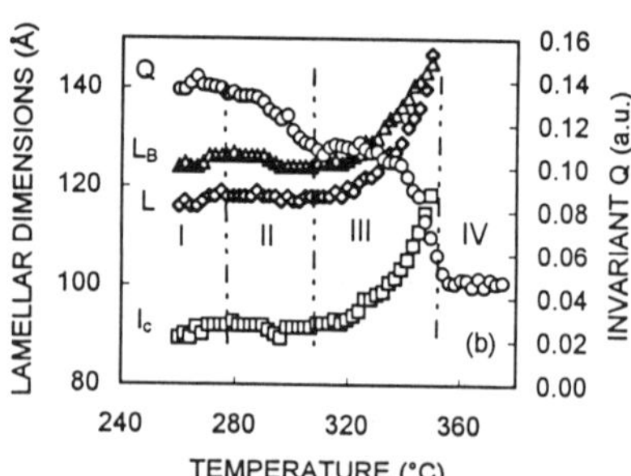

Figure 2. Changes in lamellar dimensions of PEEK|PEKK(30/70) blend during a) isothermal crystallization at 260°C and b) the subsequent melting.

increase rate eventually slows down and gradually approaches a constant value. As reported by the authors [12], for the cases of isothermal crystallization of neat polymers, where only one kind of lamellar structure is formed, both L and l_c initially decrease and then remain constant, and the invariant shows only stage I growth. The unusual changes in L, l_c and Q shown in Figure 2a is the evidence that there are two kinds of lamellar structures formed within the blend.

The changes in lamellar dimensions of PEEK|PEKK(30/70) during the subsequent melting are shown in Figure 2b. The heating rate used was 5°C/min. In region I, which corresponds to the temperature range 260-285°C, both L and l_c increase slightly with temperature, while Q shows a slight decrease. The increases in L and l_c in this region likely represent the simple thermal expansion. Region II represents the temperature range 285-310°C. In this region, L and l_c both show a small decrease with temperature and the invariant decreases more steeply, indicating the occurrence of melting of lamellae. This is consistent with the fact that the melting temperature of PEKK(30/70) is about 90°C below that of PEEK. The changes in L and l_c in this region can be thought of as the result of the competition between the effects of thermal expansion and the melting of the PEKK lamellae. As a whole, the combination of the two effects results in small decreases in L and l_c. Region III can be thought as the melting of PEEK lamellar stacks. The decrease rate of the invariant drops rapidly near the middle of this region (330°C), and becomes constant after the sample was completely melted (in region IV).

<u>PEEK|PEKK(50/50) blend</u>

Figures 3a and 3b show the lamellar dimensions of PEEK|PEKK(50/50) blend during isothermal crystallization and the subsequent melting. Unusual decreases in the long period and lamellar thickness of the blend during isothermal crystallization are clearly seen from Figure 3a. The temperatures of the specimens were kept constant throughout the process (after a short time fluctuation at the beginning). Since the long period and lamellar thickness of PEKK(50/50) are shorter and its crystallization rate is slower than PEEK, the possible reason for such decreases in L and l_c would be the appearance of the PEKK(50/50) lamellae, which shifts the observed lamellar dimensions of the blend to lower values.
The changes in lamellar dimensions and the evolution of the invariant Q of the blend during the subsequent heating are shown in Figure 3b. Similar to the case of the melting in the PEEK|PEKK(30/70) blend, the trend of the invariant Q also shows two-step decrease during

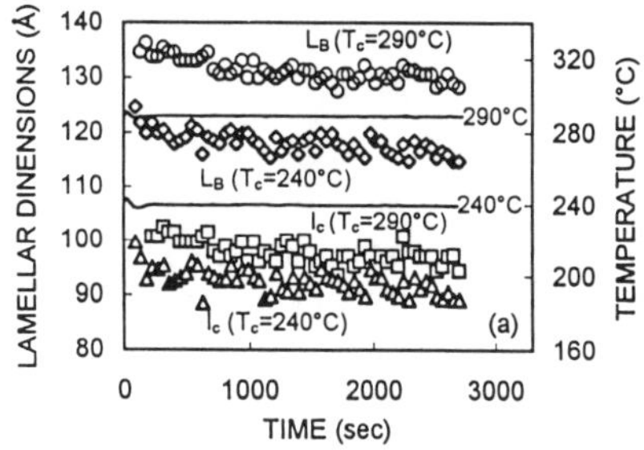
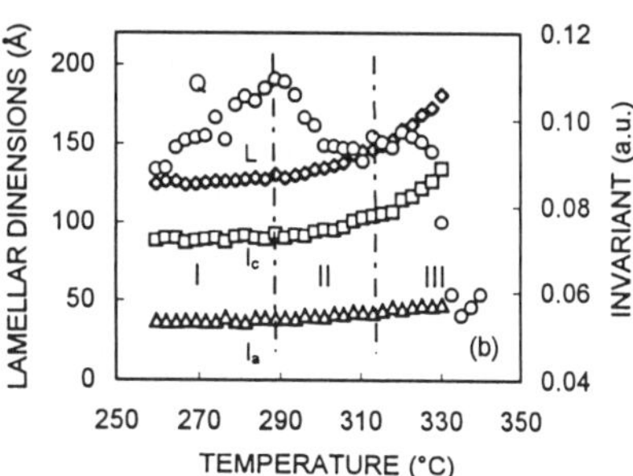

Figure 3. Changes in lamellar dimensions of PEEK|PEKK(50/50) blend during a) isothermal crystallization at 240 and 290 °C, and b) the subsequent melting.

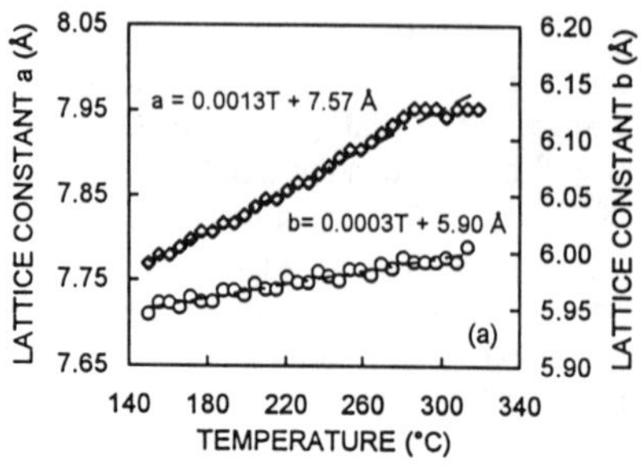
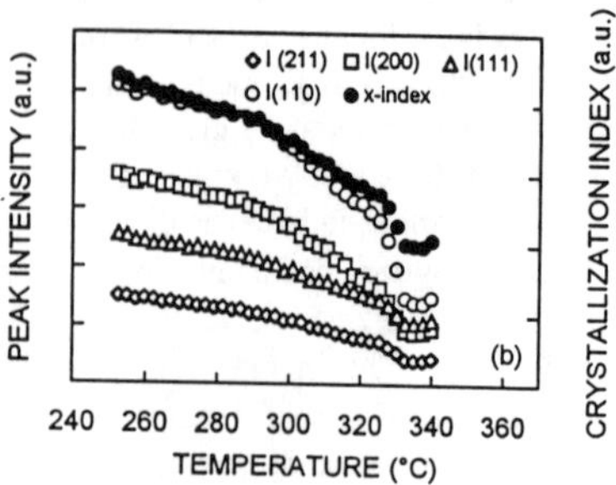

Figure 4. Changes in unit cell parameters (a) and Bragg peak heights and crystallinity index (b) of PEEK|PEKK(50/50) blend during melting.

heating. It is seen that Q first increases with temperature in region I, decreases and levels off in region II, and then shows a sharp decrease in region III. The evolution of the invariant clearly shows a two-step melting, which again indicates the existence of two phases in the blend.

Crystal structures

Figure 4a shows the changes in unit cell parameters a and b during melting. The heating rate used was 10°C/min. During heating, both a and b increase approximately linearly with temperature. The small changes in the lattice constants here are likely due to normal thermal expansion. Therefore, ignoring any possible irreversible effects, the thermal expansion coefficients of the unit cell dimensions can be estimated from these data,. Based on the data in Figures 4a, the calculated thermal expansion coefficients in a and b directions are $\alpha_a = 1.9 \times 10^{-4}°C^{-1}$ and $\alpha_b = 5.1 \times 10^{-5} °C^{-1}$. These values are also close to those of PEEK reported [14].

Figure 4b shows the changes in the heights of the Bragg peaks and the crystallinity index (defined as the ratio of the integral intensity of crystalline scattering during the process to its maximum value). Notable kinks appear on all the curves in this figure. The first kink is at approximately 295°C, and the second one is at about 325°C. Since the temperatures at which the kinks appear are very close to the starting temperatures of the melting of neat PEKK(50/50) and PEEK crystals, the first kink likely represents the beginning of the melting of PEKK, while the second one indicates that of PEEK crystals.

DISCUSSIONS

As stated above, both the SAXS and WAXS results indicate that there are two phases coexisting in the PEEK|PEKK blends. When crystallized from the melt, PEEK crystals form first, followed by PEKK crystals. For the case of PEEK|PEKK(30/70), since the L and l_c in PEKK(30/70) are thicker than those of PEEK, the appearance of PEKK(30/70) lamellar shifts the lamellar dimensions of the blend to higher values; for the case of PEEK|PEKK(50/50) blend, since PEKK(50/50) are with thinner L and l_c, the formation of PEKK(50/50) lamellae during crystallization causes the continuous decreases in lamellar dimensions. The results from WAXS also indicate the existence of two phases in the blends.

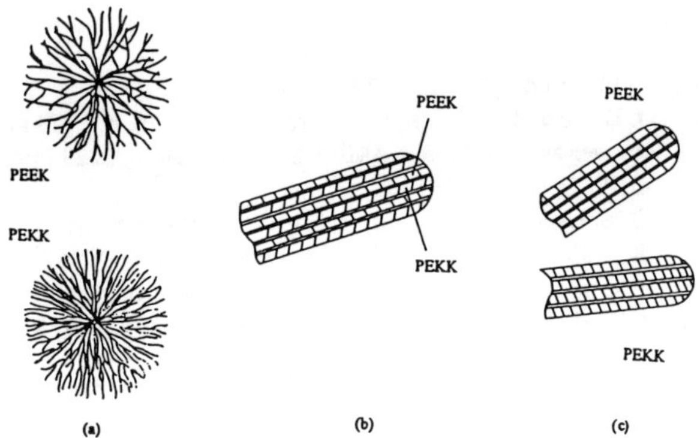

Figure 5. Possible models for the two-phase structures of PEEK|PEKK blends.

There are several possibilities for the existence of the two kinds of lamellar structures within the PEEK|PEKK blends, as illustrated in Figure 5. Figure 5a represents a model in which the two components separate completely, forming two different kinds of spherulites. If this is the case, two different kinds of spherulites should be seen during crystallization. But our results are not consistent with this either. Another possible model is that the two kinds of lamellae co-exist in the lamellar bundles, as sketched in Figure 5b. For this case, since PEKK lamellae will be melted at lower temperature, the thickness of amorphous layers should show a sharp increase during the melting of PEKK lamellae. Our results from SAXS show that l_a almost stayed constant throughout the melting process, implying that this model is also incorrect. The remaining possibility is that PEEK and PEKK lamellae reside in the same kind of spherulites but in different lamellar bundles, as shown in Figure 5c. During isothermal crystallization, the bundles contain PEEK lamellae precipitated from the melt first, forming the main frames of spherulites. PEKK lamellar bundles crystallize later between the existing PEEK lamellar bundles. Since the lamellar bundles are not formed by neat PEEK nor neat PEKK, the dimensions of lamellar stacks of the two are different from those of neat materials.

CONCLUSIONS

Crystal structures and phase behaviors of the PEEK|PEKK blends during isothermal crystallization and melting processes were investigated, and their unit cell parameters and lamellar dimensions were determined. Two-phase growth during crystallization was traced by both time-resolved SAXS and WAXS measurements. It was found that there are two kinds of lamellar structures, PEEK lamellae and PEKK lamellae, coexisting within the PEEK|PEKK blends. When crystallized from melt, PEEK lamellae form first, followed by PEKK lamellae. Possible models of the two phases within the blend were discussed, and a suitable model is proposed. It is believed that the two kinds of lamellae locate in separate lamellar bundles within a common spherulite.

ACKNOWLEDGMENTS

The authors wish to thank J. P. McKeown and R. J. Trotman for their excellent technical assistance. This work was partially supported by the National Science Foundation, under grant DMR 9115308, and by E.I. DuPont de Nemours and Company.

REFERENCES

1. D. Blundell and B.Osborn, Polymer, **24**, p. 953 (1983).
2. S. Hudson, D. Davis and A. Lovinger, Macromolecules, **25**, p. 1759 (1992).
3. J Pratt, W. Krueger and I. Chang, SAMPE Int. Symp. **34**, p. 2229 (1989).
4. I. Chang, SAMPE Q. **19**, p. 29 (1988).
5. K. Gardner, B. S. Hsiao, R. Matheson and B. Wood, Polymer, **33**, p. 2483 (1992).
6. B. S. Hsiao, K. Gardner, D. Wu and B. Chu, Polymer, **34**, p. 3986 (1993).
7. P. Damman, C. Fougnies, J. Moulin and M. Dosiere, Macromolecules, **27**, p.1582 (1994).
8. D. C. Bassett, R. H. Olley and A. Raheil, Polymer, **29**, p.1745 (1988).
9. H. Marand and A. Prasad, Macromolecules, **25**, p.1731 (1992).
10. B. S. Hsiao, K. Gardner and S. Z. D. Cheng, J. Polym. Sci. Polym. Phys. **32**, p.2585 (1994).
11. R. Ho, S. Z. D. Cheng, B. S. Hsiao and K. Gardner, Macromolecules, **27**, 2136 (1994).
12. W. Wang, J. M. Schultz and B. S. Hsiao, J. Polym. Sci. Polym Phys. in press (1996).
13. C. Vonk, J. Appl. Cryst. **6**, p. 81 (1973).
14. A. Jonas, T. Russell and D. Yoon, Macromolecules, **28**, p.8491 (1995).

MECHANICAL ALLOYING OF PET AND PET/VECTRA BLENDS

C. M. BALIK*, C. BAI*, C. C. KOCH*, R. J. SPONTAK*, C. K. SAW**
*Dept. of Materials Science and Engineering, Box 7907, North Carolina State University, Raleigh, NC 27695.
**Hoechst-Celanese Research Division, 86 Morris Avenue, Summit, NJ 07901

ABSTRACT

Mechanical alloying represents a potential method for producing finely dispersed alloys of normally incompatible polymers. In this paper, PET and blends of PET with a Vectra thermotropic copolyester have been processed via high energy ball milling at room temperature (ambimilled) and at liquid nitrogen temperatures (cryomilled). Milled powders and compacted disks have been characterized using molecular weight, density and hardness measurements, as well as DSC, WAXS, TEM and FTIR.

INTRODUCTION

Mechanical attrition — high energy ball milling of powders — is a non-equilibrium processing method that has generated great interest in recent years as a viable route to the physical synthesis of metastable inorganic materials[1]. It is convenient to divide the processes induced by ball milling into (1) "mechanical alloying" (MA), in which dissimilar elements or compounds can be alloyed at the atomic or molecular level, and (2) "mechanical milling" (MM), in which the structure/microstructure of single-component powders (e.g., intermetallic compounds) can be transformed by defects introduced through milling-induced deformation. Mechanical alloying was first developed by Benjamin and co-workers[2] in the late 1960's to produce complex oxide dispersion-strengthened (ODS) alloys. The discovery[3] in 1983 that solid-state amorphization could be achieved by MA stimulated the use of mechanical attrition as a non-equilibrium processing tool to produce metastable structures such as amorphous alloys, extended solid solutions, metastable compounds, nanocrystalline materials and quasicrystalline materials[1].

Until recently, mechanical attrition has been applied only to inorganic materials such as metals, metalloids, intermetallic compounds and ceramics. Shaw and co-workers[4-10] were the first to systematically apply mechanical attrition to polymeric materials. They studied both the effects of MM on the structure and properties of individual polymers, such as polyethylene (PE), polypropylene (PP) and a polyamide (PA), as well as the MA of mixtures of PE and PA, and PA with an acrylonitrile-butadiene-styrene (ABS) terpolymer. Both MM and MA were conducted at a milling temperature ($-150°C$) below the glass transition temperature (T_g) of the polymer components. Consolidation by hot pressing to theoretical density was accomplished at temperatures lower than possible for unmilled material. A reduction in the size scale of phase separation in binary systems was observed, along with modest improvements in mechanical properties such as hardness and ultimate compressive strength[4]. Since the early work of Shaw, several other MM and MA studies involving polymers have been reported[11-14].

For inorganic materials, mechanical attrition involves the repeated fracture and cold-welding of powder particles formed by the impact of the milling balls. The mechanism of MA of polymers is not currently understood, but may involve a complex combination of chemical reactions and physical changes that yield metastable (micro)structures. Repeated shearing and fracturing of particles during milling can induce mechanical scission of polymer chains or hydrogen abstraction, thereby creating reactive free radicals on particle surfaces and providing an opportunity for chemical bonding between particles of dissimilar polymers. Alternatively, MA may result in physical refinement of the microstructure to a scale smaller than that achievable through conventional melt processing. In this study, we present results for MM of poly(ethylene terephthalate) (PET), a thermoplastic, and MA of PET with Vectra, a thermotropic copolyester.

Mat. Res. Soc. Symp. Proc. Vol. 461 ©1997 Materials Research Society

EXPERIMENTAL

PET and Vectra (75/25 oxybenzoate/oxynaphthoate) polymers were provided by the Hoechst-Celanese Corp. (HCC). Two samples of PET were studied, one with an initially high degree of crystallinity (47%), and one with an initially low degree of crystallinity (4%). Milling was conducted in a SPEX 8000 mill by loading a vial with 4 g of polymer and 40 g of steel balls (10 balls of 7.9 mm diameter and 20 balls of 6.4 mm diameter) per vial. Samples were milled at room temperature (ambimilled) and at liquid nitrogen temperature (cryomilled) under an argon atmosphere.

DSC thermograms were obtained on a DuPont 910 at 5°C/min with an argon purge. The degree of crystallinity before scanning was calculated by subtracting the heat of crystallization from the heat of fusion and dividing by the heat of fusion for 100% crystalline PET (160 J/g)[15]. PET molecular weights were obtained from dilute solution viscosity measurements in o-chlorophenol at 25°C. As-prepared solutions were cloudy and were subsequently filtered with 0.22 μm Millipore filters to remove fine particulates. The Mark-Houwink constants[16,17] were K = 3.0 x 10^{-4} and a = 0.77.

Milled and unmilled powders were pressed into KBr pellets for infrared analysis with an Analect FX-6260 FTIR spectrometer. Milled powders were also compacted into disks approximately 13 mm in diameter and 5-8 mm high under pressure (470 MPa) at temperatures below the melting points of both polymers. Density and hardness measurements (Knoop, Rockwell M, L)[18,19] were performed on the disks. SEM micrographs were obtained with a JEOL 6400 field emission SEM in secondary and backscattered electron modes. Milled powders were pressed into disks at 200°C and into films at 300°C for cryo-microtomy at –100°C. Microtomed samples were then stained with RuO_4 (aq) vapor and imaged with a Zeiss EM902 energy-filtering TEM. Wide-angle x-ray scattering (WAXS) was performed at HCC using a Philips automatic powder diffractometer with a line source, and also at NCSU with a Rigaku Geigerflex diffractometer equipped with a graphite monochromator. Cu-Kα radiation was used in both cases.

RESULTS

<u>MM of PET</u>

The molecular weight of PET decreases for samples milled 16 hrs, as shown in Table 1. The decrease was more significant for the ambimilled sample than for the cryomilled sample. Gel permeation chromatography analysis of these samples reveals the same trend, with no obvious changes in the breadth of the molecular weight distribution.

Table 1. Effect of milling temperature on PET molecular weight.

PET sample	Intrinsic viscosity* (dl/g)	M_v (g/mol)	Percent change
As received	1.138	44,500	---
Cryomilled 16 hrs	1.015	38,300	–13.8%
Ambimilled 16 hrs	0.701	23,700	– 46.7%

*measured in o-chlorophenol at 25°C

The lower molecular weight of cryomilled PET compared to ambimilled PET is somewhat unexpected, as higher temperatures are more likely to favor chain pullout during particle fracture rather than chain rupture. If any local heating occurs, the temperature during ambimilling may even exceed the T_g of PET (T_g = 70°C). In a study[20,21] of the effect of repeatedly fracturing polystyrene (by microtoming) on the number of chain scission events per unit area (N_f), it was found that increasing the slicing temperature to near T_g reduced N_f significantly; $i.e.$ the molecular weight was less affected at temperatures near T_g. Thus, some other mechanism of chain scission must be operative during MM of PET. Sol/gel measurements for MM PET show that more than 99% of the milled powder is soluble in o-chlorophenol. The small insoluble fraction has been determined to be an inorganic additive. Thus, MM does not induce measurable crosslinking of PET.

Changes in crystallinity that occur during MM of PET are summarized in Fig. 1. DSC is the primary technique for obtaining crystallinity data, although some data have also been collected by WAXS. While a variety of samples have been analyzed, some milled at NCSU and some milled at HCC, all display similar behavior with milling time. The degree of crystallinity increases for PET samples with an initially low crystallinity, and tends to decrease for samples with an initially high crystallinity. The change is very rapid in both cases, occurring within the first hour of milling. Most of the samples tend to fall within a band of crystallinities ranging from about 26–36%, for milling times from 1 to 50 hrs. The effect of milling time on crystallinity for the low crystallinity samples appears very similar to that of annealing the sample at 190°C. Also shown in Fig. 1 are crystallinity data for PET ambimilled with Vectra, which are comparable to MM PET.

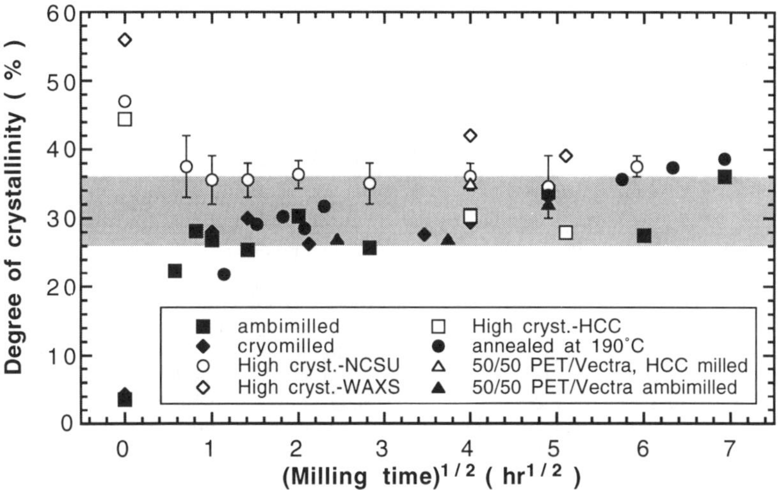

Figure 1. PET crystallinities vs. milling time for initially high- and low-crystallinity starting materials. Open symbols: samples milled at HCC, filled symbols: samples milled at NCSU.

Crystallization (T_c) and melting (T_m) temperatures are also obtained from DSC thermograms. Peak crystallization temperatures for both ambimilled and cryomilled samples decrease from about 130°C to 100°C after 2 hrs of milling for low crystallinity PET. The T_c for high crystallinity PET is observed to decrease to 100°C after 20 min of ambimilling and remain constant thereafter. Peak melting temperatures increase slightly for ambimilled samples, whereas cryomilled samples exhibit a relatively constant melting temperature with milling time.

WAXS data from milled PET provide additional insight into the effects of MM on PET at the molecular scale. Figure 2 shows WAXS data from the high crystallinity PET sample before and after ambimilling for up to 35 hrs. During the first half-hour of milling the (010) peak disappears and the (100) peak remains. These peaks correspond to intermolecular packing distances perpendicular to the chain axis. The PET unit cell is triclinic with the phenyl rings stacked parallel to each other. The structure can be described in terms of "sheets" of molecules in which the plane of the phenyl rings is nearly parallel to the sheet. These sheets are then stacked on top of each other to complete the structure. The (010) peak is related to the distance between phenyl rings within a sheet, whereas the (100) peak is related to the distance between adjacent sheets. During the early stages of milling, the order within the sheets is apparently destroyed, while the stacked parallel alignment for phenyl rings in adjacent PET molecules persists. After 35 hrs of milling, this ordering is also lost, and the molecules become rotationally disordered about the chain axis.

The (100) and (010) peaks are not visible in the WAXS spectrum of low crystallinity PET before milling (Fig. 3). However, the intensity at the position of the (100) peak increases after 0.5 hr of milling and remains constant thereafter, indicating that some chain alignment takes place during MM of this sample as well. WAXS results from cryomilled low crystallinity PET do not reveal these changes, suggesting that a certain amount of thermal energy is required for this milling-induced chain realignment to occur.

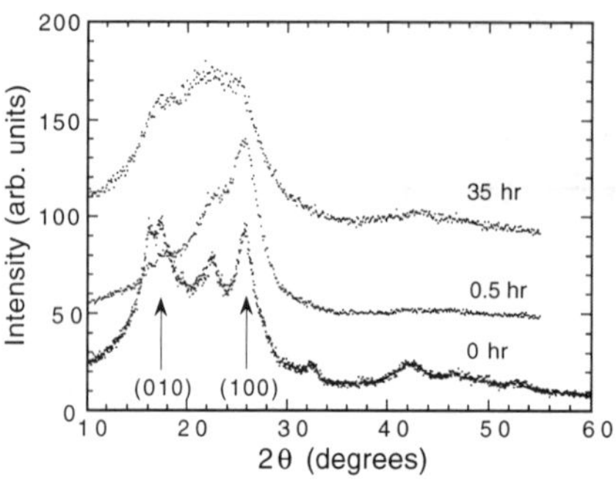

Figure 2. WAXS of high crystallinity PET ambimilled at HCC for various times.

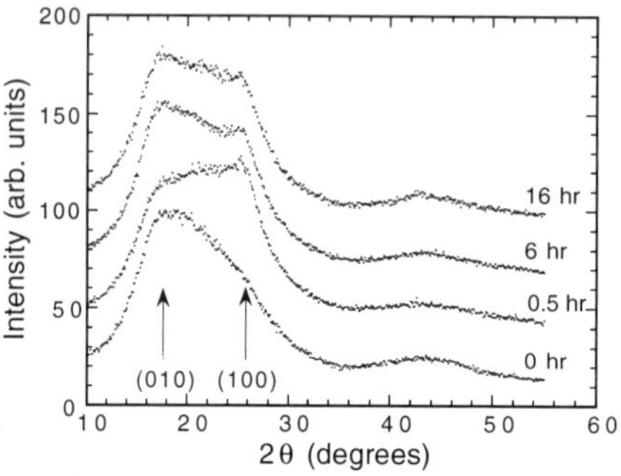

Figure 3. WAXS of low crystallinity PET ambimilled at NCSU for various times.

Particle size distributions have been obtained from SEM micrographs of PET milled for various times. The average particle size (Fig. 4) initially increases with milling time as rounded particles are flattened into flakes. After about 0.5 hr of milling, the particle size decreases monotonically with milling time and levels out at about 25 μm. The initial flattening of the particles produces biaxial stresses that can orient the PET molecules and cause stress-induced crystallization[22,23]. This could be responsible for the observed increase in crystallinity during the early stages of milling the low-crystallinity PET discussed earlier.

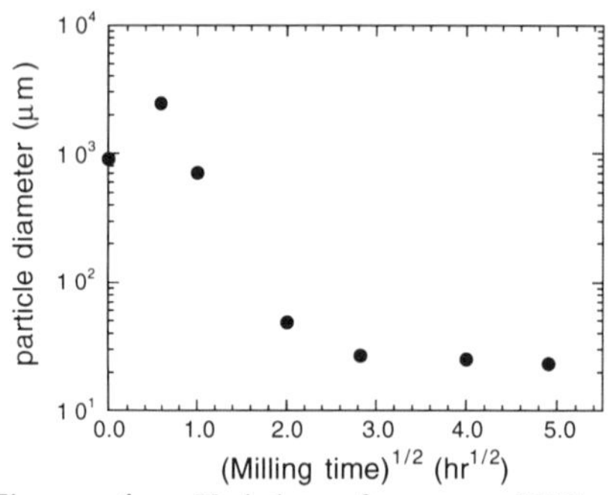

Figure 4. Variation of average PET particle size with milling time.

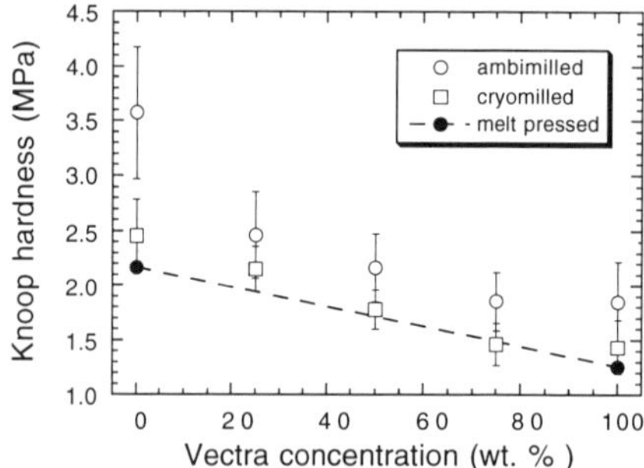

Figure 5. Knoop microhardness for compacted disks of MA PET/Vectra. The dashed line represents the rule of mixtures.

<u>MM of Vectra</u>

The properties of Vectra make the characterization of milling effects more difficult than for PET. Vectra does not readily dissolve in any solvent at room temperature, thereby precluding molecular weight and sol/gel analyses. Vectra is a nematic thermotropic copolymer with crystallites that melt at 283°C for the grade employed. Crystallinity changes that occur during MM of Vectra are complicated by the potential presence of liquid crystalline order in addition to conventional crystallinity. Upon melting, Vectra produces a weak endotherm (1.9 J/g, compared to 30–75 J/g for PET) which decreases to 1.4 J/g after 16 hrs of ambimilling. The melting temperature of Vectra is also observed to decrease to about 274°C after 16 hrs of ambimilling.

<u>MA of PET and Vectra</u>

Low-crystallinity PET and Vectra have been mixed in weight ratios of 25/75, 50/50 and 75/25 for MA studies. MA powders have been pressed into disks under a pressure of 470 MPa at temperatures from 125–225°C, and have also been melt-pressed into films. Knoop microhardness values for the disks do not depend strongly on compaction time from 5–50 min, but they increase with increasing compaction temperature up to about 160°C, then remain constant. Knoop microhardness values for the alloys are consistent with the rule of mixtures (within data scatter; see Fig. 5). The average Knoop values are consistently somewhat higher for the ambimilled samples compared to the cryomilled samples. Interestingly, significant increases in Knoop hardness are noted for ambimilled samples of pure PET and Vectra. The Rockwell L and M hardness values also follow the rule of mixtures with much less variation among the ambimilled and cryomilled samples.

The density of PET/Vectra alloys does not vary much with composition since the densities of the pure polymers are nearly identical (about 1.41 g/cm^3). Compacted disks possess densities between 95–100% of the expected values from the rule of mixtures. DSC results from MA PET/Vectra exhibited essentially the same features as pure PET since the magnitude of the thermal changes occuring in Vectra are relatively insignificant compared with PET. No chemical changes in the milled samples of PET, Vectra or their alloys have been detected by FTIR.

(a)

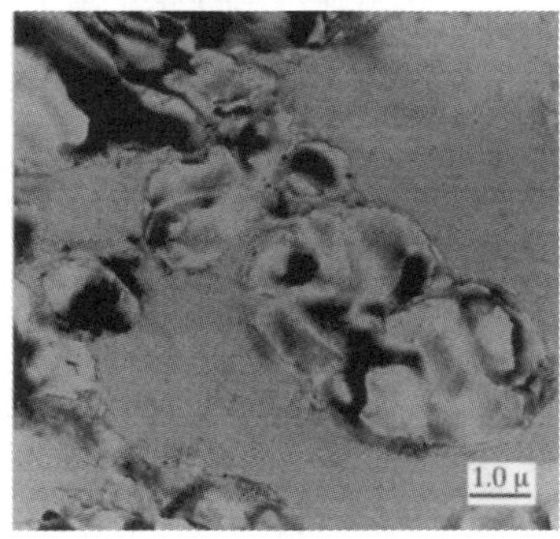

(b)

Figure 6. TEM micrographs from a MA 75/25 PET/Vectra blend cryo-microtomed and stained with RuO_4. (a) Sample from a compacted disk and (b) from a melt-pressed film.

Shown in Fig. 6 are TEM micrographs from a cryomilled 75/25 PET/Vectra blend microtomed from a compacted disk (Fig. 6a) and a melted film (Fig. 6b). As seen in Fig. 6a, there exists little phase contrast between PET and Vectra (which should be enhanced by RuO_4 staining of PET) due to poor specimen cohesion during ultramicrotomy. Conversely, the image in Fig. 6b clearly shows Vectra dispersions within the PET matrix. Such dispersions vary considerably in size, ranging from submicron-sized individual particles to micron-sized aggregates.

CONCLUSIONS

MM of PET causes a significant reduction in molecular weight, more so for ambimilled than for cryomilled samples. The crystallinity of MM PET falls in a range of 26–36% for milling times greater than 1 hr, regardless of the initial degree of crystallinity. WAXS reveals that during the early stages of ambimilling, PET molecules become aligned with their phenyl rings in a stacked parallel arrangement, but such stacking is lost upon further milling and the molecules become rotationally disordered about the chain axis. The mean PET particle diameter initially increases during the first half-hour of milling as rounded particles flatten into flakes, then the diameter decreases, and finally levels out at about 25 μm. Hardness and density values for PET/Vectra alloys have been found to follow the rule of mixtures. TEM micrographs from MA PET/Vectra indicate that micron- to submicron-scale mixing has been achieved.

ACKNOWLEDGEMENTS

This work is supported by the Hoechst-Celanese Corp. and the Kenan Foundation at NCSU.

REFERENCES

1. C.C. Koch, in *Materials Science and Technology,* Volume 15 (R.W. Cahn, ed.) VCH, Weinheim, 1991, p. 583.
2. J.S. Benjamin, *Metall. Trans.,* **1**, 2943 (1970).
3. C.C. Koch, O.B. Cavin, C.G. McKamey, and J.O. Scarbrough, *Appl. Phys. Lett.* **43**, 1017 (1983).
4. W.J.D. Shaw, J. Pan and M.A. Growler, in *Proc. 2nd Internat. Conf. Struct. Appl. Mech. Alloying* (J.J. De Barbadillo, F.H. Fores and R. Schwarz, eds.) ASM International, Materials Park, OH, 1993, p. 431.
5. J. Pan and W.J.D. Shaw, in *Proc. 24th Internat. SAMPE Conf.* (T.S. Reinhart, M. Rosenow, R.A. Cull and E. Struckholt, eds.) 24, T762 (1992).
6. J. Pan and W.J.D. Shaw, *Microstruct. Sci.* **20**, 351 (1993).
7. J. Pan and W.J.D. Shaw, *Microstruct. Sci.* **19**, 659 (1992).
8. J. Pan and W.J.D. Shaw, *J. Appl. Polym. Sci.* **52**, 507 (1994).
9. W.J.D. Shaw and M.A. Gowler, , in *Proc. 1st Internat. Conf. Mater. Prop.* (H. Henein and T. Oks, eds.) TMS, Warrendale, PA, 1993, p. 687.
10. J. Pan and W.J.D. Shaw, *Microstruct. Sci.* **21**, 95 (1994).
11. A. Ikekawa, Int. *J. Mechanochem. Mech. Alloying* **1**, 42 (1994).
12. H.L. Castricum, H. Yang, H. Bakker and J.H.Van Deursen, ISMANAM '96, Rome, May 20–24, 1996 to be published in the conference proceedings.
13. T. Ishida, *J. Mater. Sci. Lett.* **13**, 623 (1994).
14. P. Farrell, R.G. Kander and A.O. Aning, *J. Mater. Sym. Proc.* (submitted).
15. A. Mehta et al., *J. of Polymer Sci., Polymer Physics Edition* **16**, 289-296 1978.
16. I. W. Ward, *Nature* **180**, 141-142 1957.
17. M.L. Wallach, *Die Makromolekulare Chemie* **103**, 19-26 1967.
18. J. Lopez, *Polymer Testing* **12**, 437-458 1993.
19. Metals Handbook, 9th edition, Vol. 8, *Mechanical Testing-Hardness Testing.*
20. J.L. Willett, *J. Polym. Sci., Polym. Phys. Ed.* **24**, 2583 (1986).
21. P. Wool, and A.T. Rockhill, *J. Macromol. Sci.-Phys.* **B20**, 85 (1981).
22. A. Misra and R.S. Stein, *J. Polym. Sci., Polym. Phys. Ed.* **17**, 235-257 (1979).
23. S.A. Jabarin, *Polym. Eng. Sci.* **32**, 1341 (1992).

MELT-PROCESSABLE BLOCK COPOLYIMIDES

Craig L. Hunt*, Terre L. VanKirk*, Sai R. Kumar**
*New Business R&D, Naperville, IL 60566
**Polymers R&D, Alpharetta, GA 30202

ABSTRACT

We present here our probing research with segmented block copolyimdes derived on the basis of certain design principles to optimize performance and processability. We have shown that incorporation of relatively short blocks of rigid mesogenic segments along the backbone of flexible polyimide chains produce thermoprocessable polymeric systems which exhibit self-reinforcing characteristics.

INTRODUCTION

Polyimides have found well established, although limited, commercial applications because of their high performance at elevated temperatures. Processing polyimides into conventional plastic forms has remained a highly desirable goal, but extremely difficult to accomplish owing to their limited tractability. Vespel®, a commercial polyimide derived from pyromellitic dianhydride (PMDA) and oxydianiline (ODA), for example, is fabricated into certain shapes and profiles by special proprietary processing techniques that do not invole conventional melting. Even with polyimide systems capable of exhibiting tractable melting points, the melt viscosity encountered is prohibitively high to permit conventional processing. Flow enhancements have been observed in thermoplastics upon addition of small amounts of thermotropic liquid crystalline polymers[1,2]. However, such an approach for polyimides is limited to systems exhibiting tractable melting points. Moreover, inevitable phase separation between the constituent components leading to undesirable surface appearance, delamination and loss of mechanical properties in fabricated parts has precluded the commercial utilization of this approach for polyimides. Other approaches to improve the flow characteristics such as lubrication, blending, copolymerization and endcapping to control and limit the molecular mass of the polymer have typically not produced commercially successful polyimides.

We have conducted probing research with polyimides which indicates that it is feasible to realize melt processability without sacrificing their superior thermomechanical performance. We propose to do this through incorporation of short blocks of rigid mesogenic segments along the backbone of flexible polyimide chains to produce thermoprocessable polymeric systems. Owing to their inherently self-reinforcing nature of these segmented block copolymers (SBCs), we call them self-reinforcing polymer composites (SRPCs).

Several types of polymeric materials have been described variously as self-reinforced polymers in the literature: liquid crystalline polymers, semicrystalline polymers, block copolymers and even polyblends [1,3,4]. In contrast with these materials, we define SRPC as a single component SBC composed of catenated rigid and flexible polymeric segments, in which the rigid segments spontaneously organize themselves into crystalline or liquid crystalline microphases. It is important to note that the self-reinforcement occurs in these materials because of rigid segment organization and that the phase separation is limited to submicroscopic levels, corresponding to length scales of 10-200 nm. In certain cases, we may encounter spherulitic superstructural organization, but this is not to be interpreted as macroscopic phase separation, as such spherulites contain both crystalline lamellae (composed of rigid segments) and amorphous phase consisting of both type of segments. The phase separation in the amorphous phase, if any, is necessarily restricted to submicroscopic levels. Although in principle, SRPCs may be constructed with rigid and flexible segments, both of which are capable of spontaneous organization, we restrict our discussions here to SRPCs containing amorphous flexible segments.

Since melt processability is an important target property for a high performance polymeric material, we need to optimize the chain architecture and the morphology in order to accomplish superior thermomechanical performance and melt processability for the same material. We show experimental evidence in this report that a carefully designed SBC composed of concatenated sequences of rigid and flexible moieties is capapble of exhibiting respectable levels of self-reinforcement in the form of spun cast films and molded parts.

Design Principles of Segmented Block Copolymers

Since the ultimate thermomechanical performance of a SBC is governed by the characteristics of the rigid blocks, the molecular structure and the architecture of the rigid constituent should be judicously chosen. In other words, the chemical structure should be stable and rigid in the performance region of interest. The reinforcing characteristics of the rigid moiety is dependent on its aspect ratio, which corresponds to the rigid block length, for a segment with uniform width. For a copolymer obtained through equilibrium statistical reaction of monomers, the average rigid block length is governed by its molar composition. Thus we impart self-reinforcement characteristics to the rigid-flexible SBC by controlling the molar composition of the rigid constituent. Following Flory's arguments [5], we stipulate that the molar composition of the rigid constituent must exceed 80% for them to crystallize and act as reinforcement. However, in order to avoid complications arising due to macroscopic phase separation (as in the case of alloys/blends of polymers), the length of rigid sequences should not be too high, leading to undesirable aggregation/phase separation.

It is important to note that the overall performance of the SBC is critically dependent on the nature of both constituents. In particular, certain important bulk material characteristics such as solubility, deformation and rheology (processability) are sensitive to the nature of the major constituent in the SBC. Therefore, in order to get a processable SBC, the composition of the flexible component is adjusted such that it constitutes the continuous phase in which rigid reinforcing segments are distributed as the minor component. The rigid segments may exhibit crystalline or liquid crystalline order, but such an organization and hence the phase separation is necessarily restricted to submicroscopic levels. This type of microphase separation is a common feature encountered in many block copolymers [6]. The microphase separated morphology is thus obtained by selecting a compositional range corresponding to the flexible component exceeding 70% by weight (or volume).

RESULTS AND DISCUSSION

Evidence of Self-Reinforcement: Importance of Molecular Architecture

In order to demonstrate that SBCs constructed in accordance with the design principles discussed above possess self-reinforcing characteristics, we chose two polymeric systems to serve as models: the flexible portion of both copolymers was derived from a telechelic polyimideamine of desired molecular mass, synthesized from the condensation reaction of oxydiphthalic anhydride (OPAN) with calculated excess of bis(aminophenoxy phenyl propane) (BAPP). The rigid segment of the copolyimide system was derived from pyromellitic dianhydride (PMDA) and p-phenylene diamine (pPDA), whereas the copolyesterimide system consisted of rigid segments derived from p-phenylene bis(trimellitate anhydride) (PBTMA) and pPDA. The synthetic procedure used to obtain these SBCs is given in the experimental section.

We chose the monomers OPAN and BAPP for the flexible segment construction because of the tractable nature of the polyimide derived from them, i.e., solubility in N-methyl pyrolidinone (NMP) (at all molecular weights of interest) and the amorphous nature, with a glass transition temperature of 210-240°C. Neither of the rigid segments is tractable in the homopolymer form, as a consequence of which we found that the SBC in the imidized form was insoluble in the reaction medium. However, we found that test parts could be fabricated out of these SBCs through a combination of injection and compression molding at the conditions described in the experimental section. The film properties described below were obtained from free standing films fabricated from the intermediate polyamideacid solutions as described below in the experimental section.

We present experimental evidence in this section that the molecular architecture of catenated sequences of rigid and flexible segments is a necessary criterion for the material to show self-reinforcement characteristics. We illustrate this in Table I, where the tensile properties of films are compared. 16170-52 is the flexible homopolyimide obtained by stoichiometric reaction between OPAN and BAPP. The SBC, 16170-41B, was obtained through the one-pot-two-step procedure described in the experimental section, and corresponds to a rigid phase composition of 27.1 wt% (80 mol%). The random copolyimide, 16170-98B, was obtained through a single step synthetic procedure at the same composition.

TABLE I. Comparison of Properties of Copolyimide Films

Sample ID	Composition*		Tensile Properties			Comments
	Flexible Block	Rigid Block	Modulus (kpsi)	Strength (kpsi)	Elongation (%)	
16170-52	OPAN, BAPP	--	342.4	16.7	13	Homopolymer
16170-41B	OPAN, BAPP	PBTMA, p-PDA	572.8	23.2	15	SBC
16170-98B	OPAN, BAPP PBTMA, p-PDA	--	388.7	21.1	6.7	Random Copolymer

*OPAN: oxybis(phthalic anhydride); BAPP: bis(aminophenoxyphenyl) propane; p-PDA: p-phenylene diamine; PBTMA: p-phenylene bis(trimellitate anhydride)

Results shown in Table I clearly manifest the superior properties of the SBC compared to those of the flexible hompolymer and the random copolymer. Since the monomer feed composition was the same for both copolymers, the difference in performance between them is due to the segmented molecular architecture in the case of 16170-41B.

Superior thermomechanical performance of the SBC is illustrated in the dynamical mechanical thermal analysis (DMTA) curves shown in Figure 1. The storage modulus of the homopolymer falls by more than three orders of magnitude accompanying the glass transition temperature (236°C), whereas the drop in the storage modulus for the SBC is only by a factor 20 over the same temperature range. To faciliate comparison, the measured value of the storage modulus at different temperatures for the samples is shown in Table II. It is interesting to note that although there is a pronounced glass transition for both the materials (around 235°C) the retention of stiffness even at 300°C for the SBC is because of the microphase separated morphology present in it. The behavior of the random copolymer further lends support to the importance of optimizing the molecular architecture. The random copolymer films were tested three times to verify that they fail around 200°C, as shown in Figure 1.

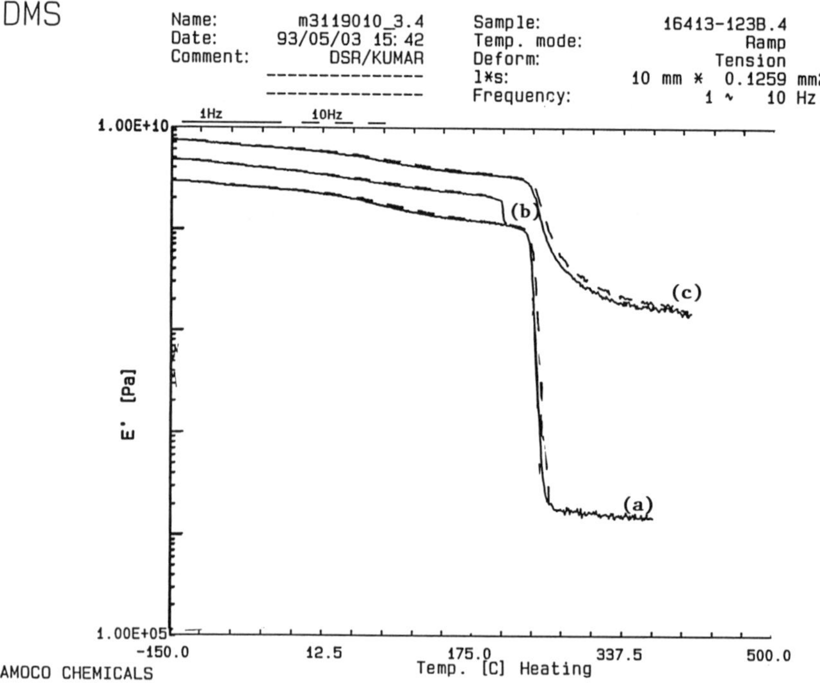

Figure 1. Comparison of DMTA Curves for (a) Homopolymer; (b) Random Copolymer, and (c) Segmented Block Copolymer

TABLE II. Thermomechanical Data Extracted from Figure 1

SAMPLES	STORAGE MODULUS, MPa*				T_g (°C)
	25°C	200°C	250°C	300°C	
HOMO	2200 (319)	1200 (174)	14 (2)	9 (1.3)	236
RANDOM	2700 (389)	FAILED	---	---	276[A]
BLOCK	5700 (826)	3500 (507)	1200 (174)	280 (40)	234

[A] by DSC   * Values in parentheses are in kpsi.

In Table III we show certain advantageous properties (the coefficient of linear thermal expansion (CLTE), moisture absorption and the chemical resistance) of the SBCs compared to those of homo and random copolymers, which once again manifest the importance of controlling the molecular architecture.

TABLE III. Comparison of Physical Properties of Polyimides

SAMPLE ID*	[a]CLTE (PPM/K)	WATER ABSORPTION (%)	TENSILE STRENGTH (kpsi)				POLYMER TYPE
			CONTROL	10% NaOH	GLACIAL ACETIC ACID	TOLUENE	
16170-137	44	1.0	12.7	12.8	9.5	9.2	HOMO
16170-138A	25	1.0	15.2	14.7	17.7	14.9	RANDOM
16170-138B	42	1.1	11.5	9.9	13.9	9.5	RANDOM
16170-139A	9	0.5	18.4	17.1	17.1	15.3	SEGMENTED
16170-139B	N.A.	1.1	20.6	19.4	18.1	18.5	SEGMENTED

[a] Measured by TMA, at 200°C   *-138A and -139A are copolyimides; -138B and -139B are copolyesterimides

The Morphology of SBCs

Several segmented block copolyimides and copolyesterimides (not discussed here) were prepared and tested in the form of films and compression-injection molded parts. The spun cast films were all optically clear, with no evidence of preferred molecular orientation (in terms of measurable birefringence) under optical microscopic examination. Since the materials show one glass transition temperature characteristic of the flexible component, a likely explanation for their thermomechanical and molding behavior is that the SBCs described here exhibit microphase separated morphology, wherein the rigid microdomains act as reinforcement. Since these domains are of submicroscopic size (below 0.5 μm), we had to examine the morphology by x-ray diffraction or other high resolution microscopic techniques. We discuss below the results from the wide-angle x-ray scattering (WAX) analysis to investigate the structural details of the domains in the SBCs in the form of films and compression molded parts.

The two dimensional WAXS patterns for the samples are shown in Figure 2 for the films and in Figure 3 for the molded parts. The results from WAX analysis of the samples are also summarized in Table IV. We infer the following from these results: (i) All the specimens are markedly crystalline. Since the OPAN-BAPP homopolymer was found to be amorphous, we may conclude that the microphase separated domains are crystalline in nature. (ii) No preferred molecular orientation due to processing was detected . (iii) Except for the sample 16170-153B, the SBCs exhibit small angle scattering as well. (iv) The Bragg reflections were considerably sharper in the case of molded parts than those for films, indicating that the domain size in the case of films is considerably smaller (estimated average domain size of 25-30 Å). (v) The estimated degree of crystallinity for the films were higher than those estimated for the molded parts, indicating the difference in crystallization kinetics between the two methods of fabrication.

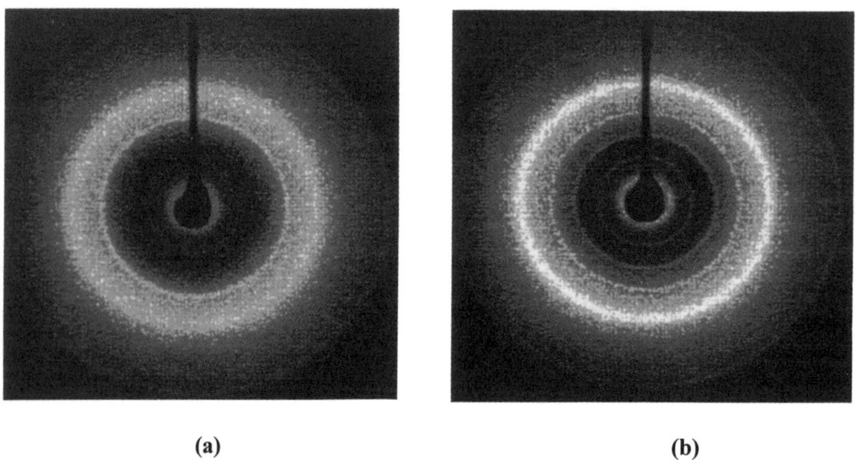

(a)   (b)

Figure 2. 2D Area Detector X-Ray Diffraction Images for the SBC Films:
(a) Copolyimide (16170-41A);
(b) Copolyesterimide (16170-41B)

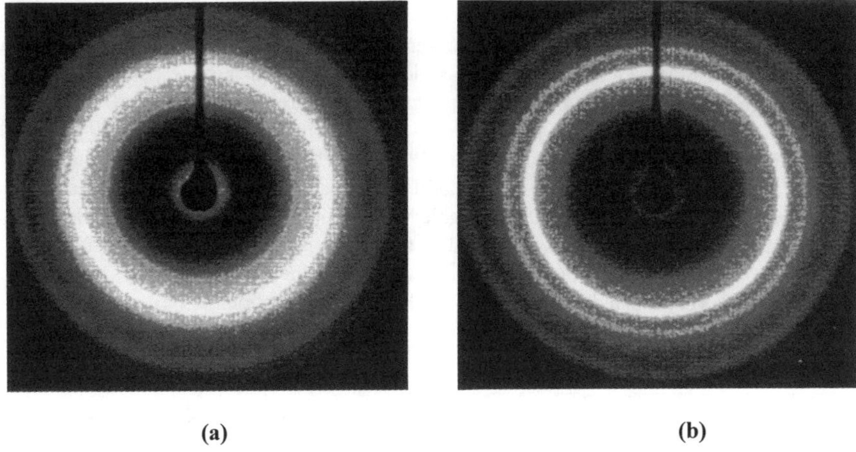

(a)   (b)

**Figure 3. 2D Area Detector X-Ray Diffraction Images for the SBC Molded Parts:
(a) Copolyimide (16170-153A);
(b) Copolyesterimide (16170-153B)**

TABLE IV. WAXS Analytical Results on Segmented Block Copolyimides

SAMPLE ID	COMPOSITION		SAMPLE TYPE*	CRYSTALLINITY (%) [+]
	RIGID	FLEXIBLE		
16170-153.A	PMDA, pPDA	OPAN, BAPP	Plaque	17(2)
16170-41.A	PMDA, pPDA	OPAN, BAPP	Film	47(12)
16170-153.B	PBTMA, pPDA	OPAN, BAPP	Plaque	22(1)
16170-41.B	PBTMA, pPDA	OPAN, BAPP	Film	64(8)
16170-52	---	OPAN, BAPP	Plaque	Amorphous
16170-52	---	OPAN, BAPP	Film	Amorphous

* Compression molded plaques and solution cast films   [+] Values in parentheses are estimated errors.

<u>How Much Reinforcement?</u>

We have shown so far that the two rigid-flexible SBC systems described here have self-reinforcing characteristics and that they exhibit microcrystalline phase separation due to rigid segment molecular organization. We now describe our systematic study of determining how much rigid phase is necessary (in terms of weight or volume) for optimal reinforcement, at a given average rigid block length. To determine this, we needed to synthesize the telechelic flexible prepolymer (i.e., OPAN-BAPP polyimideamine in this case) of varying average molecular weight.

51

We represent in Table V our systematic study of the effect of changing the rigid weight fraction, ω_r, on the tensile properties for the two SBCs under investigation. The mole fraction of the rigid phase in both cases was fixed at 0.85. The properties shown in Table V correspond to compression molded specimens. The glass transition temperature shown in column 4 of Table VI was determined from DSC measurements on as precipitated copolymer at 20°C / min. Since the flexible phase in all cases is the same (OPAN/BAPP), we see that the T_g of the copolymer is apparently the same, irrespective of the composition. The properties of the flexible homopolymer (16170-52S) is also shown in Table V for comparison. We see from Table V that optimal reinforcing characteristics, as indicated by the tensile properties, are realized around $\omega_r = 0.15$ in both cases. It is important to note, however, that the optimal properties achievable for a given SBC is very much dependent upon its chemical constitution.

TABLE V. Composition and Properties of the Segmented Block Copolyimides ($X_r = 0.85$)

[a]Sample ID:	[b]$(Mn)_f$ (g/mole)	[c]W_r	[d]T_g (°C)	Tensile Properties		
				Strength (ksi)	Modulus (ksi)	Elongation (%)
6170-76A	31225	0.05	232	13.7	225.5	24.2
-76B	14790	0.1	233	14.6	267.6	12.3
-76C	9313	0.15	233	15.1	276.4	10.5
-76D	6573	0.2	238	12.3	291.2	5.7
-76E	4930	0.25	237	12.0	307.4	5.2
16170-85B	57064	0.05	236	14.5	243.3	35
-85C	27030	0.1	234	14.6	265.9	11.2
-85D	17019	0.15	233	13.7	275.4	20.7
-85E	12013	0.20	234	6.2	303.4	2.2
-85A	9010	0.25	234	N.A.	N.A.	N.A.
16170-52S	---	---	234	14.3	158.7	49.1

[a] 16170-76 contains PMDA/p-PDA as the rigid segment; 16170-85 contains PBTMA/p-PDA as the rigid segment

[b] Theoretical number average molecular weight of the telechelic polyimide amine formed from OPAN/BAPP, based on stoichiometry.

[c] Weight fraction of the rigid segments.

[d] Glass transition temperature determined at 20°C/min by DSC.

Processability of Segmented Block Copolyimides

We briefly discuss certain issues concerning the melt processability of the SBCs described here. Although we have demonstrated that the SBC compositions can be molded into test parts by compression and minicompression-injection molding processes, we found it necessary to use significantly higher pressures (10,000 psi) than typical thermoplastics, in addition to using higher temperatures (400°C). We found that the copolymers did not exhibit good flow characteristics even at 425°C, and untraditionally high pressures were needed to achieve flow. However, we found that the molded parts can be reground and remolded with apparently no change in physical properties [7].

In order to obtain a versatile SRPC material which can be extruded into films or spun into high performance fibers, it is essential to substantially lower its melt viscosity. Since higher shear sensitivity and lower melt viscosities are particularly suitable for fiber spinning processes, we have to fundamentally incorporate physicochemical characteristics which promote such desirable processing attributes. It is a well documented fact that thermotropic liquid crystalline polymers greatly improve the processability of conventional thermoplastic materials, when blended in small amounts (5 to 10 wt%) with them [1,2]. Since the rigid rodlike segments present in the SBCs described here are mesogenic [8], we may expect improved processing characteristics compared to the homopolyimides. We found qualitative evidence for discernable flow improvement for SBCs composed of mesogenic rigid segments compared to those consisting of isotropic rigid segments (formed from PMDA and ODA, for example) during our compression-injection molding experiments.

Improved rheological behavior for the SBCs can be expected only if the crystalline rigid segments melt at accessible temperatures to form the liquid crystalline phase. As noted earlier, the rigid segments used in our study do not exhibit tractable melting points in the desired temperature range (< 350°C). Therefore, we need to modify the chemical constitution and the copolymer composition such that the rigid segments form microcrystalline domains which can melt to form the microliquid crystalline domains [9] under typical melt processing conditions. This is a prerequisite for achieving both self-reinforcement and processability for a single material.

CONCLUSION

We have demonstrated in our probing research work that the concept of assembling rigid and flexible polymeric segments to obtain SBCs with controlled molecular architecture and microphase separated morphology leads to self-reinforcement and other advantageous properties. The extent of reinforcement achievable is system dependent. Since we have shown evidence for microphase separation with formation of ordered microdomains composed of rigid segments, these copolymers have been classified as SRPCs. Although our discussion in this report is concerned with crystal-amorphous microphase separated SBCs, we believe that the design concepts described here are applicable to the construction of many other useful SRPCs. Applicability of our design approach to other useful SBC systems will be illustrated in a forthcoming publication [10].

EXPERIMENTAL

<u>Materials</u>

All monomers used in various syntheses were obtained in the highest available pure form and were used as such. The aromatic dianhydrides and diamines used in the polymerizations described below were obtained from Criskev Company. 4-chloroformyl phthalic anhydride (4-TMAC) was obtained from Aldrich or Amoco. Solvents used in the polymerization, viz., N-methyl pyrrolidinone (NMP), dimethyl acetamide (DMAc), acetonitrile, tetrahydrofuran (THF), and toluene were obtained in the anhydrous form from Aldrich.

<u>A Two-Step-One-Pot Procedure To Synthesize Polyimide Based SBCs</u>

The apparatus used was a 3-neck round bottom flask, fitted with a nitrogen inlet, mechanical stirring assembly and a Dean-Stark assembly carrying the nitrogen outlet. The apparatus was dried by heating with flowing dry nitrogen to remove moisture, and calculated excess amounts of the diamine (BAPP) and the anhydrous solvents were stirred together to get a solution. Predried dianhydride (OPAN) is then added through a powder funnel, followed by the washings. The amount of the solvent used at this stage was such that the concentration of the solution was 20% wt./vol. The polymerization was carried out usually overnight. The thermal imidization in solution was carried out by azeotropic removal of water in the usual manner.

To the resulting telechelic polyimideamine solution are added additional amounts of solvent and calculated amount of the diamine (p-PDA) to obtain a homogeneous solution. Calculated amount of dianhydride (PBTMA or PMDA) is then added to initiate the copolymerization, which is usually carried out overnight. A small portion of the resulting copolyimideamideacid solution is always used for film casting. The remaining portion is then imidized thermally in solution to get the block copolyimide. The strength of the medium during copolymerization and subsequent imidization required adjustment (varying between 5 and 15% wt./vol. depending upon the composition and the associated pseudogelation problems), in order to get a homogeneous solution prior to complete imidization. We usually observed that the copolyimide precipitated out of the medium during the final stages of the imidization. The resulting tan colored slurry is then treated with at least a 5:1 excess of water and filtered to separate the polymer. The product is then washed thoroughly twice with water and once with methanol and then dried in vacuum at 120°C overnight.

<u>Spin Casting of Polymer Solutions</u>

Intergrated Technologies Inc. model P-6000 spin coating device was modified to allow sweeping of heated nitrogen to facilitate solvent removal after casting. A polished flat glass of 4" diameter was used as the substrate, held by suction. The spinning speed is continuously variable and the conditions used were based on the polymer solution viscosity and the required film thickness. The solution viscosity was measured using a Brookfield viscometer (Model HBT), using a #7 spindle. The workable viscosity is a range from 1.2 to 600 poise, with an optimal range in the vicinity of 200-400 poise. The conditions were chosen so as to get a film thickness around 25µ, which occasionally required multiple depositions depending on the polymer concentration.

Spinning under heated nitrogen atmosphere was continued until a film, dry to the touch (and not leaving a finger print) was obtained. The glass plate is then removed and placed in a nitrogen purged oven, capable of programmed heating. A standard drying/imidization cycle included a program of 150°F for 15 minutes, 200°F for 15 minutes, a ramp to 510°F at 600°F/hour, one hour at 510°F, and another hour at 200°F. The plate was removed at this stage, edges trimmed with a sharp razor blade and placed in a hot water bath, being gently heated. After 15-30 minutes in the hot water bath, the polyimide film gets released from the substrate. The film is then removed, dried and cut into strips of required dimensions for tensile testing.

<u>Molding of Polymers</u>

The dry polymeric resin prepared in the laboratory is usually compression molded first to determine the optimal molding conditions. The imide-based SBCs were molded typically around 400°C with a compression pressure of 5-10 kpsi. The compressed plates were then machined to get tensile and flexural test bars. We then constructed a machine [12], capable of handling small quanitities of resins (about 1 g), at high temperatures (up to 425°C) and high pressures (up to 20 kpsi) by combining compression and injection processes. Tensile bars of required dimensions could directly be obtained by this miniscale molding process.

<u>Testing of Polymers</u>

All data reported here were obtained by standard analytical techniques available through our polymer physics and analytical services. The small angle x-ray scattering analysis was performed at NIST and the optical microscopic analysis, where required, was independently obtained by us.

ACKNOWLEDGMENTS

We would like to acknowledge Amoco Chemical Company for supporting the probing work and for the publication clearance. We thank members of Amoco's analytical shared services for providing x-ray and thermomechanical data included in this report.

REFERENCES

1. W. Brostow, Polymer, **31**, 979 (1990).
2. R.A. Weiss, in "<u>High Modulus Polymers</u>," A.E. Zacharides and R.S. Porter, Dekker, New York, 1988, p.45.
3. D.G. Baird and K.G. Blizard, Polym. Eng. Sci., **27**, 653 (1987); A.I. Isayev and M.J. Modic, Polym. Compos., **8**, 158 (1987).
4. G.T. Pawlikowski, Ann. Rev. Mater. Sci., **21**, 159 (1991).
5. P.J. Flory, J. Chem. Phys., **17**, 223, (1949).
6. A. Noshay and J.E. McGrath, "<u>Block Copolymers: Overview and Critical Survey</u>," Academic, New York, 1977.
7. T.L. VanKirk, Amoco Chemical Lab Notebook # 18365.
8. S.R. Kumar, unpublished work on model compounds.
9. W. Brostow, Polym. Eng. Sci., **28**, 785 (1988).
10. S.R. Kumar, manuscript in preparation.
11. T.L. VanKirk, R.E. Miller and R. Sai Kumar, Amoco Chemical Invention Disclosure, December, 1993.
12. A.P. Melissaris and J.A. Microyannidis, J. Polym. Sci. Part A: Polymer Chemistry, **27**, 245 (1989).

BLOCK COPOLYMER MICRODOMAIN MORPHOLOGIES
BY PHASE DETECTION IMAGING

Ph. LECLÈRE [1], J.M. YU [2], R. LAZZARONI [1], Ph. DUBOIS [2], R. JÉRÔME [2],
and J.L. BRÉDAS [1]

[1] Service de Chimie des Matériaux Nouveaux,
Université de Mons-Hainaut, B-7000 Mons, Belgium.
[2] Center for Education and Research on Macromolecules (CERM),
Université de Liège, Institut de Chimie, Bât B6a, B-4000 Sart-Tilman (Liège), Belgium.

ABSTRACT

Atomic Force Microscopy with Phase Detection Imaging is used to study the surface microdomain morphology of thick (i.e., ca. 2 mm) films of triblock copolymers, such as polymethylmethacrylate - *block* - polybutadiene - *block* - polymethylmethacrylate copolymers prepared by a well-taylored two-step sequential copolymerization promoted by a 1,3-diisopropenylbenzene based difunctional anionic initiator. By means of this new scanning probe microscopy technique, it is shown that the surface exhibits a segregated microphase structure, corresponding to the different types of components predicted theoretically by thermodynamic processes. We investigate the relationships between the size and characteristics of the microdomain structure as a function of the molecular parameters of the constituent polymers. Our data illustrate the interest of Phase Detection Imaging in the elucidation of surface phase separation in block copolymers.

INTRODUCTION

The increasing importance of block copolymers arises mainly from their unique properties in solution and in the solid state, which are a consequence of their molecular structure [1]. In particular, sequences of different chemical compositions are usually incompatible and therefore have a tendency to phase segregate. The microdomain formation process in the solid state is directly related to the specific molecular architecture, (di- triblock copolymers) and the chemical composition (nature of the structural monomeric units) of the copolymers. Since their discovery, block copolymers have been considered for numerous applications for instance in surface modification [2] or impact resistance [3]. Recently, self-assembled ordered structures with periodicities on the nanometer scale have become an important area of study because of their applications in nanolithography and electronic devices [4].

Structural studies of block copolymers have been based mostly on electron microscopy and X-ray scattering. These techniques indicate that segregated microphases can consist of spheres, cylinders, or lamellae. The latter tend to form a regularly-repeating order while the cylinders arrange in a bidimensional hexagonal lattice, and the spheres give rise to cubic lattices. Next to these extensively studied mesomorphic structures [5], more complex morphologies such as a bicontinuous double-diamond arrangements have also been observed experimentally [6] and rationalized theoretically [7].

Mat. Res. Soc. Symp. Proc. Vol. 461 ©1997 Materials Research Society

Progress in the understanding of surface morphology has lagged behind knowledge about the bulk, partly due to a paucity of suitable analysis techniques. Recently, Tapping Mode Atomic Force Microscopy with Phase Detection Imaging (TMAFM-PDI) has become available and promises to lead to significant progress in this field [8]. In Tapping Mode Atomic Force Microscopy (TMAFM), the cantilever is excited into resonance with a piezo driver and it only touches the surface very lightly and very briefly. The interactions between the tip and the sample lead to a modulation of the absolute value of both the resonance frequency and its amplitude; the surface morphology is then monitored by keeping this amplitude or frequency shift at a constant value. In PDI, the phase lag of the cantilever oscillation, relative to the signal sent to the cantilever piezo driver, is monitored simultaneously to the topographic response; this approach provides both the amplitude and the phase difference. The mapping of that phase during the analysis allows one to go beyond simple topographical imaging and to *detect variations in composition.*

EXPERIMENTAL

Polymethylmethacrylate - *b* - polybutadiene - *b* - polymethylmethacrylate triblock copolymers (MBM) were synthesized by a perfectly well-controlled anionic copolymerization using a difunctional anionic initiator (DIBLi$_2$) prepared by addition of *tert*-butyllithium (*t*-BuLi) to 1, 3-diisopropenylbenzene (DIB), as described elsewhere [9].

MBM triblocks with various combinations in composition and narrow molecular weight distribution (M_w/M_n < 1.15)(see Table I for details) were cast into thick films by pouring toluene/THF (40/60) solutions onto leveled glass plates that were kept away from dust as the solvent evaporated (the composition of the solvent mixture was selected on the basis of a systematic study to optimize the evaporation rate). The films were further dried under high vacuum until the samples reached a constant weight. The final film thickness was about 2 mm. As toluene and THF are good solvents for both polymethylmethacrylate (PMMA) and polybutadiene (PBD) and the evaporation rate is slow, solvent effects are not considered to be responsible for the structure of the formed films. This means that the observed microdomain structure can be considered as a thermodynamic equilibrium morphology.

Table I. Characteristics of studied samples.

Sample	M_n[a] (10^{-3})	1,2 PBD(%) [b]	Φ_{PMMA}[c]	M_w/M_n	Morphology[d]
A	50 - 100 - 50	45	0.50	1.15	Lamellae
B	16 - 65 - 16	45	0.33	1.10	Cylinder
C	6 - 85 - 6	45	0.11	1.10	Sphere

(a) Number-average molecular weight as calculated from SEC and ^{1}H-NMR data, assuming a perfect triblock structure; (50-100-10) 10^3 corresponds to a MBM with 100,000 and 50,000 M_n for PBD inner block and PMMA outer sequences, respectively.
(b) Relative content in 1, 2 microstructure, based on ^{1}H-NMR data.
(c) Relative molar composition in PMMA.
(d) Expected morphology from thermodynamic theory and triblock relative compositions [7].

All AFM images were recorded with a Nanoscope III microscope from Digital Instruments Inc. operated in air at room temperature in the Tapping Mode, using the microfabricated cantilevers provided by the manufacturer (spring constant of 30 Nm^{-1}). The system is equipped with the Extender electronics module to provide height and phase cartography. Images of each sample were taken at several locations and the time for scanning was about 5 minutes. All images were made with the maximum available number of pixels (512) in each direction. It is worth pointing out that all the images were culled from several recorded files. For image analysis, the Nanoscope image processing software was used. The images presented here were not filtered and are shown as captured. Repeated scans indicated that the observed structures were stable under the aforementioned experimental conditions.

RESULTS

Figure 1 shows a typical TMAFM height image of the surface of MBM sample A characterized by a 50,000-b-100,000-b-50,000 molecular weight. Extracting information on the microdomain morphology from this picture is not straightforward because the origin of the apparent height is not obvious in TMAFM. If the various surface elements are characterized by identical interactions with the cantilever tip, the measured surface would correspond to the actual topography. However, if the interactions are different, as can be the case for polymers with different elasticity moduli, the relation between the voltage detected by the photodetector and the physical height may not be direct.

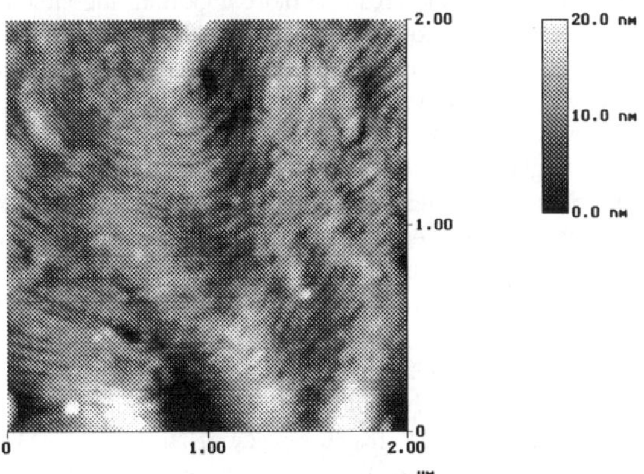

Figure 1. Tapping mode AFM height image of an MBM thick film (Sample A).

In Figure 2, we present a typical result obtained on the same sample when we map the phase difference between the piezo signal driving the oscillating tip and the response of the photodetector. The darker areas correspond to zero phase shift (the tip interacting with the surface remains in-phase with the piezo driver signal) while the white areas represent a 180° phase difference (interaction with the surface makes the tip in phase opposition to the piezo signal). It must be noted that these images were captured <u>simultaneously</u> to the height images (Figure 1). The contrast is significantly increased and the surface morphology appears more clearly. It is thus an important aspect of PDI-TMAFM that it is able to provide in real time the block copolymer morphology with a good contrast between the different phases.

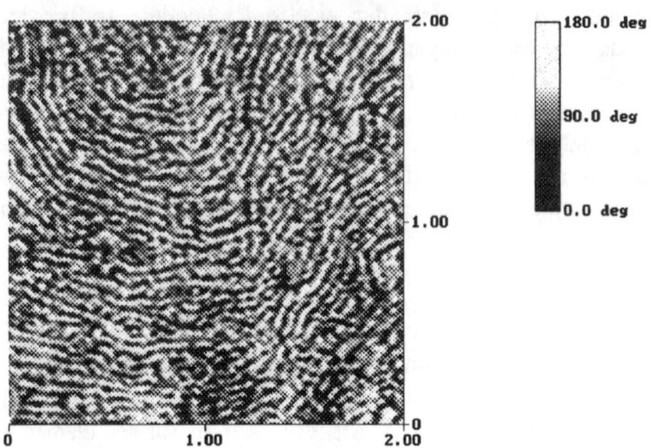

Figure 2. Tapping mode AFM phase image of an MBM thick film (Sample A).

If we consider that the origin of the phase lag is mainly due to the interaction between the sample and the tip, the brighter the area, the more important the interaction. In our view, the differences in viscoelasticity are likely responsible for the observed phase contrast, which therefore reflects the microdomain morphology. A model has been recently proposed, in which that the phase lag is taken to be directly proportional to the elasticity modulus (E) [10]. On this basis, the PBD phase gives rise to the darker areas in the PDI-TMAFM images while the PMMA domains appear brighter. This is consistent with the interpretation deduced from the topographic data and the chemical composition. While it is well-known that various types of tip-sample interactions can affect the AFM signal in general, in our case, we believe that the phase contrast is mainly governed by viscoelastic properties. Note that a similar PDI-TMAFM image obtained on a pure homogeneous polymer surface shows no phase contrast.

As mentioned before, the morphology of block copolymers essentially depends on their composition [1]. In the first system (sample A) considered in this study, the composition ($\Phi_{PMMA} = 0.50$) should correspond to a structure made of lamellae of PMMA and PBD. The average repeat distance between the lamellae was determined from the 2D power spectrum (derived from the Fourier transform) of the image to be about 30 nm. This distance is in agreement with the value deduced from electron microscopy and Small Angle X-Ray Scattering (SAXS) measurements. These SAXS results will be presented in detail elsewhere [11].

Finally, we present in Figure 3, the preliminary results obtained by PDI for the systematic study of microdomain morphology versus the PMMA content. As expected from thermodynamic theories [7], we observe PMMA spheres (diameter of about 25 nm) when $\Phi_{PMMA} = 0.11$ (sample C) and PMMA cylinders (width of about 31 nm) when $\Phi_{PMMA} = 0.33$ (sample B).

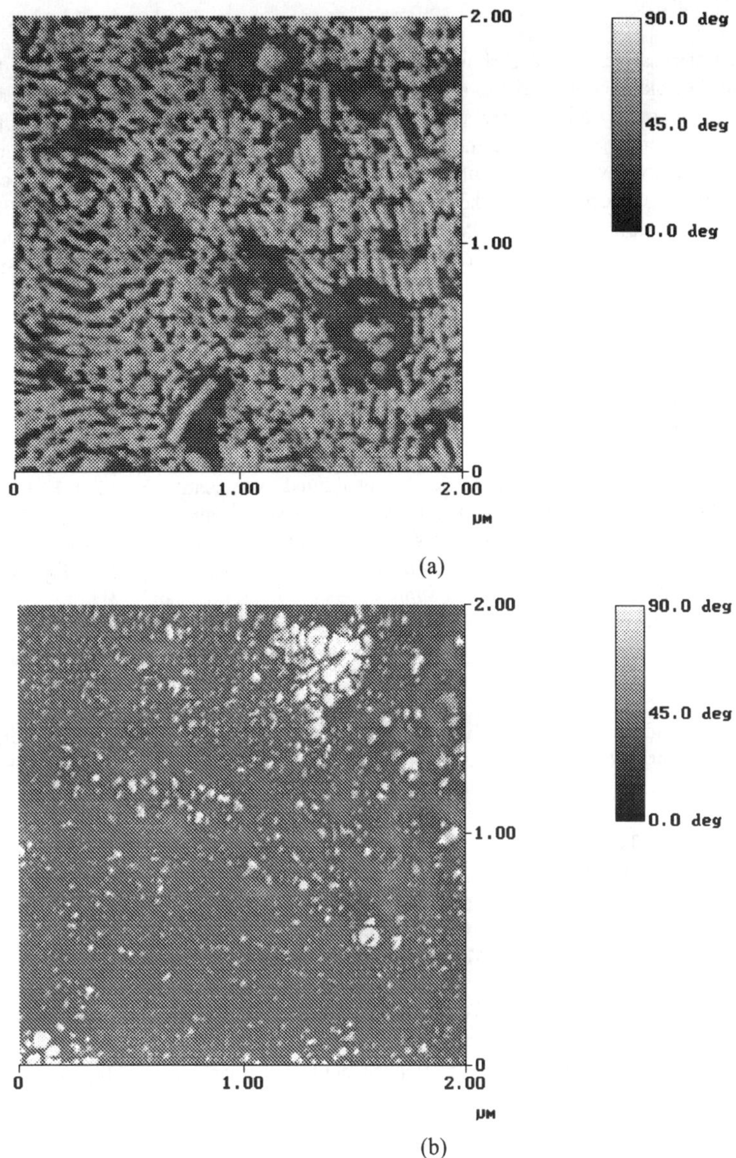

(a)

(b)

Figure 3. Tapping mode AFM phase images of MBM thick film
versus the PMMA content (Φ_{PMMA}).(a) $\Phi_{PMMA} = 0.33$; (b) $\Phi_{PMMA} = 0.11$.

CONCLUSIONS

To summarize, we have shown that the use of phase detection imaging in Tapping Mode AFM of block copolymers significantly improves the quality of the data. This leads to a much better characterization of the surface morphology as obtained on thick layers of MBM triblocks compounds. By mapping the phase difference of the cantilever oscillation during the Tapping Mode scan, we obtained information concerning the viscoelastic properties of those block copolymers. Depending of the value of Φ_{PMMA}, these preliminary PDI-TMAFM data indicate that the hard segments of PMMA form lamellar structure ($\Phi_{PMMA} = 0.50$), cylindrical ($\Phi_{PMMA} = 0.33$) and spherical ($\Phi_{PMMA} = 0.11$) microdomains dispersed in the elastomeric matrix of PBD.

ACKNOWLEDGEMENTS

The research in Mons is supported by the European Commission and Ministère de la Région Wallonne (FEDER-Objectif 1: Projet Mobilisateur *NOMAPOL*), the Belgian Federal Government Office of Science Policy (SSTC) "Pôles d'Attraction Interuniversitaires en Chimie Supramoléculaire et Catalyse", the Belgian National Fund for Scientific Research FNRS/FRFC, and an IBM Academic Joint Study. The research in Liège is supported by the SSTC "Pôles d'Attraction Interuniversitaires: Polymères". The collaboration between Mons and Liège is partly supported by the European Commission (Human Capital and Mobility Network: *Functionalized Materials Organized at Supramolecular Level*). RL and PhD are "chercheurs qualifiés" du Fonds National de la Recherche Scientifique (FNRS - Belgium).

REFERENCES

[1] G. Riess and P. Bahadur in <u>Encyclopedia of Polymer Science and Engineering</u>, edited by H.F. Mark, N.M. Bikales, C.G. Overberger, G. Menges, Wiley, New York, 1989, pp. 379-397.

[2] R.J. Angelo, R.M. Ikeda, and M.L. Wallach, Prep.-Am. Chem. Soc., Div. Org. Coat. Plast. Chem. **34**, 103 (1974).

[3] E. Helfand, in <u>Recent Advances in Polymer Blends, Grafts and Blocks</u>, edited by L. Sperling, Plenum Publishing Corp., New York, 1974, pp. 117.

[4] C. Harrison, M. Park, P. Chaikin, R.A. Register and D.H. Adamson, Polymer Preprint Proceedings of the 212th ACS Meeting, Orlando, 1996, 821.

[5] T. Hashimoto, K. Nagatoshi, A. Todo, H. Hasegawa and H. Kawai, Macromolecules **7**, 364 (1974); T. Hashimoto, A. Todo, H. Itoi and H. Kawai, Macromolecules **10**, 377 (1977); T. Hashimoto, M. Shibayama and H. Kawai, Macromolecules **13**, 1237 (1980).

[6] E.L. Thomas, D.M. Anderson, C.S. Henkee and D. Hoffman, Nature **334**, 598 (1988).

[7] F.S. Bates and G.H. Fredrickson, Ann. Rev. Phys. Chem. **41**, 525 (1990).

[8] Ph. Leclère, R. Lazzaroni, J.L. Brédas, J.M. Yu, Ph. Dubois and R. Jérôme, Langmuir, **12**, 4317 (1996).

[9] J.M. Yu, Ph. Dubois, Ph. Teyssié, and R. Jérôme, Macromolecules **29**, 6090 (1996).

[10] S.N. Magonov, private communication.

[11] R. Sobry, G. Van den Bossche, J.M. Yu, Ph. Dubois and R. Jérôme, to be published.

MOLECULAR AND TEXTURAL ORDERING OF THERMOTROPIC POLYMERS IN SHEAR FLOW

A. ROMO-URIBE[*,§], P. T. MATHER[*,†], K. P. CHAFFEE[*], C. D. HAN[**]
[*] Phillips Laboratory, Propulsion Directorate, Edwards Air Force Base, CA 93524-7680
[§] Chemistry Department, University of Southern California, Los Angeles, CA 90089-1062
[**] Department of Polymer Engineering, The University of Akron, Akron OH 44325-0301

ABSTRACT

The texture and microstructural order present in mesomorphic polymers and their relation to their macroscopic behavior has been investigated using rheological, optical and dynamic scattering (WAXS and SALS) experiments. Shear orientation is observed under constant rate-of-deformation conditions where this orientation is always parallel to the flow direction. However, the high degree of orientation suggested by optical and SALS measurements is not reflected in the degree of molecular order observed in WAXS experiments. After cessation of flow, a rapid relaxation of stress is observed, while only little microstructural relaxation is found; *i.e.*, the state of orientation is very stable.

INTRODUCTION

The mechanical properties of polymeric materials (flexible chain and mesomorphic) are closely related to their morphological structure and molecular orientation.[1,2] The need for specific data regarding their mechanical behavior has led to a great deal of interest in a better understanding of the molecular, director, and interfacial orientation phenomena. The goal is to establish more detailed relationships between flow kinematics, microstructure, and mechanical properties. Microstructural studies of polymeric fluids carried out under conditions mimicking a processing operation (*in-situ*) are of particular interest since they provide unique information on the mechanisms involved in polymer deformation.[3-7]

We have designed and built an apparatus that combines simultaneous stress measurements with optical microscopy, as well as light and x-ray scattering techniques. Additionally, the instrument can reach temperatures at least as high as 345°C, making it unique for the study of polymeric melts, both isotropic and anisotropic.

Here, we present some results of the application of in-situ microscopic probes to the monitoring of orientation/disorientation processes in thermotropic liquid crystalline polymers in the nematic phase. Our results derived from optical microscopy, small-angle light scattering, and wide-angle X-ray scattering reveal the inter-relationship of molecular and textural orientation.

EXPERIMENT

Materials For this study, semiflexible aromatic polyesters based on a triad ester mesogenic unit were synthesized. The mesogen contains an arylsulfonyl-substituted hydroquinone and has either six methylene groups as the flexible spacers (herafter referred to as PSHQ6) or ten methylene groups as the flexible spacers (hereafter referred to as PSHQ10).[7] The polymers have the chemical formula shown below.

$$\left[\!\!-O-\!\!\bigcirc\!\!-\overset{\displaystyle}{\underset{\displaystyle O}{C}}\!-O-\!\!\bigcirc\!\!-O-\overset{\displaystyle}{\underset{\displaystyle O}{C}}\!\!-\!\!\bigcirc\!\!-O-(CH_2)_m\!\!-\right]_n$$

---

† Author for correspondence

Mat. Res. Soc. Symp. Proc. Vol. 461 ©1997 Materials Research Society

Details of polymerization and characterization experiments are given elsewhere.[7] PSHQ6 was characterized by a weight average molecular weight, M_w, of 30,000, a glass transition temperature, T_g, of 100 °C, a melting point, T_m, of 130 °C, and a nematic-isotropic transition temperature, T_{ni}, of 228 °C. The PSHQ10 sample featured M_w = 45,000, T_g = 92, T_m = 110, and T_{ni} = 175 °C. As-cast specimens for *in-situ* microstructural characterization were dried in a vacuum oven at room temperature for at least three weeks and, prior to measurements, at 90°C for 48 hours to remove any residual solvent and moisture.

Shearing Cell *In-situ* rheo-microscopy measurements were performed in transmission mode using a custom built parallel-plates shear cell which has been described elsewhere.[8] This device produces shear flow by sliding one plate with respect to the other (simple shear mode) using a stepper motor, and enables microscopic measurements from room temperature to about 345°C, with temperature control better ± 0.5°C. The gap separation can be set accurately by micrometers, and the minimum gap achievable is 20 μm (± 5 μm). The instrument is computer controlled, making possible the programming of different shear protocols under well controlled conditions (steady shear, oscillatory shear, etc.). For rheo-optical experiments optical-grade glass plates were utilized, and a separation gap of 20 μm was chosen. For rheo-X-ray scattering measurements we utilized copper plates covered with a 50 μm thick polyimide (Kapton™) film. The film limits the upper working temperature to 340°C and does not leave a detectable diffraction imprint within the scattering range of interest for the polymers studied. Channel ports were drilled into the copper plates to enable the transmission of incident and scattered beam. In this case a gap separation of 600 μm was chosen.

Optical Microscopy *In-situ* rheo-optical measurements were performed by viewing the sample between crossed polarizers. The optical axis is oriented along the velocity gradient direction, therefore the plane imaged is that containing the velocity (horizontal) and vorticity (vertical) axes. Crossed polarizers are oriented at 45°-135° to the flow direction. A long working-distance lens with x25 magnification was used. Micrographs were recorded photographically.

Small-Angle Light Scattering The apparatus utilized for the *in-situ* SALS measurements is shown schematically in *Figure 1*. Components **a-e** were mounted to an optical breadboard. Components **f-k** were attached to a rail mounted vertically on the breadboard. This allowed the shear cell (**g**) to be positioned with the sample plane horizontal, a distinct advantage for samples of low viscosity. The polarized light source (**a**) used for all measurements was a 15 mW HeNe laser (λ=632.8 nm). The incident intensity was collimated and attenuated by apertures (**b,d**) and neutral density filters (**c**) respectively. The polarization direction of the vertically deflected beam was set by the $\lambda/2$ plate (**f**). The scattered light, after passing through a 80 mm diameter analyzer (**h**) and collected by the ground glass (**i**), was recorded by the color video CCD camera (**k**) (Panasonic KR222, 768 x 494 pixel density), equipped with a close focus lens (**j**). The scattering patterns were displayed on color monitor (**l**) and recorded to videotape. Frame grabbing and image analysis were performed using Global Lab Image™ software. For a typical arrangement of optical elements, the observable range of scattering angles is $0.12 < \|\vec{q}\| < 1.96$ μm^{-1}.

Wide-Angle X-ray Scattering *In-situ* X-ray scattering measurements were performed at Beamline X18-A, National Synchrotron Light Source, Brookhaven National. Symmetric transmission diffraction patterns were recorded on image plates and digitized for further analysis. A constant exposure time of 3 seconds and radiation of λ=1.127 Å were used throughout.

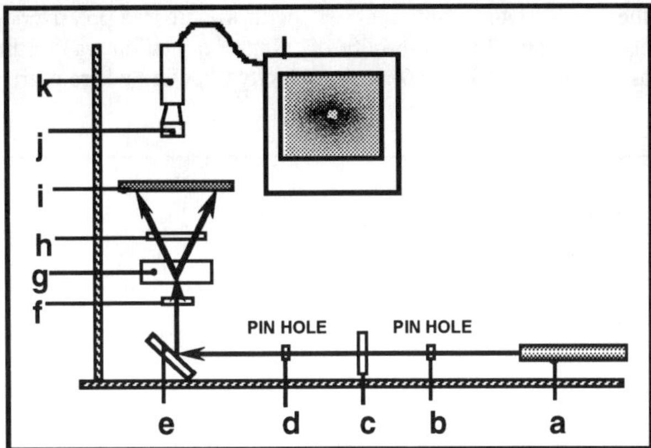

Figure 1. Schematic of Laser light scattering apparatus: a) 15 mW HeNe (632.8 nm) laser; b) and d) pin-hole apertures; c) Neutral Density Filters; e) Mirror; f) λ/2 plate; g) Shear Cell; h) Analyzer; i) Ground Glass Plate; j) Close Focus Lens; k) Color Video CCD Camera; l) Color Video Monitor.

RESULTS

The rheological characterization of PSHQ10 has been reported elsewhere,[9] whereas that of PSHQ6 is part of a future publication. It was found that, under steady shear, PSHQ10 displays a Newtonian plateau (Region II in the Onogi and Asada model[10]) at low shear rates (0.01 to 1.0 s^{-1}), whereas PSHQ6 is shear-thinning over this same range of shear rates. The typical rheo-SALS response of PSHQ10 in the nematic state, at a shear rate of 1.0 s^{-1}, is shown in *Figure 2*. It is seen that the stress-trace presents a large overshoot occurring at a strain near 2, and reaches a steady-state plateau after 20 strain units. This overshoot in stress is only observed during the first shear start-up experiment following cooling from the isotropic to nematic phase.[9] The V_V SALS pattern does not change significantly during the first several strain units (*Figure 2a-b*), while the shear stress rises dramatically to an overshoot. During this time, the sample is quite turbid and the intensity of scattered light is small due to multiple scattering. That the SALS pattern remains unchanged during such a large rise in shear stress suggests the possibility that the overshoot arises from distortional elasticity[5] of the nematic director between disclination lines of the initially fine texture. We further postulate that the stress incurred by this elastic- "distortional" - modulus ultimately achieves a yield value such that beyond the overshoot (*Figure 2c-e*), the V_V pattern evolves to show decreased turbidity while the shear stress drops precipitously by approximately a factor of three.

The sequence of SALS patterns show a slightly anisotropic pattern associated with the defect-populated texture of this material. Application of steady shearing flow produces textures yielding anisotropic patterns, with streaks oriented orthogonal to the flow direction, indicative of the texture evolution under flow. The textural evolution was further examined by studying the influence of shear rate on the SALS patterns of both thermotropic LCPs. In both cases it was found that, under steady-state conditions, the streaks shorten as the shear rate increases. The shrinkage of the SALS patterns indicate growing uniformity of the sample orientation over a length scale comparable to the laser wavelength. *Figure 3(a)* shows the optical micrograph corresponding to the steady state condition at 1s⁻¹, with the crossed polarizers oriented at 45°-135° to the velocity direction. The uniformity of texture orientation was further substantiated by rotating the crossed polarizers at 0°-90° to the velocity direction where nearly total extinction was found.

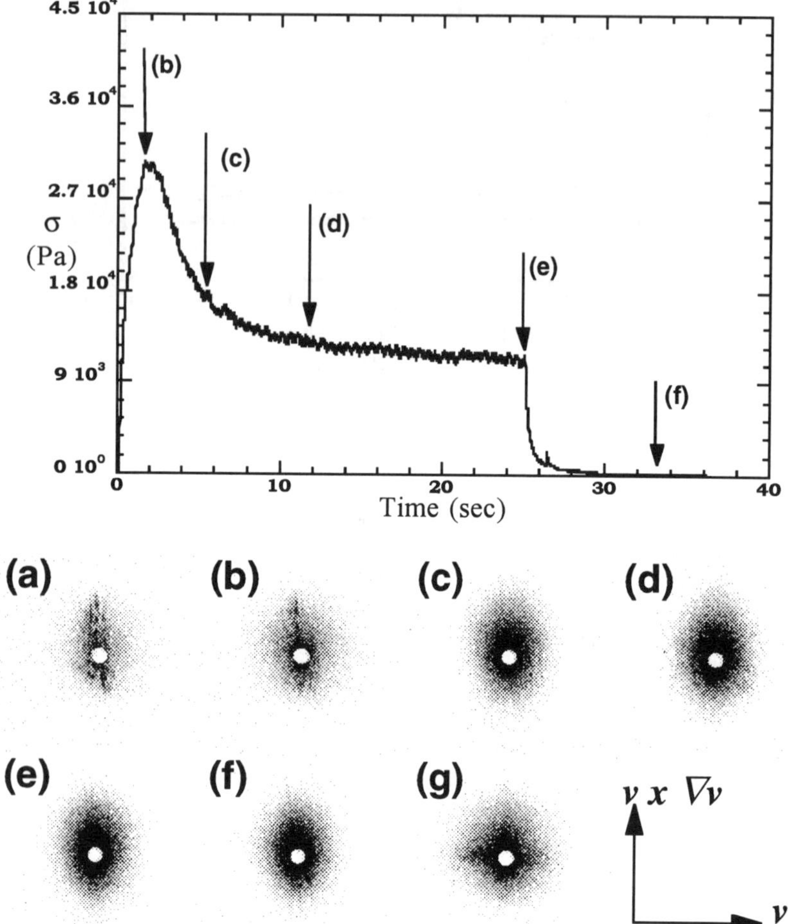

Figure 2. Stress growth trace of PSHQ10, taken at 160°C and 1.0 s-1. Arrows indicate the corresponding SALS diffraction patterns taken after: (a) 0.0; (b) 1.52; (c) 5.32; (d) 11.52 and (e) 24.77 seconds continuous shear. Relaxation was followed (f) 8.5, and (g) 215 seconds after cessation of shear. V_V polarization (polarizer and analyzer each oriented in the flow direction) was used.

Figure 2 also shows that the stress relaxation after cessation of shear takes place over only several seconds, whereas the texture shows only small relaxation within the same period of time, *Figures 2e-f*. However, after annealing the sample for several minutes, optical diffraction peaks begin to grow on the velocity axis, *Figure 2(g)*, indicating that a banded texture, akin to that observed in other high molecular weight LCPs[11], has formed. *Figure 3(b)* shows the corresponding micrograph of this banded texture, observed after following an identical thermo-mechanical history.

The degree of molecular order was investigated by using wide-angle X-ray scattering measurements. *Figure 4* shows diffraction patterns taken (a) in the quiescent condition, before shear was applied, and (b) during steady shearing flow at 1.0 s^{-1}. The angular range observed covers $0.42 < \|\bar{q}\| < 3.1$ Å^{-1}. Initially the diffraction pattern looks slightly anisotropic, indicative of the mechanical history associated with the loading of the sample between the shear plates. The amorphous nature of the equatorial reflection (at about 1.4 Å^{-1}) indicates that the material is indeed molten. Application of shearing flow produces a concentration of intensity on the vorticity axis, indicating that the polymer chains are, on average, aligned in the flow direction. The azimuthal spread of the reflections, however, show that the degree of alignment is rather modest. Calculation of the orientation parameters[12] gives $<P_2>=0.61$ and $<P_4>=0.21$. We note that increasing the shear rate does not produce a significant increase in the degree of molecular alignment, whereas the texture does coarsen further. In other words, molecular orientation appears to saturate at a lower shear rate than does textural coarsening. Experiments on PSHQ6 showed that the degree of molecular alignment is practically the same as found in PSHQ10, and in both cases the same limit of orientation ($<P_2>\sim0.6$) was found. Studies on orientation relaxation after cessation of steady shear in PSHQ6 or PSHQ10 showed that this modest level of molecular chain ordering remains relatively unchanged even after five minutes has elapsed.

(a)   **(b)**

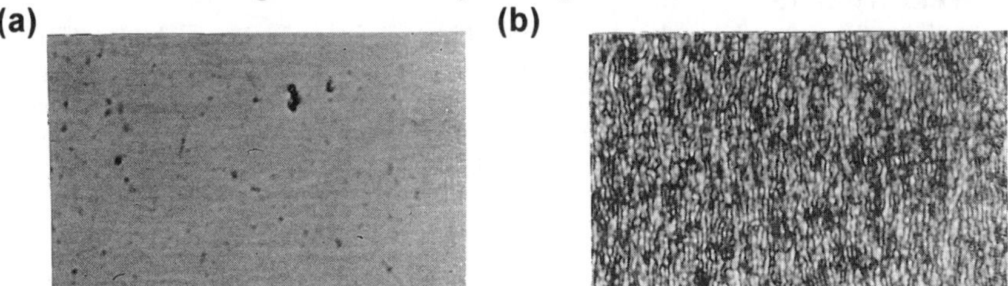

Figure 3. Optical micrographs of thermotropic PSHQ10 taken at 160°C. Crossed polarizers oriented at 45°-135° to flow (horizontal) direction. (a) Steady state shear at 1s-1. (b) Micrograph taken 215 seconds after cessation of shear.

CONCLUSIONS

We have begun to examine the microstructural response of two semiflexible thermotropic LCPs using several probes: small-angle light scattering, polarizing optical microscopy, and wide-angle x-ray diffraction. It has been shown in this preliminary study that such observations reveal, qualitatively, a high degree of orientation in sheared PSHQ10 (and PSHQ6) samples when examined with small-angle light scattering and polarizing optical microscopy, provided the shear rate is sufficiently high. Surprisingly, however, quantitative examination of wide-angle x-ray scattering data of samples sheared under identical conditions indicates a limiting orientation parameter, $<P_2>$, of only ~0.6. A possible explanation for this is that for PSHQ polymers the local nematic director (between defects) is characterized by a relatively low order parameter (at the temperatures studied) so that shear-induced coarsening of the defect texture can only raise the macroscopic orientation parameter to values approaching this modest local nematic order parameter. Infrared dichroism measurements yielding the orientation parameter of various groups within the polymer chains would be enlightening in this regard. Of additional interest is the relationship of the stress-growth trace to the evolution of the V_v SALS patterns (*Figure 2*) which for low strain values suggests the possibility that the large stress overshoot results from "distortional elasticity." We are currently examining this possibility in greater detail by investigating the relationship between the textural length scale prior to shear start-up and the magnitude of the stress overshoot.

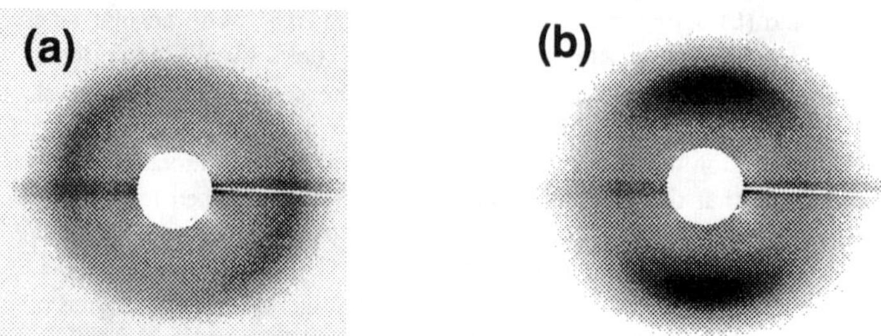

Figure 4. In-situ wide-angle X-ray scattering patterns of PSHQ10, at 160 °C. (a) Quiescent conditions, before any shear was applied. (b) Steady state at 1 s^{-1}. Synchrotron X-ray radiation of λ=1.127 Å was used.

ACKNOWLEDGMENTS

This research was supported by the Air Force Office of Scientific Research, Division of Chemistry and Life Sciences, and the Phillips Laboratory, Propulsion Directorate. Discussions with Prof. Wesley Burghardt of Northwestern University regarding x-ray scattering measurements were very helpful. Mr. Victor Ugaz of Northwestern University kindly helped us with image processing of wide-angle x-ray scattering data.

REFERENCES

1. A. Ciferri, 1987, *Developments in Oriented Polymers - 2*, edited by I.M. Ward (Elsevier Applied Science) Chapter 3.
2. A. M. Donald and A. H. Windle, *Liquid Crystalline Polymers*, Cambridge University Press, Cambridge, 1992.
3. W. Richtering, J. Läuger and R. Linemann, Langmuir, 10, 4374 (1994).
4. S. Okamoto, K. Saijo and T. Hashimoto, Macromolecules, 27, 5547 (1994).
5. P. T. Mather, D. S. Pearson and W. R. Burghardt, J. Rheol., 39, 627 (1995).
6. A. Romo-Uribe and A. H. Windle, Macromolecules, 28, 6246 (1996).
7. S.S. Kim and C.D. Han, Polymer, 35, 93 (1993).
8. P.T. Mather, H.R. Stüber, K.P. Chaffee, T.S. Haddad, A. Romo-Uribe, and J.D. Lichtenhan, *MRS Proceedings: Liquid Crystals for Advanced Technologies*, edited by S. Chen and T. Bunning (Mater. Res. Soc. Proc., New York, NY, 1996).
9. S.S. Kim and C.D. Han, J. Rheol., 37, 847 (1993).
10. T. Asada, T. Koda, and S. Onogi, Mol. Cryst. Liq. Cryst., 68, 231 (1981).
11. C. Viney, A.M. Donald, and A.H. Windle, Polymer, 26, 870 (1985).
12. W. Haase, Z.X. Fan, and H.J. Müller, J. Chem. Phys., 89, 3317 (1988).

PHASE BEHAVIOR OF TRIBLOCK COPOLYMERS UPON INCORPORATION OF NON-PARENT, MIDBLOCK-ASSOCIATING ADDITIVES

J.H. Laurer*, J.F. Mulling*, R. Bukovnik**, R.J. Spontak*

*Department of Materials Science & Engineering, North Carolina State Univ., Raleigh, NC 27695
**Telecom Division, Raychem Corporation, Fuquay-Varina, NC 27526

ABSTRACT

Addition of a block-selective homopolymer to a microphase-ordered block copolymer is known to result in preferential swelling of the chemically compatible microdomain. In this work, we examine the miscibility between a triblock copolymer and a relatively low-molecular-weight, chemically dissimilar, midblock-associating homopolymer and demonstrate that the homopolymer molecules residing in the swollen midblock matrix self-assemble to avoid repulsive interactions with neighboring microdomains. We extend this investigation to include systems composed of a very low-molecular-weight, midblock-associating additive (an oil). At high oil concentrations, the glassy copolymer endblocks micellize, resulting in the formation of a thermoplastic elastomer gel.

INTRODUCTION

As an attractive alternative to tailored *chemical* synthesis, the morphological characteristics and corresponding properties of ordered block copolymers can be controllably modified through the *physical* incorporation of either a miscible (i) parent or block-selective homopolymer [1-3] or (ii) a second copolymer of differing composition [4-6] or molecular weight [7,8]. Previous experimental efforts exploring blends with an added parent homopolymer have focused primarily on determining the dependence of blend morphology on factors such as homopolymer molecular weight (M_h), copolymer molecular weight and volume-fraction concentration (ϕ_h) [1,2,9-11]. The presence of an added homopolymer impacts the effective thermodynamic repulsion (characterized by the Flory-Huggins interaction parameter, χ) of the blend [12], as well as the efficiency of interfacial chain packing [13]. It has been shown that the incompatibility between the homopolymer and non-host block decreases as M_h is reduced. Thus, if M_h is significantly less than that of the host block, the added homopolymer behaves as a preferential solvent. At sufficiently high concentrations of a low-M_h homopolymer (i.e., within the disordered copolymer regime), the compatible (host or parent) blocks are highly swollen, whereas the incompatible (non-host) blocks typically organize into various microstructural elements to reduce the propensity of energetically unfavorable contacts. Spheroidal [14] micelles, as well as complex bilayered membranes [15,16], have been observed in numerous binary copolymer/homopolymer blends varying in M_h.

In the events that (i) M_h is much smaller than the molecular weight of the host copolymer block and (ii) the copolymer is a triblock composed of glassy endblocks and elastomeric midblocks (e.g., styrene-diene-styrene copolymers), micellization of the glassy endblocks induced by extensive midblock swelling results in the formation of a thermoplastic elastomer gel (TPEG). The characteristic features of TPEGs are schematically depicted in Fig. 1. A fraction of the swollen midblocks forms bridges between endblock-rich micelles; these bridges serve as physical crosslinks and are responsible for stabilizing the gel network. Within a reasonable composition range, the low-M_h additive is completely bound within the midblock matrix, which consists of biconformational midblocks that are either bridged or looped. Reynaers and co-workers [17,18] have examined copolymer gels via small-angle neutron scattering (SANS), and provide evidence that is consistent with the illustration displayed in Fig. 1. Their results have been recently verified by transmission electron microscopy (TEM) [19] of comparable gels (see Fig. 2). The thermomechanical properties of TPEGs are expected *a priori* to be particularly sensitive to processing due to variation in the fraction of network-forming bridged (or viscosity-enhancing looped) midblocks.

Mat. Res. Soc. Symp. Proc. Vol. 461 © 1997 Materials Research Society

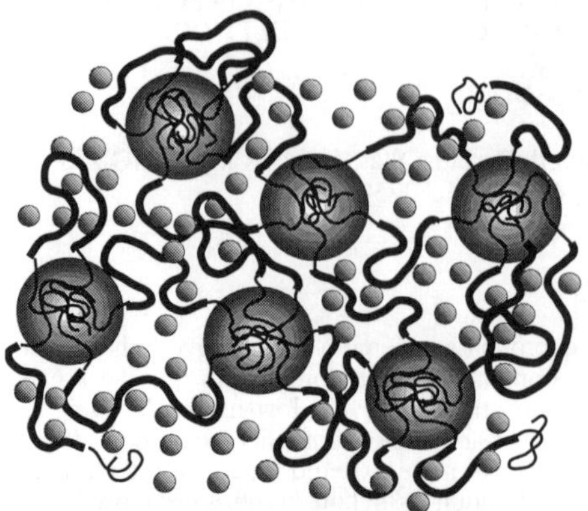

Fig. 1. Schematic illustration of the micellar network formed in a TPEG that is stabilized by a triblock copolymer with glassy endblocks.

Fig. 2. Transmission electron micrograph of a SEBS/oil TPEG containing 90 wt% oil. The S-rich micelles (dark) are stained with RuO_4.

While several prior studies have examined the morphological characteristics and properties of block copolymers in the presence of a single, miscible nonparent homopolymer [11,20-22], the present work focuses on a series of triblock copolymer/homopolymer systems in which ϕ_h and M_h are varied (with the latter remaining small relative to the molecular weight of the host copolymer block) so that morphological and property similarities between blends and TPEGs can be identified.

EXPERIMENT

Gels and blends were prepared using a single poly[styrene-*b*-(ethylene-*co*-butylene)-*b*-styrene] (SEBS) triblock copolymer, Kraton G-1654, obtained from the Shell Development Co. According to the manufacturer, this copolymer possesses a molecular weight of about 150,000 g/mol, a polydispersity index of less than 1.04 and a styrene (endblock) fraction of 31 wt%. Two EB-associating additives were selected for this study. The first was a homopolyisoprene (hI), prepared via living anionic polymerization with a *sec*-butyllithium initiator in cyclohexane. The hI molecular weight and polydispersities were 15,000 g/mol and 1.05, respectively. The second additive was a short-chain aliphatic oil (Hydrobrite 380PO, from Witco Corp.) with a molecular weight of 468 g/mol (as discerned from ASTM D2502) and a viscosity of 75 cSt at 40°C. Blends possessing 70, 80 and 90 wt% oil were prepared via mechanical mixing at 180°C under vacuum in a Ross double planetary mixer. The role of gel composition on morphological and property development was examined by pressing the mixed gels at 180°C, followed by slow cooling to ambient temperature. Solution casting was used here to produce bulk films of the neat SEBS copolymer and a 70/30 w/w SEBS/hI blend. Solutions were prepared in cyclohexane (at a concentration of 4% wt/v), cast into Teflon trays and dried over the course of three weeks. Resultant films were heated to 90°C for 4 hrs under vacuum to remove residual solvent and facilitate microstructural equilibration.

Thin (electron transparent) sections of the resultant films were obtained by cryosectioning at −100°C in a Reichert-Jung Ultracut-S cryoultramicrotome and, in the case of the TPEGs, stained with the vapor of RuO_4 for 5 min at 25°C to enhance phase contrast by selectively staining the S-rich micelles. The SEBS/hI blend was stained with a combination of RuO_4 and OsO_4 vapors to stain the S copolymer blocks and hI chains, respectively. Images were obtained on a Zeiss EM-902 electron spectroscopic microscope, operated at 80 kV and an energy loss (ΔE) of 50-100 eV. The stress response of the TPEGs to dynamic oscillatory shear strain was measured at ambient temperature on a Rheometrics Mechanical Spectrometer (RMS-800) equipped with 25 mm parallel plates separated by a 1 mm gap. Stress relaxation was also recorded at ambient temperature by monitoring the force required to maintain a probe 5 mm within each of the TPEGs analyzed.

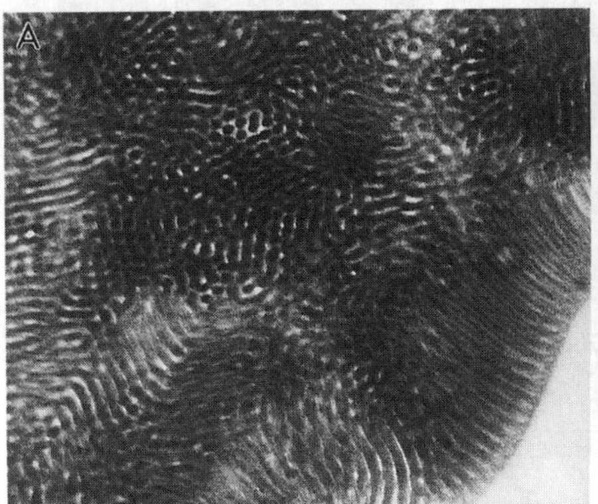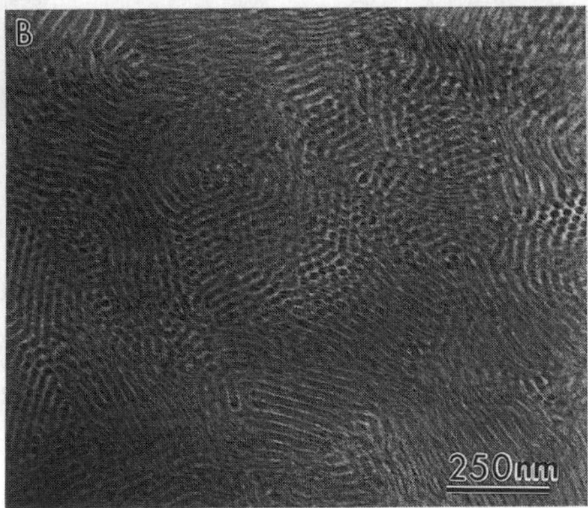

Fig. 3. Electron micrographs of the 70/30 w/w SEBS/hI blend stained with (a) RuO_4 (styrene appears dark) and (b) OsO_4 (isoprene appears dark). Not seen in these micrographs are the macrophase-separated hI-rich domains (stainable only with OsO_4) that are prevalent in this blend.

RESULTS

All of the gels and blends investigated in this work have been prepared with a single SEBS triblock copolymer, which exhibits a cylindrical morphology (not shown here). If incorporated entirely within the SEBS copolymer (the molecular ratio of hI to EB midblock is 0.14), the added hI is expected to swell the EB matrix, eventually resulting in an order-order transformation from styrenic cylinders to spheres. As seen in the electron micrograph displayed in Fig. 3, a blend composed of 30/70 w/w SEBS/hI exhibits substantial intramicrodomain mixing. In Fig. 3a, the styrene-rich microdomains of the SEBS copolymer are preferentially stained with RuO_4 and appear dark in transmission. Light regions therefore correspond to either EB or hI. On the other hand, the hI molecules are stained with OsO_4 in Fig. 3b. Upon comparing these images and noting that macroscopic domains of hI (not shown) are also present in this blend, it is clear that the SEBS and hI partially mix at this blend composition. Partial miscibility of the two components, even with a low molecular weight ratio of hI to EB, reflects their intrinsic chemical incompatibility. Close examination of the micrographs in Fig. 3 reveals that the hI molecules form spherical or cylindrical microstructural elements within the SEBS matrix. Since the neat SEBS copolymer orders into a cylindrical morphology, this observation implies that the hI molecules, forced to occupy the EB-rich matrix region between neighboring cylinders, minimize repulsive S-I interactions by adopting an interstitial morphology much in the same fashion as do ordered ABC triblock copolymers [23]. It must be remembered, however, that this morphological characteristic is only a consequence of the chain length of hI incorporated in the SEBS copolymer, but is not representative of the overall blend composition (due to SEBS/hI macrophase separation).

Since the affinity of hI (at 15,000 g/mol) and EB is clearly insufficient to produce a TPEG, a short-chain aliphatic oil is employed in the next series of blends. An electron micrograph of the SEBS/oil system that is comparable to the SEBS/hI blend (each contains 70 wt% additive) is provided in Fig. 4. Unlike the TPEG presented in Fig. 2, this gel has been prepared via mechanical mixing. It nonetheless exhibits a uniformly-distributed population of S micelles, each measuring about 20 nm in diameter. [Solvent casting has been found to yield a similar morphology, but with fewer well-oriented regions.] As depicted in Fig. 1, the glassy S micelles stabilize the TPEG by serving as physical crosslinks for highly extended bridged EB midblocks. An increase in the concentration of oil would swell the EB-rich matrix (and midblocks) further,

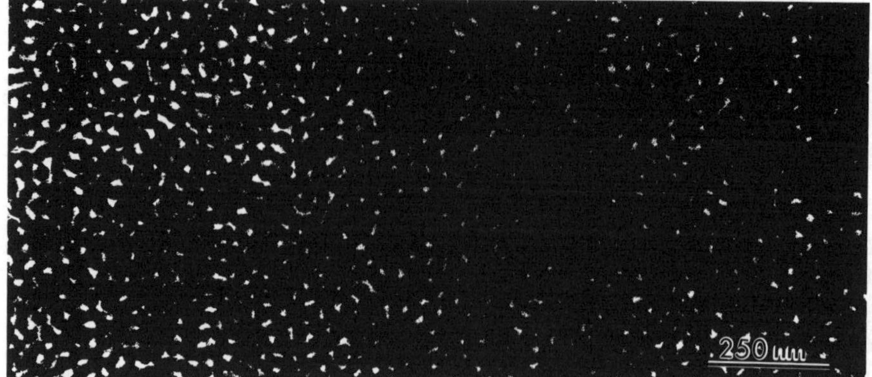

Fig. 4. Electron micrograph of a TPEG composed of 70 wt% oil. As in Fig. 2, the styrenic micelles, which serve as physical crosslinks to stabilize the gel network, are stained with RuO$_4$.

thereby reducing the extent to which the micelles connect via bridged EB blocks into a three-dimensional network. The impact of oil concentration is evident in Fig. 5, which shows the dependence of the dynamic elastic (G') and viscous (G") shear moduli on strain amplitude (γ, Fig. 5a) and oscillatory frequency (ω, Fig. 5b). In both cases, G' appears to be nearly independent of γ and ω over most of the ranges examined. Note that, as the concentration of oil is increased, both G' and G" decrease markedly (by over an order of magnitude in some cases). The reduction in G', as well as the variation in tan δ (=G"/G'), with increasing oil content is presented in Fig. 6. As the fraction of oil is increased from 70 to 90 wt%, G' (a measure of the network elasticity) is observed to decrease by over two orders of magnitude. In contrast, tan δ is, within experimental error (the shaded region in Fig. 6), independent of gel concentration over the range explored.

Another means by which to discern the effect of gel composition on morphological and property development is through stress (σ) relaxation. Shown in Fig. 7 are stress relaxation curves obtained from two TPEGs differing in gel content (one 70 wt% oil and the other 90 wt% oil). If the TPEG

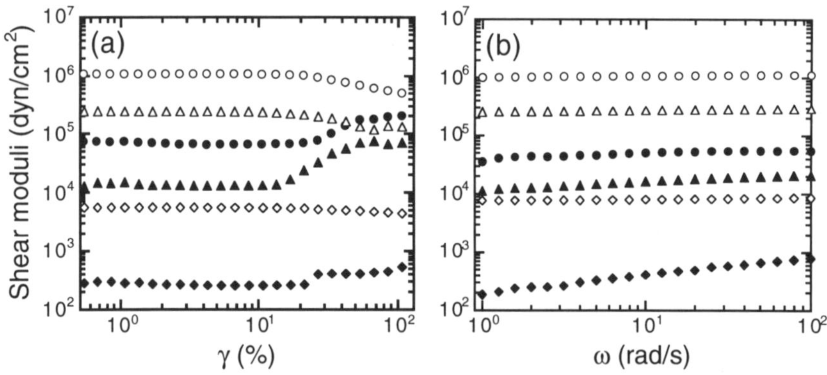

Fig. 5. Shear moduli G' (open symbols) and G" (filled symbols) obtained from three TPEGs varying in composition (in wt% oil): 70 (circles), 80 (triangles) and 90 (diamonds). The frequency (ω) was held constant at 10 rad/s in (a), while the strain amplitude (γ) was fixed at 1 % in (b).

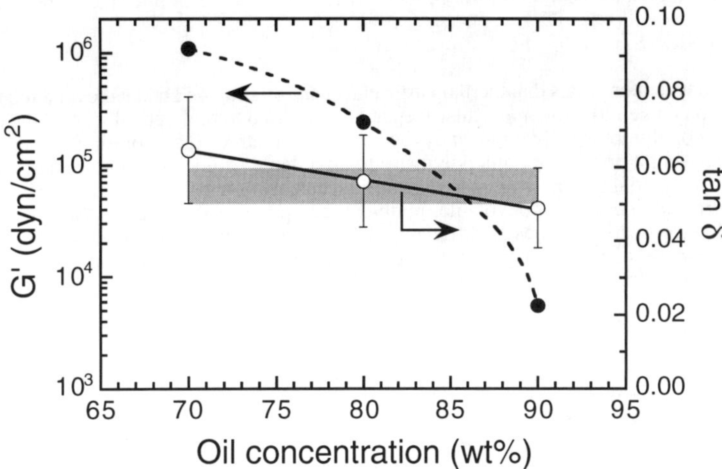

Fig. 6. Variation of G' (●), the elastic shear modulus (measured in the linear viscoelastic regime), as well as tan δ (O), with oil concentration for the three TPEGs investigated in this work.

consists of a fraction that relaxes slowly (Φ_s) and another that relaxes relatively quickly (Φ_q) with corresponding relaxation times of τ_s and τ_q, respectively, stress relaxation can be modeled as

$$\frac{\sigma(t)}{\sigma(t=0)} = \sum_{i\,=\,s,\,q} \Phi_i \exp\left(-t\,/\,\tau_i\right) \tag{1}$$

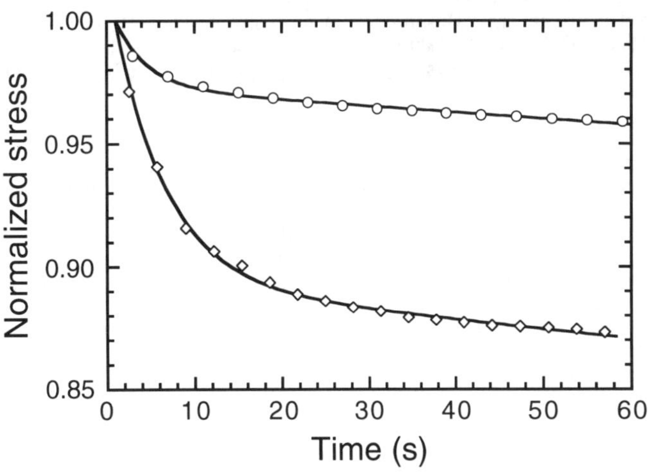

Fig. 7. Normalized compressive stress relaxation as a function of time for the same TPEGs described in the caption of Fig. 5. The solid lines correspond to least-squares regressions of Eq. 1.

If Eq. 1 is fitted to the data provided in Fig. 7, the long relaxation times are found to decrease with increasing oil content: 63 min for the 70 wt% oil TPEG *versus* 38 min for the 90 wt% TPEG.

CONCLUSIONS

This study demonstrates that thermoplastic elastomer gels can be obtained over a relatively broad composition range if the low-molecular-weight additive used to preferentially swell the midblock of a triblock copolymer is fully imbibed by the copolymer matrix. Incorporation of a relatively low-molecular-weight nonparent homopolymer into an ordered triblock copolymer has been found to result in only partial miscibility due to chemical incompatibility between the homopolymer and the copolymer midblock. If a nonvolatile, midblock-compatible oil is used to swell the copolymer, endblock micellization ensues at sufficiently high oil concentrations. We have shown that these micelles can be directly imaged with TEM, and that the morphological characteristics and mechanical properties of these gels are very sensitive to the oil content.

ACKNOWLEDGMENTS

This work has been supported by the Raychem Corporation and the Shell Development Co. We thank Dr. S.D. Smith for providing the hI sample and Dr. S.A. Khan for the use of his rheometer.

REFERENCES

1. K.I. Winey, E.L. Thomas, L.J. Fetters, J. Chem. Phys. **95**, 9367 (1991).
2. K.I. Winey, E.L. Thomas, L.J. Fetters, Macromolecules **25**,422 (1992); **25**, 2645 (1992).
3. M.W. Matsen, Macromolecules **28**, 5765 (1995).
4. A.D. Vilesov, G. Floudas, T. Pakula, E. Yu. Melenevskaya, T.M. Birshtein, Yu. V. Lyatskaya, Macromol. Chem. Phys. **195**, 2317 (1994).
5. J. Zhao, B. Majumdar, M. F. Schulz, F. S. Bates, K. Almdal, K. Mortensen, D. A. Hajduk, S. M. Gruner, Macromolecules **29**, 1204 (1996) .
6. R.J. Spontak, J.C. Fung, M.B. Braunfeld, J.W. Sedat, D.A. Agard, L. Kane, S.D. Smith, M.M. Satkowski, A. Ashraf, D.A. Hajduk, S.M. Gruner, Macromolecules **29**, 4494 (1996).
7. T. Hashimoto, K. Yamasaki, S. Koizumi, H. Hasegawa, Macromolecules **26**, 2895 (1993).
8. L. Kane, M.M. Satkowski, S.D. Smith, R.J. Spontak, Macromolecules (in press).
9. H. Tanaka, H. Hasegawa, T. Hashimoto, Macromolecules **24**, 240 (1991).
10. R.J. Spontak, S.D. Smith, A. Ashraf, Macromolecules **26**, 956 (1993).
11. M.F. Schulz, F.S. Bates, in <u>Physical Properties of Polymers Handbook</u>, edited by J.E. Mark (AIP Press, New York, 1996), p. 427.
12. S. Koizumi, H. Hasegawa, T. Hashimoto, Macromolecules, **27**, 4371 (1994).
13. M.W. Matsen, F.S. Bates, Macromolecules, **29**, 7641 (1996).
14. D.J. Kinning, E.L. Thomas, L.J. Fetters, J. Chem. Phys. **90**, 5806 (1989).
15. D.J. Ponchan, S.P. Gido, S. Pispas, J.W. Mays, A.J. Ryan, J.P.A. Fairclough, I.W. Hamley, N.J. Terrill, Macromolecules, **29**, 5091 (1996).
16. J.H. Laurer, J.C. Fung, J.W. Sedat, D.A. Agard, S.D. Smith, J. Samseth, K. Mortensen, R.J. Spontak, Langmuir (submitted).
17. N. Mischenko, K. Reynders, K. Mortensen, R. Scherrenberg, F. Frontaine, R. Graulus, H. Reynaers, Macromolecules **27**, 2345 (1994).
18. N. Mischenko, K. Reynders, M.H.J. Koch, K. Mortensen, J.S. Pedersen, F. Frontaine, R. Graulus, H. Reynaers, Macromolecules **28**, 2054 (1995).
19. J.H. Laurer, R. Bukovnik, R.J. Spontak, Macromolecules **29**, 5760 (1996).
20. P.S. Tucker, J.W. Barlow, D.R. Paul, Macromolecules **21**, 1678 (1988); **21**, 2794 (1988).
21. T. Hashimoto, K. Kimishima, H. Hasegawa, Macromolecules **24**, 5704 (1991).
22. R. Xie, B. Yang, B. Jiang, Macromolecules **26**, 7097 (1993).
23. R. Stadler, C. Auschra, J. Beckmann, U. Krappe, I. Voight-Martin, L. Leibler, Macromolecules, **28**, 3080 (1995).

EFFECT OF ADDITIVE CONSTRAINT ON THE MORPHOLOGICAL AND MECHANICAL PROPERTIES OF TRIBLOCK COPOLYMER BLENDS

L. Kane[*], D.A. Norman[*], S.A. White[**], and R.J. Spontak[*]
[*]Department of Materials Science & Engineering, North Carolina State Univ., Raleigh, NC 27695
[**]Becton Dickinson Research Center, Research Triangle Park, NC 27709

ABSTRACT

While numerous studies have addressed the morphological characteristics of diblock copolymer blends either with a second copolymer or a parent homopolymer, relatively few have examined comparable blends containing a triblock copolymer. In this study, we investigate the role of midblock bridging on the morphological and physical characteristics of blends composed of a poly(styrene-b-isoprene-b-styrene) (SIS) triblock copolymer with either an unconstrained homopolyisoprene (hI) or an end-grafted SI diblock copolymer. Blend compositions and molecular weights of the hI, as well as the I-block of the copolymer, have all been systematically varied to elucidate the effect of additive constraint on the extent of nonideal intramicrodomain mixing. Blend morphologies are characterized using transmission electron microscopy, while blend properties have been measured by dynamic mechanical analysis.

INTRODUCTION

Block copolymers are composed of long contiguous sequences ("blocks") of chemically dissimilar repeat units. When the blocks are sufficiently incompatible, these materials microphase-order into a variety of periodic nanoscale microstructures differing in interfacial curvature and area [1,2]. In this work, only the alternating lamellar (planar) morphology is considered. Although previous attempts to manipulate the microstructural characteristics, as well as the corresponding bulk properties, of these materials have relied extensively on chemically tailored copolymers, physical blending at the molecular level (i.e., within the ordered microdomain structure) is also a viable means of producing morphologies with controlled characteristics and mechanical properties. Efforts exploring this potential materials synthesis route have focused principally on microphase-ordered diblock copolymers in the presence of either a parent homopolymer [3-7] or a second copolymer [8-14]. In lamellar diblock copolymer/homopolymer blends, homopolymer molecules tend to localize near the center of their host microdomains to minimize repulsive interactions. The extent of localization is dependent on the homopolymer molecular weight, decreasing with a reduction in molecular weight. If the additive is a second diblock copolymer of chemically identical repeat units, the covalently linked junctions of both copolymers are restricted to interfacial regions, in which case each microdomain consists of a bidisperse mixture of grafted chains [15].

Triblock copolymer molecules, on the other hand, are of more applied and fundamental interest due to their biconformational midblocks, which serve to enhance the physical properties of such copolymers. Unlike single-grafted endblocks, which impart strength only through entanglements and which fully comprise diblock copolymers, double-grafted copolymer midblocks are capable of adopting either a looped or bridged conformation, the latter of which effectively strengthens linear multiblock copolymers through the formation of a physically crosslinked network. The addition of a midblock-associating additive to a microphase-ordered triblock copolymer is expected *a priori* to hinder the ability of the midblock to form bridges, thereby resulting in an increase in the population of midblock loops (in which both ends of the midblock reside in the same interfacial region) and a noticeable reduction in mechanical properties. In this work, we explore the differences derived by adding two series of midblock-associating additives, one consisting of homopolyisoprenes (hIs) varying in molecular weight and the other of poly(styrene-b-isoprene) (SI) diblock copolymers differing in I-block length, to a single lamellar poly(styrene-b-isoprene-b-styrene) (SIS) triblock copolymer and, more specifically, examine the effect of additive constraint on the morphological and property characteristics of the resultant blends. While the added hIs are anticipated [16] to localize along the microdomain midplane much in the same manner as in diblock copolymer/homopolymer blends (and frustrate midblock bridging), the constrained diblock molecules must compete with the bridged and looped triblock molecules for available interfacial area.

Mat. Res. Soc. Symp. Proc. Vol. 461 © 1997 Materials Research Society

EXPERIMENTAL

Six triblock copolymer blend series were prepared using the SIS triblock copolymer and the hI and SI additives listed in Table 1. Predetermined quantities of the blend constituents were dissolved in toluene at 5% w/v, and the solutions were cast into Teflon molds. Films measuring approximately 1 mm thick were allowed to dry over a period of three weeks and then annealed under vacuum at 110°C for 4 hrs to remove residual solvent. Specimens for dynamic mechanical analysis (DMA) were obtained by cutting the films into strips measuring 35 mm long by 5 mm wide. The dynamic response of each blend to uniaxial tension was recorded using a Rheometrics solids analyzer (RSA II). Tensile storage (E') and loss (E") moduli were measured (at ambient temperature) as functions of frequency (ω) at a strain amplitude of 0.01% (within the linear visco-elastic regime). Electron-transparent specimens for transmission electron microscopy (TEM) were produced via cryosectioning in a Reichert-Jung Ultracut-S ultramicrotome maintained at −100°C, and were subsequently stained with the vapor of OsO_4(aq) for 90 min. Images were collected on a Zeiss EM902 electron spectroscopic microscope operated at 80 kV and an energy loss of 50 eV.

Table 1. Material characteristics of the copolymers and homopolymers employed in this study.

Designation	w_I [a]	M_{total} [b] (kg/mol)	M_I (kg/mol)	R_I
Triblock copolymer				
SIS	0.46	111	51	1.00
Homopolyisoprenes				
hI8	1.00	7.5	7.5	0.31
hI15	1.00	15	15	0.59
hI30	1.00	30	30	1.18
Diblock copolymers				
SI15	0.30	50	15	0.59
SI30	0.46	65	30	1.18
SI48	0.64	75	48	1.89

[a] Isoprene weight fraction, measured by ^{1}H NMR.

[b] Number-average molecular weight, measured by GPC (polydispersities were <1.05).

RESULTS

Shown in Fig. 1 are representative TEM micrographs of the neat triblock copolymer (Fig. 1a), a SIS/hI30 blend (Fig. 1b) and a SIS/SI30 blend (Fig. 1c). In these micrographs, the isoprene-rich lamellae appear electron-opaque (dark) due to the preferential OsO_4 stain. Note that the blends displayed in Fig. 1 contain isoprene additives of nearly identical molecular weight. Within most of the composition ranges discussed in this work, the lamellar morphology characteristic of the neat SIS copolymer is retained. Higher additive concentrations typically result in transitions to non-lamellar morphologies due to changes in interfacial packing and, hence, curvature [17,18]. As shown elsewhere, addition of a preferential homopolymer [3-5] or a block copolymer differing in either composition [9-12] or molecular weight [8,13,14] to a lamellar diblock copolymer results in microphase swelling or contraction, depending on the ratio of the molecular weight of the additive (hI or I-block of the SI copolymer) to that of the host block. Here, only half of the I-midblock of the SIS copolymer is considered, since midblocks exist as either loops or bridges. Values of this ratio, designated R_I, are included in Table 1 for the blends examined here. Two of the hI materials (hI8 and hI15) are expected to remain fully miscible within the microphase-ordered morphology,

Figure 1. Representative transmission electron micrographs of (a) the neat SIS copolymer, (b) the 96.3/3.7 w/w SIS/hI30 blend and (c) the 50/50 w/w SIS/SI30 blend. Due to OsO_4 staining, the isoprene-rich lamellae appear dark.

whereas the hI30 specimen may, at sufficiently high concentrations, macrophase-separate from the triblock copolymer due to its relatively high molecular weight. Such macrophase separation has, in fact, been observed in the SIS/hI30 blend series. Since the SI copolymers employed here are all ordered and possess blocks of comparable length to those of the SIS copolymer, macrophase separation is not expected. In compositionally symmetric $(SI)_\alpha/(SI)_\beta$ copolymer blends, phase demixing is observed [8,14] and predicted [19] when the block length ratio exceeds a critical value.

The effect of hI15 concentration on a mechanical property (i.e., the tensile storage modulus, E') of the triblock copolymer is seen in Fig. 2. All of the data presented here correspond to an average of at least five different runs. Since bridged midblocks are responsible for the physically crosslinked network, it follows that E' should provide a measure of the extent to which the network exists and, consequently, to which the SIS midblocks adopt the bridged conformation. The open symbols in Fig. 2 are indicative of the lamellar morphology, while the filled symbol designates a nonlamellar morphology. Note the difference in the functionality of E'(ω) between the lamellar blends and the nonlamellar blend, for which E' is less strongly dependent on ω. While

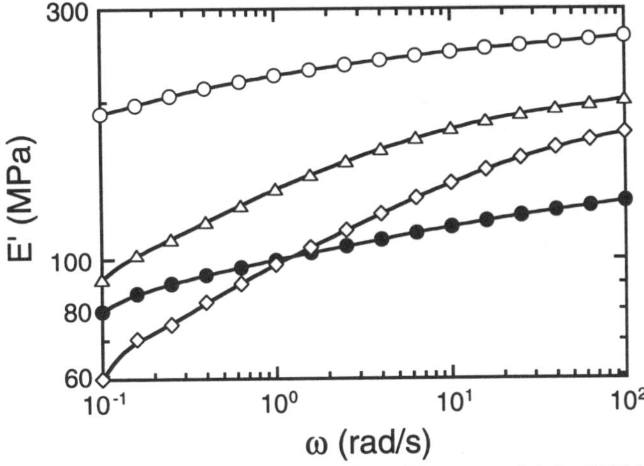

Figure 2. Dynamic tensile modulus (E') as a function of frequency (ω) for SIS/hI15 blends with the following bulk isoprene compositions (in wt%): 46 (O), 48 (Δ), 56 ($\Diamond$) and 58 ($\bullet$).

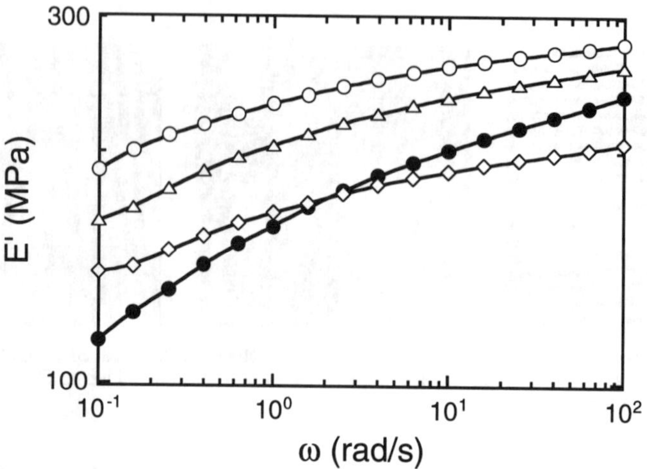

Figure 3. Double-logarithmic representation of E'(ω) for SIS/SI15 blends with the following SI concentrations (in wt%): 0 (O), 40 (Δ), 80 ($\Diamond$) and 100 ($\bullet$).

only a few data sets are included in Fig. 2, intermediate blend morphologies follow similar trends, with their E'(ω) curves lying between the curves corresponding to the 48 and 56 wt% isoprene blends in Fig. 2. Shown for comparison in Fig. 3 are E'(ω) curves obtained from SIS/SI15 blends of comparable isoprene composition. It should be remembered that as the concentration of SI15 increases in the blend, the total isoprene fraction decreases. To compare blends with similar isoprene content, Fig. 4 shows E'(ω) for two sets of blends from the SIS/hI30 and SIS/SI48 series. In these series, the molecular weights of the hI and I-block (of the diblock) are not vastly different, and an increase in either hI or SI serves to increase the isoprene content of the blend. From Fig. 4, it is evident that the SIS/SI blends have a higher storage modulus at low ω, whereas

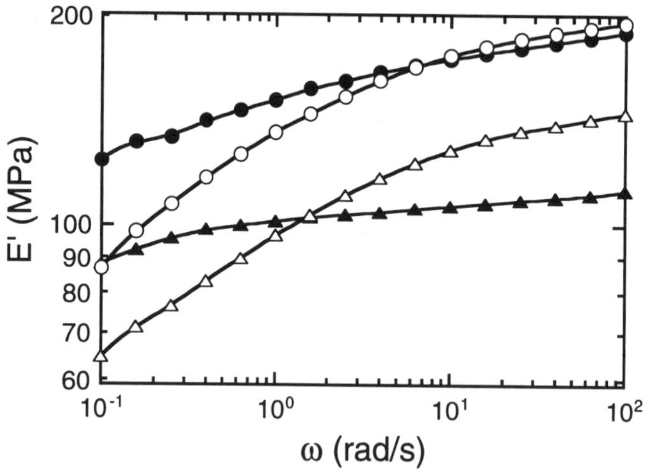

Figure 4. Variation of E' with ω for SIS/hI30 (open symbols) and SIS/SI48 blends (filled symbols) at two bulk (overall) isoprene compositions (in wt%): 50 (circles) and 60 (triangles).

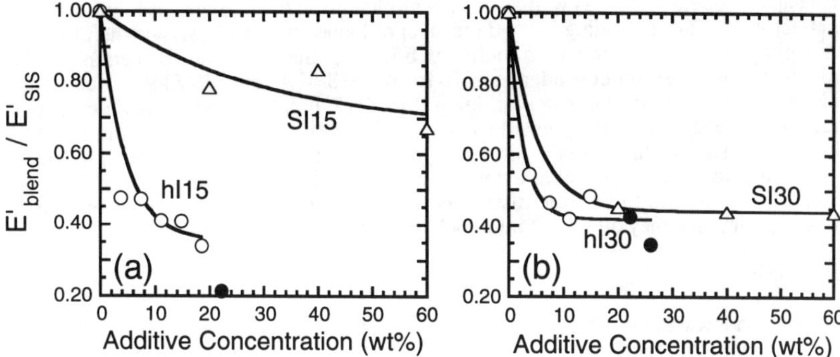

Figure 5. E'(ω=0.1 rad/s) relative to that of the neat SIS copolymer, E'$_{SIS}$(ω=0.1 rad/s), as a function of composition for (a) 15 and (b) 30 kg/mol isoprenic additives. Shown are hI (O) and SI ($\triangle$) data. Filled symbols denote nonlamellar morphologies, and solid lines are guides for the eye.

the SIS/hI blends exhibit higher E' at high ω. Note that the shapes of the curves from each series also differ. To discern the effect of additive constraint on the E' data acquired from each of these SIS/hI and SIS/SI blends, we elect to compare E' from each blend at the lowest measured frequency of 0.1 rad/s since low frequencies probe relatively larger microstructural dimensions. Shown in Fig. 5 are E' values normalized to that of the neat triblock copolymer as a function of isoprene content for blends with either hI15 and SI15 (Fig. 5a) or hI30 and SI30 (Fig. 1b). In Fig. 5a, it is apparent that the added hI15 induces a substantial reduction in E', whereas the added SI15 causes a more gradual reduction, as the concentration of additive is increased. A similar, but less pronounced, trend is observed in Fig. 5b. These results suggest that the unconstrained hI, residing near the lamellar midplane, hinders bridging more effectively than an interfacially constrained block of comparable chain length. Comparison of Figs. 5a and 5b also reveals that, while hI15 and hI30 yield similar results, addition of the SI30 copolymer results in a more dramatic decrease in E' than the SI15 copolymer, indicating that the long isoprene tails of the SI30 copolymer are more successful than the SI15 tails at inhibiting midblock bridging. Moreover, the SIS/SI30 blends exhibit a lower E' than that of the neat SI30 copolymer. This observation implies

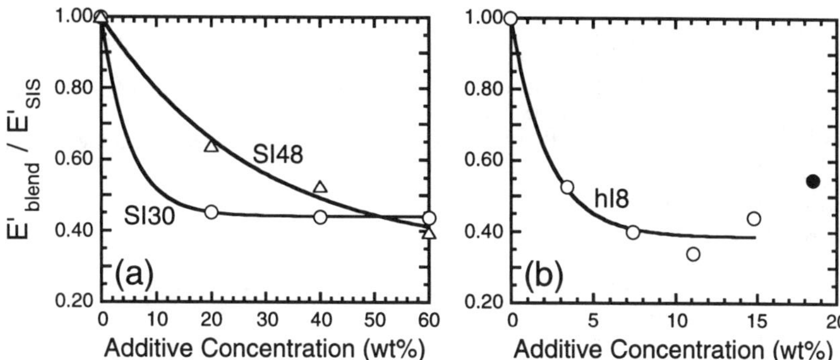

Figure 6. Dependence of the normalized storage (elastic) modulus on additive content for (a) two SI copolymers differing in isoprene block length (SI30 $vs.$ SI48) and (b) hI8. Filled symbols denote nonlamellar morphologies. Solid lines represent guides for the eye.

that, since the SI30 tails extend to nearly the same extent as midblock loops [20] and since loops are not capable of entangling as well as tails, the SI30 tails in the presence of loops cannot entangle very effectively. Tails extending beyond the layer of looped midblocks should therefore be capable of entangling, as well as eliminating midblock bridging. Such behavior is seen in Fig. 6a, which compares the composition dependence of E' for the SIS/SI30 and SIS/SI48 series. Thus, in the blends examined here, SIS midblock bridging (SIS/SI15, Fig. 5a) is found to yield greater storage moduli than SI endblock entanglement (SIS/SI48, Fig. 6a). Figure 6b shows the effect of adding the low-molecular-weight hI8 to the SIS copolymer. This additive is not expected to localize along the lamellar midplane to the same extent as hI15 and hI30. Upon comparing Figs. 5 and 6b, however, there appears to be little difference in E' among the three hI grades employed in this work, suggesting that they all inhibit midblock bridging to similar extents.

CONCLUSIONS

This study addresses the effect of additive constraint on the properties of SIS triblock copolymer blends with isoprene-associating additives. Addition of unconstrained hI consistently yields the greatest reduction in E' and, by inference, the midblock bridging fraction of the blends examined here. Blends containing constrained SI copolymer molecules exhibit a gradual reduction in E' with increasing SI content if the I-block of the copolymer is either relatively short (to form a thin layer of SI tails along the interface) or relatively long (so that the SI tails undergo entanglement).

ACKNOWLEDGMENTS

We are indebted to Dr. S.D. Smith for the provision of model materials, and we thank Mr. J.H. Laurer for technical assistance. L.K. was supported by a GAANN Fellowship.

REFERENCES

1. F.S. Bates and G.H. Fredrickson, Annu. Rev. Phys. Chem. **41**, 525 (1990).
2. M.F. Schulz and F.S. Bates, <u>Physical Properties of Polymers Handbook</u>, edited by J.E. Mark (American Institute of Physics, New York, 1996) Chap. 32.
3. K.I. Winey, E.L. Thomas, and L.J. Fetters, Macromolecules **24**, 6182 (1992).
4. K.I. Winey, E.L. Thomas, and L.J. Fetters, Macromolecules **25**, 422, 2645 (1992).
5. T. Hashimoto, H. Tanaka, and H. Hasegawa, Macromolecules **23**, 4378 (1990).
6. M.W. Matsen, Macromolecules **28** 5765 (1995).
7. D.A. Hajduk, P.E. Harper, S.M. Gruner, C.C. Honeker, G. Kim, E.L. Thomas, and L.J. Fetters, Macromolecules **27**, 4063.(1994).
8. T. Hashimoto, K. Yamasaki, S. Koizumi, and H. Hasegawa, Macromolecules **26**, 2895 (1993).
9. A.D. Vilesov, G. Floudas, T. Pakula, E. Yu Melenevskaya, T.M. Birshtein, and Yu. V. Lyatskaya, Macromol. Chem. Phys. **195**, 2317 (1994).
10. J. Zhao, B. Majumdar, M.F. Schulz, F.S. Bates, K. Almdal, K. Mortensen, D.A. Hajduk, and S.M. Gruner, Macromolecules **29**, 1204 (1996).
11. R.J. Spontak, J.C. Fung, M.B. Braunfeld, J.W. Sedat, D.A. Agard, L. Kane, S.D. Smith, M.M. Satkowski, A. Ashraf, D.A. Hajduk, and S.M. Gruner, Macromolecules **29**, 4494 (1996).
12. M.W. Matsen and F.S. Bates, Macromolecules **28**, 7298 (1995).
13. E.K. Lin, A.P. Gast, A.-C. Shi, J. Noolandi, and S.D. Smith, Macromolecules **29**, 5920 (1996).
14. L. Kane, M.M. Satkowski, S.D. Smith, and R.J. Spontak, Macromolecules (in press).
15. T.M. Birshtein, Yu. V. Lyatskaya, and E.B. Zhulina, Polymer **31**, 2185 (1990).
16. C.C. Han, D.M. Baek, J. Kim, K. Kimishima, and T. Hashimoto, Macromolecules **25**, 3052 (1992).
17. M.W. Matsen and F.S. Bates, Macromolecules **29**, 7641 (1996).
18. E.L. Thomas and R.L. Lescanec, Phil. Trans. Royal Soc. A (Lond.) **348**, 149 (1994).
19. M.W. Matsen, J. Chem. Phys. **103**, 3268 (1995).
20. H.S. Gulati, C.K. Hall, R.L. Jones, and R.J. Spontak, J. Chem. Phys. **105**, 7712 (1996).

THE EFFECT OF CHEMICAL FUNCTIONALITY ON ADHESION HYSTERESIS: A STUDY USING THE JKR METHOD

SOOJIN KIM, GUN YOUNG CHOI, JEFF NEZAJ, ABRAHAM ULMAN*
Department of Chemical Engineering, Chemistry, and Materials Science and the NSF MRSEC
for Polymers at Engineered Interfaces, Polytechnic University, Brooklyn, New York 11201

CATHY FLEISCHER
Material Science and Engineering Division, Eastman Kodak Company, Rochester, NY 14650

ABSTRACT

The adhesion of crosslinked PDMS surfaces to self-assembled monolayers with different chemical functionality was investigated using the JKR method, the contact mechanics of solids spreading their interfacial area under load. Interfacial H-bonding was shown to be an important chemical interaction causing significant adhesion hysteresis. The number of H-bonds between PDMS and silanol groups on SiO_2/Si surfaces increased with time of the contact under a constant load, indicating pressure-induced reorganization of the PDMS network near the interface. The interaction between PDMS and carboxylic acid groups showed somewhat smaller hysteresis which suggests weaker H-bonding strength. The interaction between PDMS and functionalized biphenyl groups exhibited small hysteresis which is believed to be caused by dipolar interaction, whereas that between PDMS and nonpolar perfluorocarbon groups showed negligible hysteresis. The distinction in the behavior of the unloading data between H-bonding related interaction and dipolar interaction seems to indicate the difference in the nature between non-specific (van der Waals, dipolar) and specific (donor-acceptor, H-bond, acid-base) interactions.

INTRODUCTION

The ability to control surfaces and interfaces is critical for the development of new technologies and the modification of existing ones in the materials research and engineering. Yet, there is little understanding of the relationship between surface chemistry and adhesion. We have launched a program to address this issue, whereby we study the adhesion of cross-linked poly(dimethylsiloxane) (PDMS) lenses to molecularly engineered surfaces, using the JKR method.

The JKR theory is a continuum contact mechanical model developed by Johnson, Kendall, and Roberts [1] that considers the effect of surface energy on the properties of an elastic contact and has been widely used by Chaudhury [2-5], Kramer [6], Brown [5,7], and Tirrell [8], for direct estimation of the surface free energy and the work of adhesion using functionalized PDMS surfaces. The radius of contact, a, between an elastic semispherical surface and a flat nonelastic surface at equilibrium is described by the equation:

$$a^3 = \frac{R}{K}\left[P + 3\pi WR + \left(6\pi WRP + (3\pi WR)^2\right)^{1/2}\right] \tag{1}$$

where R is the radius of curvature of the spherical elastomer, P is the applied load at the contact interface, K is the elastic constant of the elastomer, and W is the thermodynamic work of adhesion between two surfaces. The JKR apparatus and experimental procedures have been described by previous authors. [2,6]

We have prepared model PDMS networks using low polydispersity PDMS polymers synthesized from hexamethyl cyclotrisiloxane. [9-11] The narrow molecular weight distribution resulting from this particular polymerization method allowed us to create a model network containing almost no defects and only a tiny amount of unconnected chain ends; hence, the

Mat. Res. Soc. Symp. Proc. Vol. 461 ©1997 Materials Research Society

possibility of viscoelastic processes in the JKR experiment can be minimized. The amount of tetrafunctionalized hydrosilylation crosslinker was accurately controlled (somewhat more than stoichiometric ratio) in order to yield more complete crosslinking and thus optimal elastic behavior, as in the process developed by Patel et al. [12] PDMS lenses were prepared by allowing the polymer droplets to crosslink in a heated vacuum desiccator for 3 days as described by Perutz et al. [13]

Self-assembled monolayers (SAMs) prepared on the surface of gold from functionalized thiols were used as molecularly engineered surfaces with different chemical functionality. These thiol monolayers on gold are especially suited for studies of adhesion because of the possibility of fine-control of diverse surface functional group structure and concentration since it is relatively easy to functionalize and purify thiols compared to other precursors such as chlorosilane derivatives. In addition, SAMs of thiols on gold have well-characterized homogeneous and close-packed structure with strong chemical and thermal stability. In this study, we have used SAMs of $HS(CH_2)_2(CF_2)_9CF_3$, $HS(CH_2)_{15}COOH$, $HS(C_6H_4)_2CH_3$, and $HS(C_6H_4)_2F$ on gold following the procedure described elsewhere [14-16].

EXPERIMENTAL RESULTS AND DISCUSSION

Many authors have investigated the origin of adhesion hysteresis in the JKR experiments and attributed them to different effects such as irreversible chemical interaction between surface molecular groups [2], reorganization, entanglement, or interdigitation of surface molecular chains [8,17], viscoelastic energy loss due to plastic deformation of the elastomer [18], heat dissipation, and contact history [17]. In our effort aimed at decoupling different sources of hysteresis (whether chemical, mechanical, or kinetic, etc.) we first carried out the JKR experiments of self-adhesion of PDMS using two identical surfaces in order to eliminate the chemical effect on the adhesion so that observed hysteresis can be related to other effects. When continuous loading and unloading were carried out, the system was hysteretic as shown in Figure 1. However, if loading and unloading were carried out in steps, allowing the system to approach "quasi-equilibrium" after each step (for 5 min), the hysteresis apparently disappeared. We believe that the process responsible for the observed hysteresis in the continuous loading experiment was the incomplete (and thus dissipative) mechanical relaxation of the PDMS crosslinked network because it takes a finite length of time for the change of mechanical potential energy (applied by the load) to be balanced against those of elastic energy and surface adhesion energy. This experiment provided a procedure in which we can eliminate any mechanical effect on the observed hysteresis; all our experiments utilized this step-wise loading and unloading protocol. When the JKR equation (1) was used to fit the equilibrium data of PDMS self-adhesion in order to estimate the parameters W and K using the least square method, W came out to be 42mJ/m^2 which is approximately twice the surface free energy of PDMS commonly determined from the contact angle measurements and in adequate agreement with previous investigators' works [2, 19].

One example for the chemical interaction attributed for adhesion hysteresis is the formation of H-bonds. We investigated time-dependent adhesion hysteresis when H-bonding exists between the PDMS lens and oxidized silicon wafer surface. Silicon wafers were cleaned in H_2SO_4/H_2O_2 solution, rinsed with H_2O and dried under N_2. Three experiments were carried out with different waiting periods between maximum loading and the start of unloading. As shown in Figure 2, the initial hysteresis (0 min waiting) is very significant and much larger than that observed in other experiments with continuous loading (Figure 1). The hysteresis can be attributed to the H-bonding between the naturally-present surface Si-OH groups on silicon wafer and the Si-O-Si groups of PDMS (Figure 3). The fact that the hysteresis increases with the

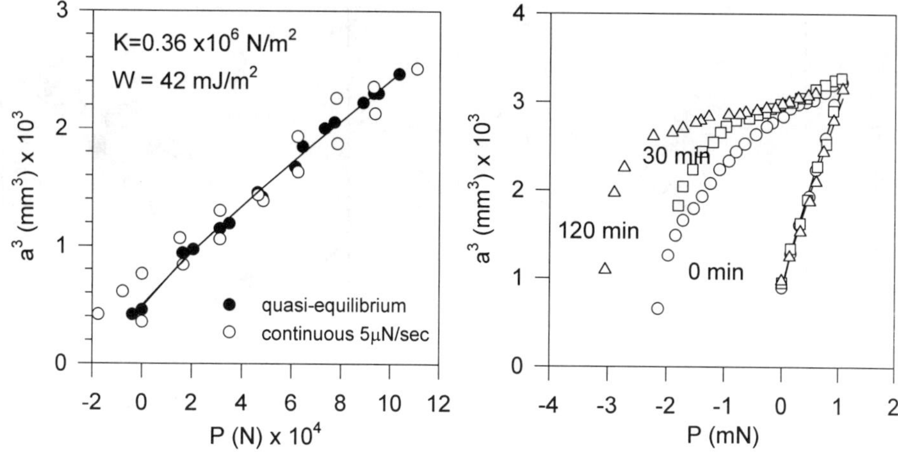

Figure 1. Radius of contact, a^3 vs. load, P in self-adhesion of two PDMS lenses.

Figure 2. H-bonding interaction between PDMS and SiO_2/Si surface with different waiting time between maximum loading and the start of unloading.

increasing time of contact indicates that there is a time-dependent and pressure-induced reorganization of the PDMS network near the interface that increases the number of H-bonds. Notice that a^3 does not increase with time, indicating that these H-bonds are not strong enough to deform the hemisphere near the contact line. Hence all the observed change in hysteresis results from chain reorganization at the PDMS-Si/SiO_2 interface. Interestingly, when the silicon wafers were cleaned using Ar plasma, very little hysteresis was observed. This is probably because the outermost layers of the native SiO_2 have been removed, thus leaving a surface with no Si-OH group.

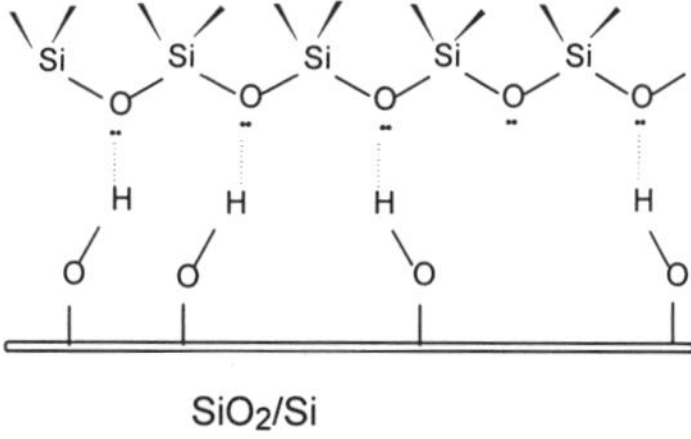

Figure 3. H-bonding between PDMS and silanol groups on SiO_2/Si surfaces.

The interaction between PDMS and $HS(CH_2)_{15}COOH$ monolayer (shown in Figure 4) also demonstrates the same effect of the H-bonding between the PDMS Si-O-S groups and the carboxylic acid -OH groups. The effect of H-bonding on adhesion hysteresis shows a marked distinction when compared to another type of chemical interaction such that between PDMS and CH_3- or F- biphenyl SAMs on gold (shown in Figures 5 and 6) which contain no source of H-bonding. The slopes of the H-bonding-related unloading curves (Figures 2 and 4) increase as the separation advances whereas the unloading curves from the interaction between PDMS and the functionalized biphenyl thiol SAMs (Figures 5 and 6) are almost straight lines with constant slopes. We believe that the hysteresis observed in Figures 5 and 6 is caused by dipolar interaction. The interaction between PDMS and nonpolar $HS(CH_2)_2(CF_2)_9CF_3$ surface (shown in Figure 7) exhibits very small hysteresis since there is no irreversible chemical interaction. In addition, the rigidity of the perflurocarbon chains and of the biphenyl groups precludes the

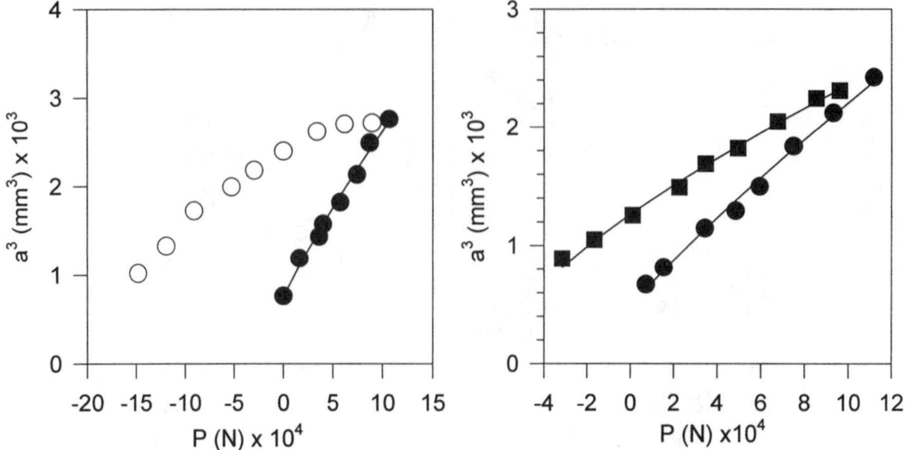

Figure 4. H-bonding interaction between PDMS and SAM of -COOH.

Figure 5. Surface interaction between PDMS and CH$_3$-biphenyl thiol SAM.

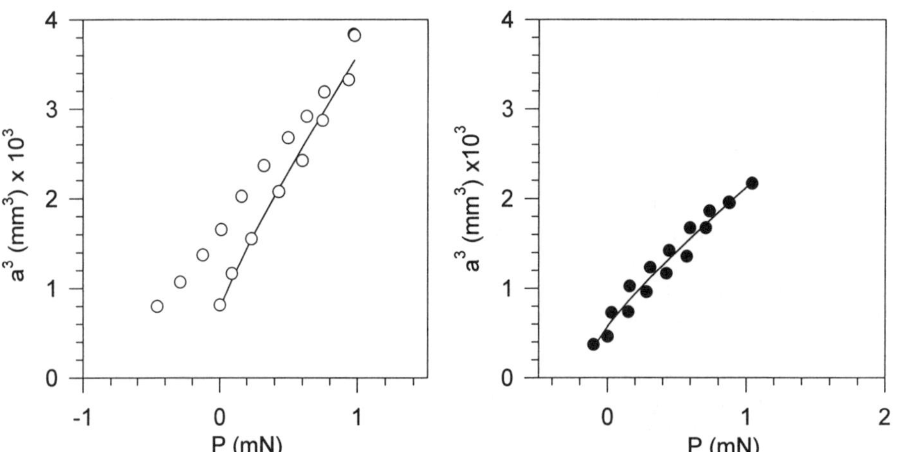

Figure 6. Surface interaction between PDMS and F-biphenyl thiol SAM.

Figure 7. Surface interaction between PDMS and CF$_3$(CF$_2$)$_9$(CH$_2$)$_2$SH SAM.

possibility of surface reorganization of the monolayers which can be another source of hysteresis. Currently we are in the process of investigating van der Waals and dipolar interactions using a series of biphenyl SAMs with different chemical functionality.

This distinction in the unloading curvature seems to indicate the difference in the nature between non-specific interaction (van der Waals, dipolar) and specific interaction (donor-acceptor, H-bond, acid-base). Furthermore, the strength of H-bond is generally much greater than the other type of interaction, and thus it is likely during the unloading process the PDMS at the crack tip is extensively stretched by the attempts to pull apart strongly interacting surfaces until the bonds break. In such case a good amount of adhesion energy is consumed in plastic dissipation at the crack tip, and the unloading data are in a significant departure from the JKR regime.

In the viewpoint of fracture-mechanics, the contact experimental data can be expressed in terms of the strain energy release rate, G as a function of contact radius, a [20]:

$$G = \frac{(P_H - P)^2}{6\pi K a^3} \tag{2}$$

where P_H is the apparent Hertz load ($P_H = a^3 K / R$) which would produce the radius of contact, a, in the absence of surface forces. G is defined as [20]:

$$G = \left(\frac{\partial U_E}{\partial A} + \frac{\partial U_P}{\partial A} \right) \tag{3}$$

where A is the contact area, and U_E and U_P are the stored elastic energy and the potential energy of the load, respectively. Figures 8 and 9 show the calculated values of G plotted with respect to the contact area, πa^2. The elastic constant, K, was estimated for each experiment by fitting the loading data to the JKR equation (1), assuming that the loading part followed the JKR behavior. For the loading data the value of G is mostly constant near the value of W for all the experiments, indicating that the loading proceeded more or less near equilibrium. On the other hand, for the unloading data, G increases continuously as the separation increases for all the experiments except for the case of the perfluorocarbon SAM. This indicates $\partial G/\partial A<0$ which means unstable equilibrium where the separation can spontaneously increase toward a new stable equilibrium. The plots show highest values of G of unloading for the H-bonding interaction between PDMS and silanol groups and somewhat lower for that between PDMS and carboxylic acid groups. The unloading G for the dipolar interaction (PDMS and biphenyl functional groups) is much lower than that for H-bonding interaction. Detailed studies are underway to further elucidate these issues.

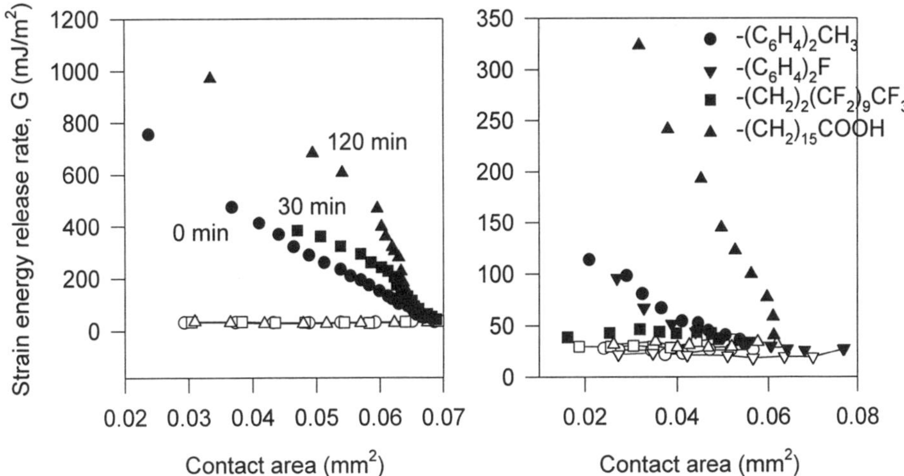

Figure 8. Strain energy release rate vs. contact area for the H-bonding interaction between PDMS and Si-OH on SiO₂/Si. Hollow and filled symbols indicate loading and unloading data, respectively.

Figure 9. Strain energy release rate vs. contact area for the interaction between PDMS and various SAMs. Hollow and filled symbols indicate loading and unloading data, respectively.

ACKNOWLEDGMENTS

Funding from Eastman Kodak Company is greatly appreciated. Many useful discussions with Prof. E. Kramer, as well as his help in our initial research stages are greatly appreciated.

REFERENCES

1. K.L. Johnson, K. Kedall, and A.D. Roberts, Proc. R. Soc. London Ser. A **324**, 301 (1971).

2. M.K. Chaudhury and G.M. Whitesides, Langmuir **7**, 1013 (1991).

3. M.K. Chaudhury and G.M. Whitesides, Science **255**, 1230 (1992).

4. M.K. Chaudhury, J.Adhes.Sci. Technol. **7**, 669 (1993).

5. B. Zhang Newby, M.K. Chaudhury, and H.R. Brown, Science **269**, 1407 (1995).

6. P. Silberzan, S. Perutz, E. Kramer, and M.K. Chaudhury, Langmuir **10**, 2466 (1994).

7. H.R. Brown, Annual Rev. Mater.Sci. **21**, 463 (1991).

8. W.W. Merill, A.V. Pocius, B.V. Thakker, and M. Tirrell, Langmuir **7**, 1975, (1991).

9. C.L. Lee, C.L. Frye, and O.K. Johnson, Polym. Prepr., Am. Chem. Soc. Div. Polym. Chem. **10**, 1361 (1969).

10. C.L. Lee and O.K. Johnson, J. Polym.Sci.: Polym. Chem. Ed. **14**, 729 (1976).

11. C.L. Lee, U.S. Patent No. 3 445 426 (1969).

12. S.K. Patel, C. Malone, J.R. Cohen, J.R. Gilmore, and R.H. Colby, Macromolecules **25**, 2541 (1992).

13. S. Perutz, J. Wang, C.K. Ober, and E.J. Kramer, 1996 ACS Meeting Proceedings, 45 (1996).

14. A. Ulman, S.D. Evans, Y. Shnidman, R. Sharma, J.E. Eilers, and J.C. Chang, J. Am. Chem. Soc. **113**, 1499 (1991).

15. A. Ulman, Ultrathin Organic Films: From Langmuir-Blodgett to Self Assembly (Academic Press, Boston, 1991).

16. A. Ulman, Chem. Rev. **96**, 1533 (1996).

17. Y.L. Chen, C.A. Helm, and J.N. Israelachvili, J. Phys. Chem. **95**, 10736 (1991).

18. R.G. Horn, J.N. Israelachvili, and F. Pribac, J. Colloid Interface Sci. **115**, 480 (1987).

19. M.K. Chaudhury, Materials Science and Engineering **R16**, (19) No. 3, (1996).

20. D. Maugis and M. Barquins, J. Phys. D: Appl. Phys. **11**, 1989 (1978).

INTERFACIAL REACTION AND ELECTROCHEMICAL ACTIVITY OF THIOKOL RUBBER/CONJUGATED POLYMER COMPOSITES

GONG KE-CHENG*, MA WEN-SHI

*Polymer Structure & Mod. Res. Lab. , South China University of Technology, Guangzhou, 510641, China

ABSTRACT

The highly electroactive thiokol rubber (TR)/ conjugated polymer (eg. polyaniline (PAn) or polypyrrole (PPy)) composite films were prepared by electropolymerization deposition via one-step process in the electrolytic solutions containing aniline or pyrrole and TR oligomer. The electrocatalysis of PAn or PPy for the electrodepolymerization (reduction)- electropolymerization (oxidation) reaction of TR in the interface between PAn or PPy and TR is determined by cyclic voltammograms. The differeme between the oxidation potential and the reduction potential is 0.05V and 0.36V or less for TR/PAn and TR/PPy composite films, respectively. The chemical bands between the nitrogen atoms of PAn or PPy and the mercaptan groups of TR (oligomer) are formed in the electropolymerization process that is indicated by XPS. The conductivities of TR/PAn and TR/PPy composite films and the stability of the cells consisting of those films are remarkably improved after electrochemical reduction with addition of a suitable conducting carbon black.

INTRODUCTION

All-solid-state energy-storage system have attracted worldwide attention and are currently being pursued by a number of battery developers. To date. the majority of positive electrode materials used in these systems have been intercalation compounds, but their limited rate capability and low utilization of cathode capacity have hindered their practical application. Disulfide compounds have recently been proposed as candidates for high energy cathodes in lithium batteries or lithium ion batteries. A series of compounds having-SH groups in the molecules are thought to serve as the energy storage material, where by energy exchange occurs according to a reversibe polymerization-depolymerization process$(2SH\leftrightarrow S\text{-}S)$[1,2]. The theoretical energy content of these materials for exceeds that of conventional battery materials as well as those of intercalation compounds[3] and conducting polymers.

To get better charge-discharge performame at ambient temperature, it has already been reported that polyaniline can be acted as an effective electrocatalyst to speed up the relatively slow redox of the DMcT[4,5]. The authors have already reported that the highly electroactive thiokol rubber(TR)/conjugated polymer (eg. polyaniline (PAn) or polypyrrole (PPy)) composite films were prepared by electropolymerization deposition via one-step process in the electrolytic solutions containing aniline or pyrrole and TR oligomer[6].
In this work, we report the interfacial reaction and electrochemical activity and microstructure of thiokol rubber/conjugated polymer composites.

EXPERIMENTAL

1. Chemicals

Acetonitrile (MeCN) and trifluoroacetic acid (TFA) and trichloroacetic acid (TCA) were synthetic grade. Aniline, pyrrole and propylene carbonate (PC) were used after distillation under reduced pressure. Liquid thiokol rubber (JLK — 121, Mn=1,000) and conducting carbon black were used as received.

2. Composite film electrodes and composite films preparation

The TR/PAn or TR/PAn/C and TR/PPy or TR/PPy/C composite films were prepared by galvanostate eletrochemical polymerization on a platinium plate electrode of 0.25 cm^2 (the surface of the platinium electrode, as the working electrode, was polished with alumina powder on a microcloth wetted by twice-distillation water and rinsed with excess water and acetone before use.) in a MeCN solution containing 0.5 mol dm^3 aniline, 2.0 mol dm^3 TCA, 2.0 mol dm^3 TFA, 0.125 mol dm^3 structural unit of TR oligomer, 0 or 10g dm^3 conducting carbon black, or 0.2 mol dm^3 pyrrol, 2.0 mol dm^3 TCA, 0.05 mol dm^3 structural unit of TR, 0 or 10g dm^3 conducting carbon black as an electrolyte, respectively. The current density was 4mA/cm^2, but the film thickness was controlled by passing charges during electropolymerization. After polymerization, the resulting films were washed thoroughly with distillation water to be used in cyclic voltammograms tests or to be tried in vaccum for 12 hours.

3. Electrchemical measurement

A standard three-electrode, two-compartment electrochemical cell was used for all the electrochemical experiments. Saturated calomel electrode (SCE) and the composite film electrode were used as the reference electrode and the working electrode in the MeCN electrolytic solutions. Cyclic voltammograms were obtained on a normal potentiostate ECO-553 instrument.

4. Characterization

The conductivities of the composite films were measured by a 4-probe method. The vacum-dried composite film samples were mounted onto a standard sample holder by using double -sided adhesive tape. For XPS analysis, a Perkin-Elmer PHI Model 5600 was used with a Al Ka X-ray source (1486.6 eV).

RESULTS AND DISCUSSION

1. The electrochemical properties of the composite films

A typical cyclic voltammogram for PAn in MeCN solution is shown in Fig. 1a. In the voltammograms, two oxidation peaks and two reduction peaks can be observed. Moreover, the cyclic voltammogram curves did not change with the scan films. To identify the reversibility of PAn, the scanrate was from 50 mV/sec to 100 mV/sec, and the oxidation potential and the reduction potential did not change with the scan rate changed, but the peak current increased. These results indicated that PAn had a good reversibility and stability. A cyclic voltammogram for TR oligomer on the platinium electrode in MeCN solution is shown in Fig. 1b. In its voltammogram, the different between the oxidation potential and the reduction potential is 0.5 V or more, and its value increase with the scan rate increasing. These results imply that the redex reversibility of TR oligomer was poor. However, in the case of using a conbination of PAn and TR oligomer, (containing about 48 wt%TR), the difference between the oxidation potential and the reduction potential is reduced to 0.05 V or less (shown in Fig. 1c.), and its value does not chang with the scan rate changing. In the TR oligomer which is combined with PAn, the electrode reaction is promoted and a higher current density at room temperature is obtained on electrolysis, i.e., on charging or discharging and $\left| i_{pa}/i_{pc} \right| \approx 1$. When the electrode material is subjected to electrolytic oxidation, PAn is oxidized at first and the resulting oxidized form of PAn oxidizes the reduced type of TR oligomer. Thus, the oxidized form of PAn returns to the reduced form and an oxidized form of TR is generated. As described above, PAn has a function for promoting the movement of the electrons in the oxidation-reduction reaction. Therefore, the interfacial contact state between PAn and TR has played an

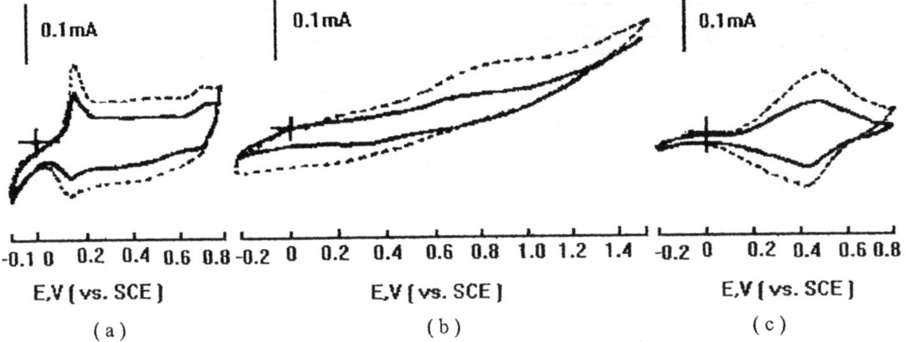

Fig. 1   Cyclic voltammograms of (a) PAn, (b) TR and (c) TR~48wt%/PAn film in the MeCN solution
containing 2.0 mol dm^{-3} TCA and 2.0 mol dm^{-3} TFA. Room temperature.
(——) 50 mV /sec. scan rate, 1~5 times. (········) 100 mV/sec. scan rate.

important role in the electrocatalysis of PAn for the electrodepolymerization-electropolymerization reaction
of TR. The results of the cyclic voltammogram indicated that the TR/PAn composite film electrode had a
highly interfacial electrochemical activity.

The cyclic voltammograms for PPy, TR oligomer, TR/PPy composite film in MeCN solution are shown
in Fig. 2a,b,c respeatively. The redox reaction rate, redox reversibility and the charge or discharge current
density of these electrode materials are much smaller than that of PAn and TR/PAn composite film. In the
case of using a combination fo PPy and the TR oligomer, the difference between the oxidation potential and
the reduction potential is smaller than that of the TR oligomer alone. The several electrochemical properties
of the composite electrodes and TR oligomer in the different situation are shown in table 1.

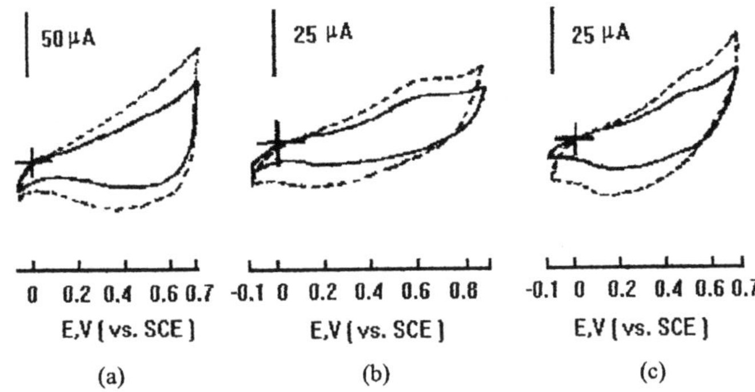

Fig. 2   Cyclic voltammograms of (a) PPy film, (b) TR on the Pt electrode, (c) TR~ 5.53wt% PPy film
in the MeCN solution containing 2.0 mol dm^{-3} TAC and 0.2 mol dm^{-3} LiClO$_4$. Room temperature.
(——) 50 mV/sec. scan rate, 1~5 times,   (········) 100 mV/sec. scan rate.

Electrode	Electrolyte Solution	E_{pa}	E_{pc}	ΔE_p	$|i_{pa}/i_{pc}|$
TR	TCA-TFA-MeCN	0.720	0.225	0.495	> 1
PAn/TR-48[a]	TCA-TFA-MeCN	0.480	0.430	0.050	≈ 1
TR	TCA-LiClO$_4$-MeCN	0.630	0.163	0.440	> 1
PPy/TR-5.53[b]	TCA-LiClO$_4$-MeCN	0.560	0.203	0.357	> 1

Table 1   Several electrochemical parameters for the composite films and TR

* a: PAn/TR film having about 48 wt% TR   b: PPy/TR film having about 5.53 wt% TR

2. Interaction between TR oligomer and PAn or PPy

As from the above cyclic voltammograms studies, there are some interaction between TR oligomer and PAn or PPy, and the interaction gives rise to the fast electrochemical reaction of TR oligomer in a PAn or PPy matrix. These have been supported by another research groups[4,7]. In the present work, we intend to examine the chemical reaction between TR oligomer and PAn or PPy. Fig.3a,b exhibit high-resolution XPS examination of a $S_{(2p)}$ region of the TR/PAn and TR/PPy composite films, respectively. The outline of the two figures are extremely similar, but the binding energy of S_{2p} electron in the TR/PAn composite is higher 0.3 eV than that of S_{2p} electron in the TR/PPy composite. Therefor, the chemical environment of the sulphur atoms in the two composites is very similar. Because of producing self-rotation-orbit coincindence effect, sulphur atom has two state electrons $S_{2p1/2}$ and $S_{2p3/2}$, and their binding energies are 165.1 eV and 163.8 eV in the Fig. 3a and 164.8 eV and 163.5 eV in the Fig.3b, respectively. The pesk at binding energy 164.6 eV or 164.3 eV is assigned to the sulphur in the sulphur-nitrogen band combination. The result implies that the chemical reaction between PAn or PPy and TR oligomer occured in the one-step electrochemical composite process. This can be supported by the $N_{(1s)}$ XPS spectra of TR/PAn or TR/PPy composite(shown in Fig. 4a,b). The investigation of the $N_{(1s)}$ peaks indicates the presence of -NH$_2$-, -NH-, -NH-, N-S, -N= groups in the composites.

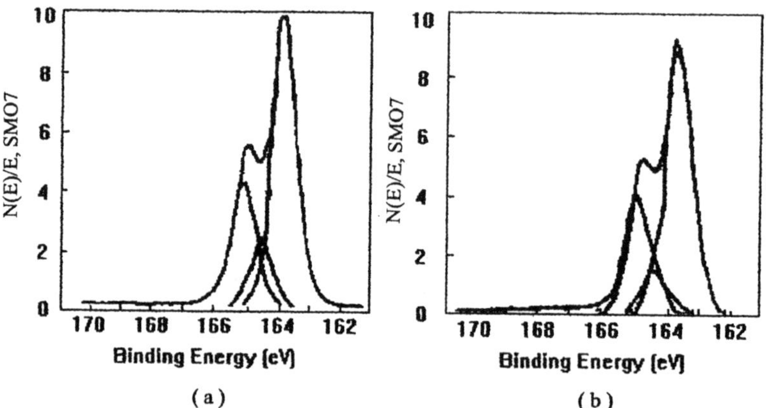

Fig.3 S_{2p} XPS spectra of the surfaces with the solution side of (a) TR/PAn and (b) TR/PPy film

3. Modified TR/PAn and TR/PPy composite films by conducting carbon black

The TR/PAn and TR/PPy composite films have a good electrochemical reversibility, but the stability of the electrolytic cells consisting of those films is poor. When the cell is electrolytic reduced, PAn or PPy molecular chains of the composite electrodes change first from oxidation form to reduction form, and its conducting ability swiftly desend and can not be controlled. The phenomena is more clear at the high current density (Fig. 5). While the cells consisting of the TR/PAn/C or TR/PPy/C composite film prepared by one-step electrochemical process from the solution containing aniline, TR oligomer and conducting carbon black ($10g\ dm^3$) were electrolytic reduced, the reduction current density can be accurately controlled, and can be fast reduction or oxidation at high current density (Fig. 5). The comparision of the properties of the composite films between oxidation form and reduction form is shown in table 2.

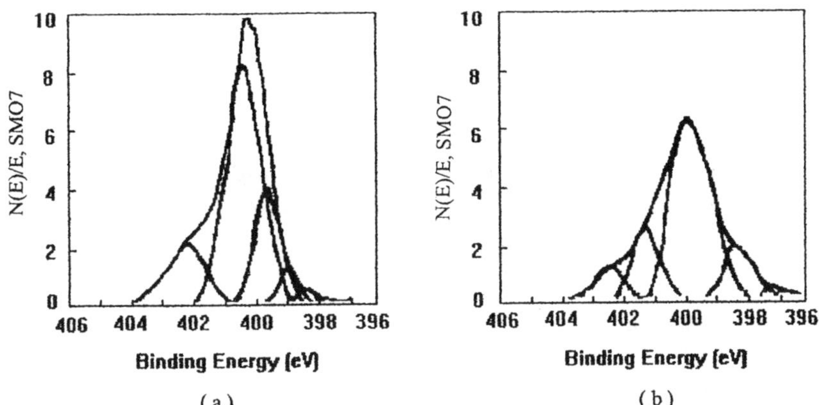

(a)   (b)

Fig. 4 N_{1s} XPS spectra of the surface with the solution side of (a) TR/PAn and (b) TR/PPy films.

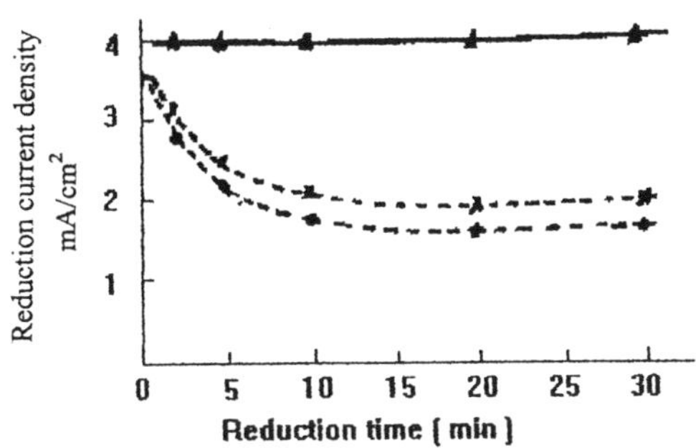

Fig. 5   The relationships of reduction current density and reduction time of the composite films in the electrolytic cell.   (Δ) PPy/TR/C or PAn/TR/C film, (×) TR/PAn film, (o) TR/PPy film.

Table 2   The comparison of the properties of the composite

Film	Appearance density	Visual appearance	Conductivity (S cm^{-1})	
			oxidation form	reduction form*
PPy/TR	0.527	rough, flexible	5.8	$<10^{-4}$
PPy/TR/C	0.21	rough, brittle	1.72	7.6×10^{-2}
PAn/TR	0.50	rough, flexible	0.21	$<10^{-4}$
PAn/TR/C	0.32	rough, brittle	2.3×10^{-2}	2.0×10^{-2}

* The reduction current density and reduction time were 4 mA/cm^2 and 30 min., respectively.

CONCLUSION

The electrocatalysis of Pan or Ppy for the electrodepolymerization-electropolymerization reaction of TR in the interface between TR and Pan orPPy is very clear. The difference between oxidation potential and reduction potential is 0.05 V and 0.36 V or less for TR/PAn and TR/PPy composite films, respectively. The chemical bands between the nitrogen atoms of Pan or Ppy and the mercaptan groups of TR are formed. The conductivities of the TR/PAn and TR/PPy composite film and the stability of the cells consisting of these films are remarkable improved after electrochemical reduction with addition of a suitable conducting carbon black.

REFERENCES

1. S. J. Visco, C. C. Mailhe, L. C. De Jonghe, and M. B. Armand, J. Electrochem. Soc., 136,661 (1989)
2. S. J. Visco and L. C. De Jonghe, Mater. Res. Symp. Proc., 135, 553 (1989)
3. M. Ue, S. J. Visco, and L. C. De Jonghe, Denki Kagaku, 61, 1409 (1993)
4. K. Naoi, M. Menda, H. Doike, and N. Oyama, J. Electroanal. Chem., 318, 395 (1991)
5. T. Sotomura, H. Uemachi, K. Takeyama, K. Naoi, and N. Oyama, Electrochem. Acta, 37, 1851 (1992)
6. Gong Kecheng and Ma Wenshi, Abstract of MAS Fall Meeting, T.9.2, P.554, 1995
7. S. Ye and D. Belanger, J. Electrochem. Soc., 141, L49 (1994)

PHASE SEPARATION KINETICS DURING DRYING

R.F. SARAF (1), S. OSTRANDER (2), R.M. FEENSTRA (3)

(1) IBM Corp., T.J. Watson Research Center, Yorktown Hts., NY 10598

(2) Dept. of Material Sci., Alfred University, Alfred, NY 14802

(3) Physics Dept., Carnegie Mellon University, Pittsburgh, PA 15213

Abstract

Kinetics of phase separation at air/polymer interface in a binary polymer mixture on evaporation of common solvent is studied. The lateral dimension of the highly anisotropic, pancake-like, minority phase increases with a growth exponent of 2/3 identical to 'late-stage' growth under (classical) thermal-quench at interface. In contrast to the thermal-quench, during drying the kinetics depends on the initial condition (i.e., initial concentration, c_0) that is rescaled to obtain a master-curve.

Introduction

Recently, there has been a great deal of interest in surface influenced phase separation (SIPS) phenomena in polymer and monomeric systems upon thermal-quenching (1-6). We will refer to SIPS due to thermal quench as classical SIPS to distinguish it from our case of SIPS on drying. The characteristic length of one of the phases during separation increases with time as, t^a where, a is the growth-exponent. There are distinctions from the classical 3D (three dimensional) system where the growth-exponent, $a=1/3$ can increase to 1.0 due to coalescence (7). Some general observations from the study of classical SIPS phenomena are as follows: (i) Due to preferential wetting of one phase over the other, there is a low energy interfacial layer of the latter at the interface as predicted by Cahn (8). The growth kinetics of the characteristic length is also influenced due to the differences in the wetting tendencies of the phases at the interface. (ii) According to Tanaka's model (5), at an early stage the lateral growth occurs mainly due to 2-dimensional (2D) capillary instability with the more-wetting phase transported to the interface leading to a growth exponent of a=3/2. This is experimentally observed for a polymeric system where due to slow dynamics the early-stage could be probed (4). (iii) The late-stage kinetics, observed for small-molecule fluids indicated a slower exponent of a=2/3 (1). (iv) Tanaka's model (5) further indicates that the growth exponent for the surface layer thickness should be 1.0 (at least) during the early-stage regime. This seems consistent with the small-angle neutron scattering results for binary polymer mixture (3). (v) An interesting observation is recently reported by Cumming et al., where a can range from 1 to 1.5 depending on the quench depth with respect to the coexistence curve (9).

Here we report our studies on phase separation kinetics at the surface in a binary polymer blend as the cosolvent evaporates. Essentially, in the experiment we achieve a 'concentration-quench', where the solvent is instantaneously dried (compared to the slow phase separation kinetics of the polymeric system) in the top layer. This layer continues to phase separate until all the solvent from the bulk layer is evaporated. Since the glass transition tem-

Mat. Res. Soc. Symp. Proc. Vol. 461 © 1997 Materials Research Society

peratures (Tg's) of the two polymers chosen are >80 °C above the drying temperature the system is frozen after solvent evaporation (10). No measurable difference in the lateral size is observed when the sample is left over 60 mins. beyond the evaporation time, τ (10). By controlling the evaporation time various stages of phase separation is obtained. The characteristic lateral size of the discreet phase in the top layer is measured by Atomic Force Microscopy (AFM). AFM is chosen because, this way we only probe the phase separation in the top layer avoiding the morphology in the bulk. The topography from the deeper layers in the bulk are filtered by a special image analysis described below. We first discuss the system chosen to achieve the proper concentration-quench. Then describe the growth curve that is dependent on the initial solvent concentration. Next the observations are discussed in terms of a phenomenological model. Finally, a master curve is obtained from the various growth curves at different initial solvent concentrations.

Experimental

We start with a one phase, ternary mixture of two polymers, acid (PAA) and ester (PAETE) precursor of poly(pyromellitic dianhydride-oxydianiline) (PMDA-ODA) in a cosolvent N-methyl pyrrolidinone (NMP). The weight average molecular weights and polydispersity index for PAA and PAETE is 152,000 and 88,000, and 2.2 and 2.1, respectively. In this study the relative weight fraction of PAA with respect to PAETE, defined as ϕ, is fixed at 0.2. The weight fraction of NMP in the solution is c_0. The c_0=0.950, 0.936 and 0.920 for this study.

Films from solutions at given c_0 are spin coated on ~0.2 mm thick single crystal Si wafer with ~1 nm native oxide. During spin coating at room temperature, insignificant amount of solvent evaporates (<0.5%/hour at 30 °C). The resultant thin film is immediately placed on a hot plate at T_{dry}=70 °C under dry forming gas blanket. As the films dries, distinct thickness fringes are observed as a result of solvent evaporation. The fringes and color change ceases after the evaporation time, τ. Since the glass transition temperature of both PAA and PAETE is above 150 °C, no phase separation is expected after the evaporation of NMP. By comparing films made under identical conditions but dried for τ and $\tau+800\tau$, it was confirmed that phase separation process indeed cease after τ sec. (10).

For each c_0, the thickness is varied from ~1 to ~20 μm to obtain broadest possible variation in τ. The lower thickness is limited by absence of any (observed) phase separation below τ~5 sec., under AFM (11). In films thicker than 20 μm, the morphology of the bulk layers (below DTL) begin to superimpose on the surface topography. The final thickness after drying is, L=100-1000 nm. A regression straight line is fit between L versus τ to calculate the actual τ reported in the growth-curve. The surface topography of the dried film is measured using a commercially available Digital Instrument's Nanoscope III Atomic Force Microscope (AFM). The images were taken at ~50 nN contact force and scan rate of 1 Hz.

Phase Diagram

Fig. 1 shows the ternary phase diagram for the system studied. The cosolvent NMP forms a stable complex with PAA and PAETE as shown by points D and E. The corresponding solvent fractions are c_D=0.1915 and c_E=0.0946 (11,12). Line DE is the locus of concentrations where all the solvent is complexed with either PAA or PAETE, i.e., there is no 'free-solvent'. The only physical method to decomplex the NMP is by imidizing the polymer. Since, in this study we are only concerned with the unimidized polymers, viz, PAA and PAETE, the phase diagram at solvent concentrations, c above DE is of interest. The solvent weight fraction, c on DE as a function of q (= weight fraction of PAA / weight fraction of PAETE) is given by,

$$c\Big|_x = \frac{2c_D c_E - (1-p)c_D - (1+p)c_E}{(1+p)c_D + (1-p)c_E - 2} \tag{1}$$

where, p=(1-q)/(1+q) and subscript, x defines any particular point on CD, such as D', I, E'.

We define ϕ as the weight fraction of PAA relative to PAETE component. No phase separation was observed in films made from solution with ϕ between 1 to 0.85 and 0.03 and 0. Thus the spinodal resides between CM and CN shown in Fig. 1. The onset of phase separation in PAA-rich and PAETE-rich solution defined as D' and E' respectively are given by the intersection of segments CM and CN with DE. Using eq. (1), and q=17/3 and 3/97 for CM and CN respectively, $c_{D'}$=0.1783 and $c_{E'}$=0.0978.

The onset of phase separation for ϕ=0.8, 0.2 and 0.1 is measured by observing turbidity. With these three points, D' and E', a nominal spinodal curve is drawn in Fig. 1. For the concentration of interest for this study, at ϕ=0.2, spinodal solvent weight fraction, c_s=0.89. The final equilibrium concentration after all the free solvent is evaporated is point I. The solvent concentration, c_i=0.1158 at x=I, calculated from eq. (1) for q=1/4. As will be seen later, for this study an exact shape of the spinodal is not required. The important points in phase diagrams are S, I, D', E'. Furthermore, due to high molecular weights, high solution viscosity makes the measurement beyond c=0.85 difficult.

Concentration Quench

We start with, a one-phase film of ϕ weight fraction of A (i.e., PAA) in (1-ϕ) parts of B (i.e., PAETE) in c_0 weight fraction of solvent relative to the total polymer is cast on a substrate (point, R). In this study, ϕ=0.2 and c_0>0.89. As the film is bought to drying temperature, T_{dry}, the solvent begins to dry from initial concentration, c_0 to the final, equilibrium concentration, c_i (point, I). For the polymer composition, ϕ=0.2, the film begins to phase separate at solvent concentration, c_s (point S, on the spinodal curve). The final concentration is given by point I in Fig. 1. The phase separated morphology probed by AFM is seen in Fig. 2. The mounds are the discreet PAA-rich, minority phase.

Unlike temperature induced separation phenomena where the driving force (proportional to the magnitude of the temperature quench) can be controlled fairly uniformly throughout the sample, in drying induced process, the sample will intrinsically have a concentration gradient of the solvent at the surface. However, under proper conditions, analogous to thermal-quench, a 'concentration-quench' can be achieved at the film/air interface. The resultant morphology can then be probed by a surface sensitive technique, such as AFM for this study.

To achieve proper 'concentration-quench', two conditions must be satisfied: (i) The solvent should change from c_0 to c_i instantaneously compared to the phase separation kinetics time scales. (ii) The final solvent concentration should be fairly uniform compared to the phase separation driving force, c_s-c_i in the film.

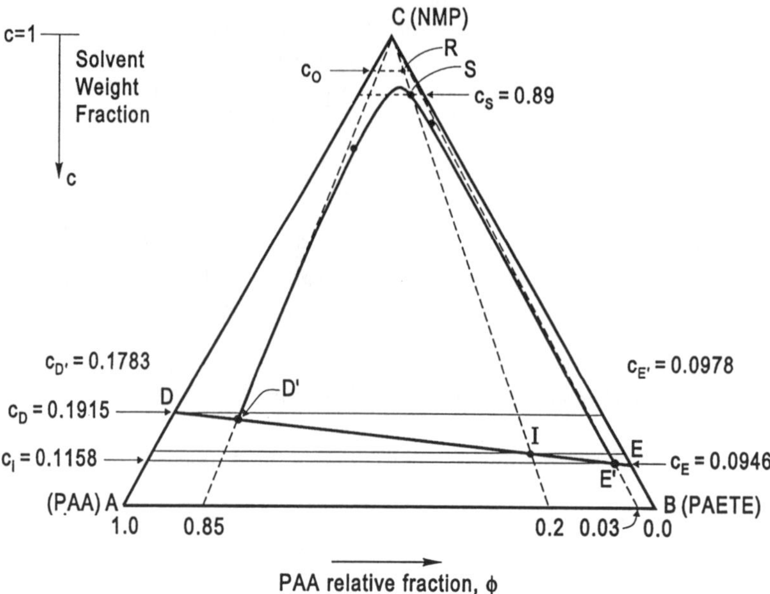

Figure 1: Phase diagram of polymers PAA, PAETE and cosolvent, NMP at 70 °C. All compositions are in weight percent.

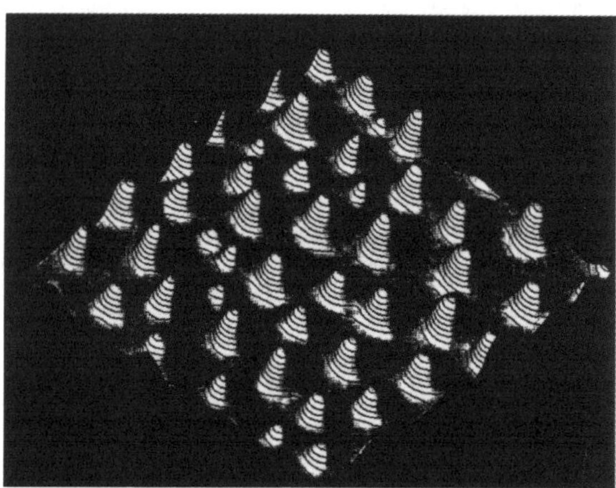

Figure 2: The AFM topograph is obtained on Digital Instrument's Nanoscope III model at contact force of ~50 nN range. The 400x400 pixel image is reproduced using Galaxy on RS6000 machine. The range in x-y plane is 4 μm. The z-range corresponding to the highest elevation, (measured) evaporation time, τ and $<D_\xi>$ for the sample made from c_0=0.916 solutions is, 27.3 nm, 28.3 sec., and 473±13 nm, respectively.

Consider condition (i). From thermal diffusivity of ~0.895 and >0.0017 cm^2/s. for Si and polyimide respectively, the film will attain hot plate temperature of 70 °C in <0.5 sec. under the experimental conditions chosen. This is instantaneous compared to the induction time of t_i~5 sec. before any lateral phase separation is observed by AFM. Thus in <0.5 s, the surface will attain $c_f=c_i$. In t_i, the top layer defined by the penetration depth of the solvent will attain concentration close to c_i. This layer is subsequently referred to as the 'dry-top-layer' (DTL). Assuming Fickian diffusion, for t_i=5 sec. (time before phase separation) and solvent diffusion coefficient, D~10^{-9} cm^2/s at 70 °C (13), the DTL thickness, l ~1.5$(Dt_i)^{1/2}$/S ~170-200 nm for c_0 in 0.920-0.950 range. S is the shrinkage given by $\{[(1-c_0)/\rho]+[c_i/\rho_s]\}/\{[(1-c_0)/\rho]+[c_0/\rho_s]\}$, where ρ~1.2 and ρ_s=1.033 gm/cm^3 are densities of the polymer and NMP. Thus condition (i) is satisfied in the top 170-200 nm layer.

Next consider condition (ii). The concentration in DTL will be nominally linear ranging from $c=c_i$ at the surface to $c=c_l$ at depth of l (14). The non-dimensional solvent concentration profile, $\theta=(c_0-c)/(c_0-c_i)$, in the DTL will range from $\theta=1$ to 0.97 at the surface and l deep from the the surface, respectively. Thus the solvent concentration variation in DTL will be, c_l-c_i=0.03(c_0-c_i). To satisfy condition (ii), $(c_l-c_i)<<(c_s-c_i)$, i.e., $0.03(c_0-c_i)/(c_s-c_i) << 1$. Thus to satisfy condition (ii) we choose a system such that c_0 is close enough to c_s relative to the driving force, (c_s-c_i). Since, for the largest c_0=0.95 (studied), $0.03(c_0-c_i)/(c_s-c_i)$=0.032 <<1, condition (ii) is satisfied. A value of 0.032 implies that the variation in concentration profile is only 3.2% of the driving force, (c_s-c_i). Visually, in the phase diagram, c_i=0.1399 to 0.1408 for c_0=0.92 to 0.95 is very close to point I (below c_D and $c_{D'}$). Hence, we obtain concentration-quench conditions in the DTL.

The DTL will continue phase separate until there is sufficient solvent flux from the layers below. After this time, defined as the evaporation time, τ, the system phase separation process will cease and the system will be frozen, since the Tg of PAA and PAETE is more than 90 °C above the drying temperature. Thus by increasing the film thickness (for given ϕ and c_0), various stages of phase separation in the DTL at increasing τ can be frozen-in for ex-situ analysis. To restrict the observation to surface morphology we probe the structure by AFM.

Image Analysis

The characteristic length scale of the phase separated morphology probed by AFM, is defined by the diameter, D_ξ of the PAA-rich mounds seen in Fig. 2. As expected, and shown later, D_ξ increases with τ. The z-scale is highly exaggerated compared to the x-y scale. The actual shape of the mounds shown are 'pan-cake'. Typically, D_ξ is ~80-fold larger than the corresponding mound height. Such an anisotropic shape of phase separated domains close to the interface is consistent with classical SIPS experiments (1).

The average lateral size, $<D_\xi>$ of the PAA-rich regions is computed by analyzing the 512x512 pixel AFM image. From the cross-sectional view, the mounds are shaped close to inverted paraboloids (see Fig. 2 and (10)). Thus, by plotting the curvature of the topography (defined as d^{2}z/dx^2 + d^{2}z/dy^2 for topography z(x,y)), an image is achieved where the PAA-rich regions are uniformly higher than the background. An example of this is shown in Figs. 3(a) and (b), for a film made from c_0=0.934 solution with τ=26.2 sec. A histogram of pixel values for a curvature such as Fig. 3(b) reveals two peaks, one corresponding to the PAA-rich regions

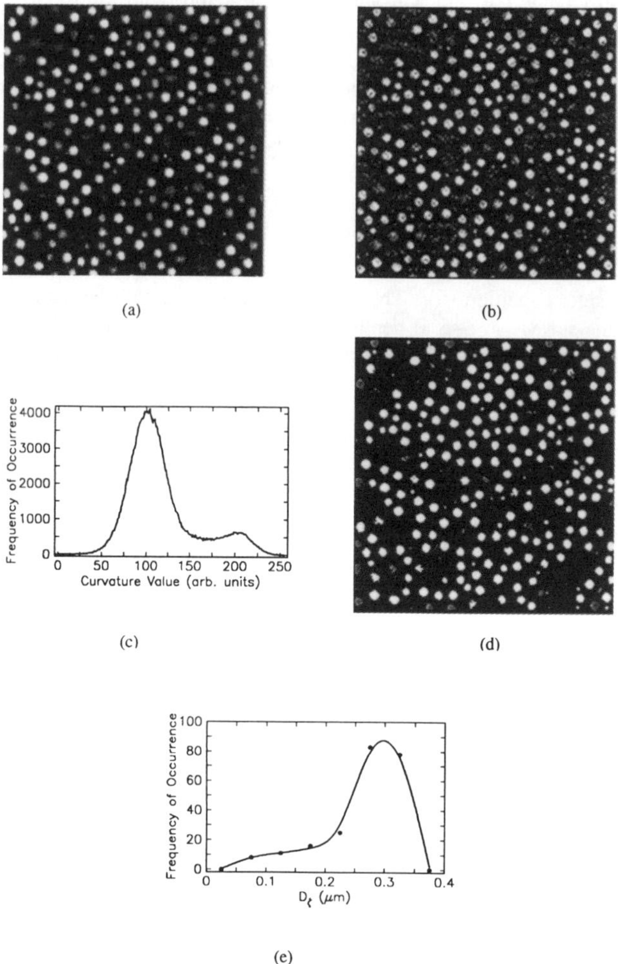

(a)

(b)

(c)

(d)

(e)

Figure 3: An image analysis consisting of five steps is performed to obtain the average size, $\langle D_\xi \rangle$ of the PAA-rich phase. 3(a) is a (raw) 10x10 μm AFM image, the grey level corresponds to height (i.e., topography). In 3(b), the image is replotted with pixels corresponding to curvature, defined in the text. 3(c) shows the curvature distribution of the pixels. For a discriminator curvature value of 152 (defined in text), image 3(d) is obtain. The pixels above and below 152 are shaded white and black respectively. Fig 3(e) shows the size distribution of the discreet white (PAA-rich) regions in 3(d) corresponding to, $\langle D_\xi \rangle$=343±5 nm at τ=26.2 sec.

and the other to the background mixed phase, as illustrated in Fig. 3(c). Defining a discriminator level at the midpoint between the two peak positions in the histogram allows a measure of the diameter of the PAA-rich regions. Fig. 3(d) shows pixels with curvature above (below) the discriminator value are shaded white (black). PAA-rich regions which lie on the edge of the image, or are non-spherical (with ratio of principal moments of inertia outside the range 0.5-2) are excluded, and are shown by the grey regions in Fig. 3(d). Thus both the edge effect and features from phase separation at deeper layer are identified and neglected from the computation. Finally, the D_ξ of the PAA-rich regions is obtained by the area A of the white regions in Fig. 3(d), and using $D_\xi=2(A/\pi)^{1/2}$. Fig. 3(e) shows the distribution curve corresponding to Fig. 3(d).

The typical size distribution seen in Fig. 3(e) (is likely to) occur due to time lags in the initiation of the unstable fluctuations. Although the quench depth in the surface layer is fairly uniform, (as discussed in the next paragraph), the small concentration gradient in solvent concentration will cause phase separation at greater depths to initiate later. This leads to a distribution of lateral size, D_ξ, of the PAA-rich mounds. This implies that the larger domains have better correspondence with the total time available, τ, for phase separation. Thus, the characteristic lateral size scale, $<D_\xi>$, for a given τ, is computed by averaging on the large domains. Specifically, the images are split into four quadrants, and the largest size PAA-rich region in each quadrant is found. The mean $<D_\xi>$ and standard deviation of these largest regions are computed.

Growth Law

Fig. 4 shows the $<D_\xi>$ versus τ for films made from solution with $\phi=0.2$ as a function of c_0. The growth law in the late-stage can be written as, $<D_\xi>=K(t/\kappa)^a$. Unlike classical case, where the scaling factors, K and κ, for correspondence condition (and universality) depend only on the temperature difference between the final (in the unstable region) and the phase separation on-set temperatures, in the drying case they depend on initial condition (i.e., c_0) also. The c_0 variation is rescaled using $\kappa^{-1}\sim c_0$ and $K\sim(1-c_0)^2$ (15). The rescaled curves are shown in Fig. 4.

The visual inspection of data indicates that the growth exponent, a=2/3, except for the lowest c_0. The reasonable superposition of curves as different c_0 indicates good correspondence-law. The good correspondence-law and excellent agreement with the late-stage growth exponent for classical SIPS (1) suggests that the 2/3 growth exponent may be a universal behavior. We note that although the time-range is small (limited by the thickness discussed earlier), the slope is distinctly different from 1/3, 1, and 3/2 exponent lines drawn to aid the eye.

Interestingly, the films made from $c_0=0.9$ indicate the classical 1/3 power law. One explanation may be that for thickness' above 400 nm, the bulk behavior contributes significantly to the surface topography. Thus the $<D_\xi>$ measured corresponds to a 3D bulk geometry rather than the 2D surface.

Summary

We have measured the kinetics of phase separation at an air/film interface during drying. The time sequences of phase separation were obtained by controlling the evaporation time, τ,

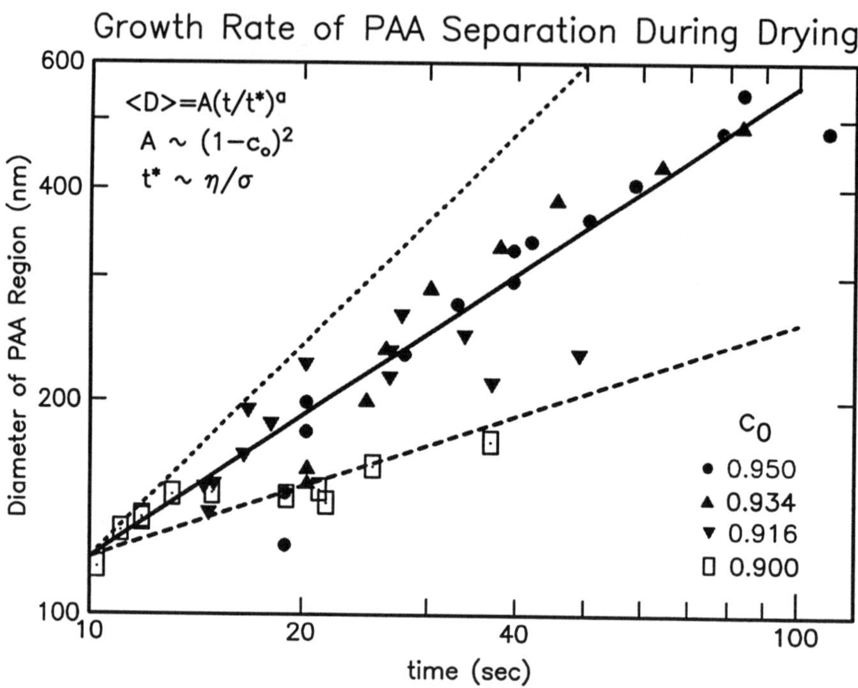

Figure 4: Master growth-curve made from solutions of c_0=0.900, 0.916, 0.934 and 0.950. The vertical shift and horizontal shifts of the the other two curves made relative to the curve for c_0=0.950 are discussed in the text. The lines are just to guide the eyes with respect to expected scaling-laws.

and allowing the phase separation to occur only during this drying time. Taking advantage of slow dynamics in polymeric-fluids, an ideal concentration-quench was ensured by achieving deep concentration quench relative to the concentration variation in DTL formed by solvent evaporation within an induction period where no lateral phase separation is observed. The growth law, $<D_\xi> \sim \tau^{2/3}$ is consistent with late-stage surface influenced phase separation due to thermal quench (1). The growth curves at different initial solvent concentration, c_0 can be re-scaled to superimpose forming a master curve. The consistent rescaling parameters to obtain the master curve and identical late-stage growth exponent as the classical SIPS process, suggests that the observed behavior may be universal.

References

(1) Guenoun, P., Beysens, D., Robert, M., Phys. Rev. Lett., **1990**, <u>65</u>, 2406.

(2) Wiltzius, P., Cumming, A., Phys. Rev. Lett., **1991**, <u>66</u>, 3000.

(3) Jones, R.A.L., Norton, L.J., Kramer, E.J., Bates, F.S., Wiltzius, P., Phys. Rev. Lett. **1991**, <u>66</u>, 1326.

(4) Bruder, F., Brenn, R., Phys. Rev. Lett., **1992**, <u>69</u>, 624.

(5) H. Tanaka, Phys. Rev. Lett., **1993**, <u>70</u>, 2770.

(6) H. Tanaka, Phys. Rev. Lett., **1993**, <u>70</u>, 53.

(7) Siggia, E.D., Phys. Rev. A, **1979**, <u>20</u>, 595.

(8) Chan, J.W., J. Chem. Phys., **1977**, <u>66</u>, 3667.

(9) Shi, B.Q., Harrison, C., Cumming, A., Phys. Rev. Lett., **1993**, <u>70</u>, 206.

(10) Saraf, R.F., Macromolecules, **(1993)**, <u>26</u>, 3623.

(11) Brenker, M.J., Feger, C., J. Polym. Sci., Polym. Chem. Ed., **1987**, <u>25</u>, 2005.

(12) Stoffel, N.C., Kramer, E.J., Volkeson, W., Russell, T.P., Polymer, **1993**, <u>34</u>, 4524.

(13) The reported diffusion constant of water in PMDA-ODA polyimide film is $\sim 5 \times 10^{-9}$ cm^2/sec. (Li, S.Z., Pak, Y.S., Adamic, K., Greenbaum, S.G., Lim, B.S., Xu, B., Xu, G., Nowick, A.S., J Electrochem. Soc., **1992**, <u>139</u>, 662) at room temperature. Since size of NMP molecule is larger than water, $D > 10^{-9}$ cm^2/sec. However, the films in these studies are not cured to rigid polyimide structure. Thus, conservatively, at 70 °C, we assume $D \sim 10^{-9}$ cm^2/sec. (at most).

(14) Crank, J., **The Mathematics of Diffusion**, 2nd ed., Oxford University Press, Oxford **(1975)**, p.50, Fig. 4.1.

(15) Saraf, R.F., Feenstra, R.M., Ostrander, S., Langmuir, submitted.

CHARACTERIZATION OF THIN POLYMERIC NANOFOAM FILMS BY TRANSMISSION ELECTRON MICROSCOPY AND SMALL ANGLE NEUTRON SCATTERING

R. M. BRIBER[+], J. S. FODOR[+], T. P. RUSSELL[++], R. D. MILLER[++], K. R. CARTER[++], J. L. HEDRICK[++]
[+]Department of Materials and Nuclear Engineering, University of Maryland, College Park, MD 20742
[++]IBM Research Division, Almaden Research Center, 650 Harry Rd., San Jose, CA 95120

ABSTRACT

Thin film polymer nanofoams are produced from triblock copolymers of a fluorinated polyimide, 3F/PMDA (derived from pyromelletic dianhydride (PMDA) and 1,1-bis(4-aminophenyl)-1-phenyl-2,2,2-trifluoroethane (3F)) as the center block and polypropylene oxide (PO) as the end blocks. The nanofoam is produced using a three step process: 1.) spin casting the triblock copolymer onto a silicon substrate, 2.) thermal treatment in an Argon atmosphere to imidize the center block and 3.) thermal treatment in air to degrade the PO domains and form nanoscale voids. This process was characterized using both transmission electron microscopy (TEM) and small angle neutron scattering (SANS). For the TEM studies, Ruthenium tetroxide staining was used to enhance the contrast between the polyimide (PI) matrix and the PO microdomains or voids, which permitted a more detailed view of the microstructure of both the foamed and unfoamed materials. From the two dimensional Fourier transform of the micrographs the spatial correlation between the PO microdomains in the unfoamed material and between the voids in the foam were found. An interdomain separation distance of ~37 nm was observed. SANS was performed to follow the imidization and foaming processes both *in-situ* and on a Si substrate. The SANS results indicated that the films are homogeneous when spun from solution and that the microphase separation of the PO domains occurs during the imidization step. The subsequent foaming step leaves the morphology generally intact with the PO domains converting to voids. A peak was found in the SANS curve with a spacing of about 26 nm, which is qualitative agreement with the TEM data.

INTRODUCTION

Recently, a novel scheme for producing high temperature polyimide based nanofoams has been developed [1-4]. These nanofoams are produced by spin casting uniform films of triblock copolymers, consisting of a polyimide precursor center block (poly(amic ester)) and two labile polypropylene oxide end blocks onto a substrate. Thermal treatment under an oxygen free atmosphere effectively converts the polyimide precursor block into a stable polyimide. Subsequent thermal treatment in air results in decomposition of the PO rich domains into low molecular weight products that diffuse out of the polyimide matrix. This

Mat. Res. Soc. Symp. Proc. Vol. 461 © 1997 Materials Research Society

process is illustrated in Figure 1. The resulting materials contain voids typically on the order of 5-80 nm in size [1-4] with pore size and morphology highly dependent on the relative size of the endblocks, changes in the processing procedure, substrate/polymer interactions and overall film thickness.

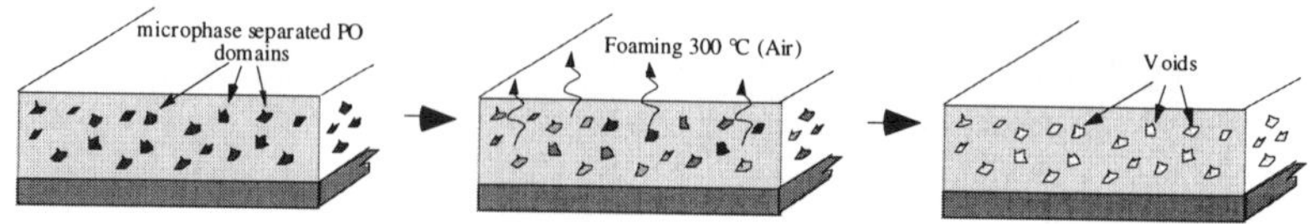

Figure 1: Process to form polymer nanofoam materials

The polyimide centerblock forms the matrix material in these nanofoams and is desirable for a number of reasons. Polyimides exhibit ease of processing (relative to inorganic materials) yet, unlike most polymeric materials, can withstand high temperatures, are resistant to solvents and have lower coefficients of thermal expansion (CTE) [5,6]. By incorporating nanometer length scale voids into a polyimide matrix a unique "foamed" material is obtained, while maintaining the desirable properties inherent to the polyimide matrix. Such a material is expected to contribute greatly in many technological areas. With continued research they are expected to lead to advances in microelectronics packaging, storage cells and high temperature polymeric membranes. These materials also have the potential to exhibit reduced dielectric constants, due to the incorporation of voids (dielectric constant (ε) = 1) into an already low dielectric constant polyimide matrix ($\varepsilon \sim 2.0 - 4.0$). The decrease in the dielectric constant is greater than what is predicted by a simple linear weighting of the dielectric constants of the two phases (polyimide/void) [7]. Initial studies [3] on these nanofoams have indicated a significant reduction in their bulk dielectric constant.

From a purely fundamental perspective, the size, distribution and connectivity of the voids is an important and interesting physical problem, since it greatly influences the mechanical stability, solvent resistance and defines the porosity of the foam. Characterization of both the imidized and foamed material and the imidization and foaming process are important to assess final film properties and better control and optimize the thermolysis of the labile blocks. Characterization of the unfoamed and foamed material as well as the effects of imidization and foaming on morphology and structure is a key issue in the development and improvement of these materials.

EXPERIMENTAL

The chemical structure of the polymers is shown in figure 2.

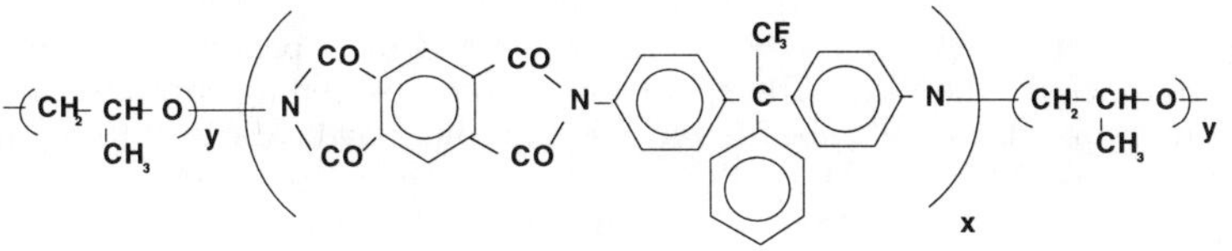

Figure 2: Chemical structure of the triblock copolymer (polyimide form).

The PO end blocks of the sample used for the TEM study had a molecular weight of 5.6 kg/mol, and the total PO content was 24 wt. % as determined by NMR. The triblock was solution cast to form a 10 µm thick film from a 50/50 vol. % dimethylformamide (DMF) / 2-ethoxyethyl acetate mixture onto a clean glass slide. The film was then placed in an inert (oxygen free) environment, heated at 5 °C/min. to 300 °C and held at that temperature for 1 hour to imidize the center block. The film described above was then cut into approximately 1 mm x 6 m m films. Several of these pieces were set aside and left unfoamed. The remaining pieces were foamed by placing them in an air environment, heating them at 5 °C/min. to 240 °C and holding at that temperature for 5 hours. The 1 mm x 6 mm films (both foamed and unfoamed) were embedded in epoxy and microtomed at room temperature using a diamond knife mounted in a Reichert-Jung Ultracut. The sections were stained using Ruthenium tetroxide. A study on the homopolymers indicated that RuO_4 stains PO much more rapidly than PI. Transmission electron microscopy was performed on both stained and unstained samples using a JEOL 2000 FX II electron microscope, with an accelerating voltage of 200 kV at magnifications ranging from 20k to 80k.

The polymer used for the SANS studies was a triblock copolymer with 8.5 kg/mol PO endblocks and a 25% PO content. The films for SANS were spun onto 1″ diameter Si wafers as poly(amic ester) material (i.e. prior to imidization) from cyclohexanone. The imidization and foaming processes were then studied *in-situ* using the 30m NG-7 SANS instrument at the NIST Cold Neutron Research Facility [8].

RESULTS AND DISCUSSION

Figure 3 shows TEM micrographs of microtomed sections of various nanofoam films. Figure 3a is a film which has been imidized (but not foamed) and stained with RuO_4. A microphase separated structure is clearly seen in the micrograph with the darker regions corresponding to the stained PO domains. Figure 3b is the same film which has been foamed and then stained. The morphology observed is essentially identical to that in the unfoamed material. This indicates that the morphology is unchanged during the foaming process with the PO domains being replaced by voids. It is clear that the RuO_4 stains the voids, which is probably due to either a higher reactivity of the void surface or incomplete degradation of the PO. Figure 3c is a micrograph of a film similar to that shown in 3b but without staining. The density of visible voids is not as high as in the stained film but on close inspection of the micrograph it is clear that the voids which are most prominent in 3c are ones which penetrate the top and bottom surfaces of the film. There are many voids exhibiting lower contrast in the micrograph which are embedded in the film (but may be difficult to see in the figure). Figure 3d is a plot of the radially averaged power spectra of the 3 micrographs. The two micrographs from the stained samples give essentially identical power spectra which have a clearly defined peak at a spacing of 37 nm. In the unstained sample the voids in the micrograph do not show a clear spatial correlation. This is probably because most of the voids are not imaged with sufficient contrast without staining.

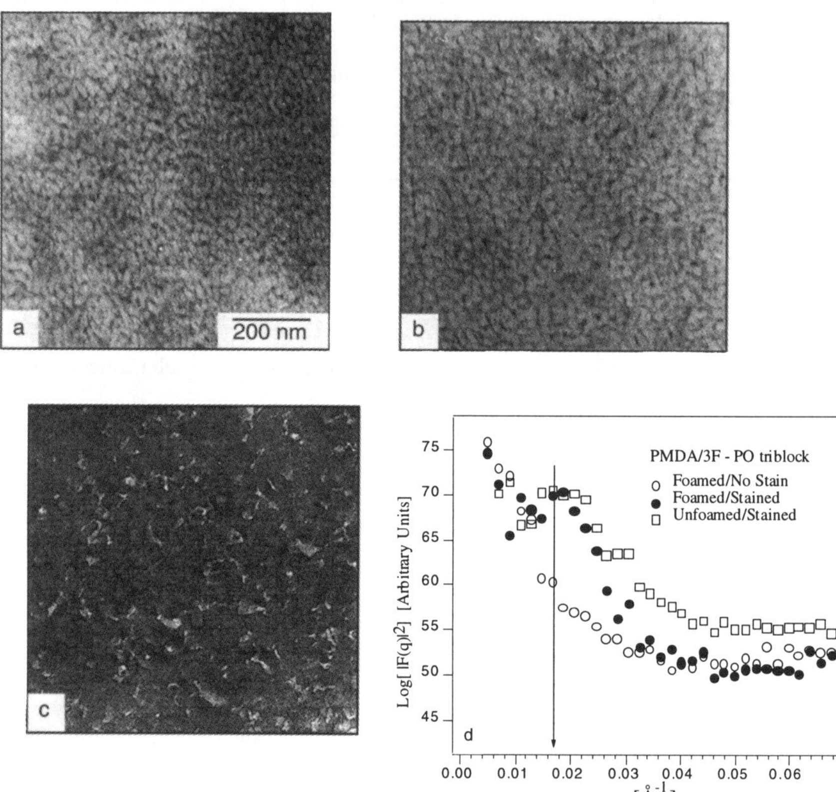

Figure 3: a.) TEM micrograph of imidized but unfoamed sample stained with RuO_4. b.) TEM micrograph of same polymer as in a.) but foamed and stained. c.) Foamed and unstained micrograph. d.) Circularlly averaged power spectra (2d-FFT) calculated from figures a, b and c. All micrographs are the same magnification.

SANS experiments performed during the foaming process while the samples were on a Si wafer substrate. The absorption of neutrons by a typical silicon wafer is negligible and the wafer doesn't contribute to the small angle scattering. Figure 4a shows SANS data from a 25% PO content (8.5k Mw PO block) copolymer film for the poly(amic ester) (PAE) (i.e. preimidization), the cured polyimide and the foamed material. The most interesting aspect of the is that the small angle scattering is almost flat for the poly(amic ester) film and upon imidization a pronounced peak develops with a spacing of about 26 nm. This peak remains at approximately the same q value and intensity after foaming. The data implies that the poly(amic ester) film as spun from solution is *not* microphase separated and that microphase separation occurs upon imidization. Calculation of the neutron scattering length densities (SLD) for the PAE, PI and PO, indicate that contrast between the PO and PAE is only slightly less than for PO/PI and if microphase separation was present in the initial PAE films it should be readily observed by SANS. The relatively small change in the scattering upon foaming is due to the fact that the SLD for PO is close to zero (i.e. almost the same as a void) and hence the change neutron contrast with foaming is negligible. In figure 4b SANS data during imidization is shown. A 2.34 µm thick film of PAE was heated in an Argon atmosphere from 25-300°C. Data was collected for 1/2 hour intervals as the temperature was increased. The scattering was constant at temperatures up to 250°C. At 250°C a peak develops during the first 1/2h (indicating the onset of microphase separation) but stays constant in position and intensity for the second 1/2h at 250°C. At 300°C the peak increases in intensity and moves to lower q. The peak continues to shift to lower q during the second 1/2h at 300°C. The time dependence of the scattered intensity for the 2.34µm film probably indicates that the change in morphology which occurs during imidization may be controlled by diffusion of ethanol (which is released during imidization reaction) to the free surface of the film.

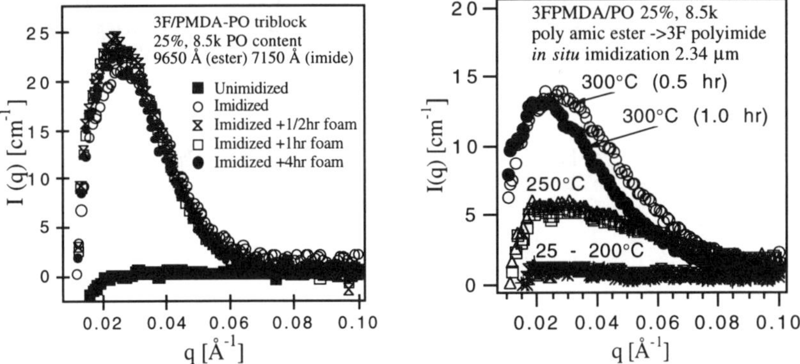

Figure 4: a.) *In-situ* SANS curves for a thin film showing the scattering from the poly(amic ester) (unimidized), the imidized and the foamed material; b.) SANS scattering during the imidization process for a relatively thick film.

CONCLUSIONS

TEM and SANS has been used to follow the formation of nanofoam materials from triblock copolymers containing a PI center block and PO endblocks. The films were spun from solution where the centerblock was still in the PAE form. The films as spun appear not to be microphase separated. Microphase separation then occurs during the imidization step (300°C in Argon). During the subsequent foaming step (300°C in air) the morphology which develops during the imidization step is preserved with the PO domains being converted to voids.

REFERENCES

1. Carter, K. R.; Labadie, J. W.; DiPietro, R. A. Sanchez, M. I.; Russell, T. P.; Swanson, S. A.; Auman, B. C.; Lakshmanan, P.; McGrath, J. E. *Polymer Preprints (Am. Chem. Soc., Div. Polym. Mater. Sci. Eng.)* **1995**, *72*, 383.

2. Charlier, Y.; Hedrick, J. L.; Russell, T. P.; DePietro, R. *Polymer Preprints (Am Chem. Soc., Div. Polym. Mater. Sci. Eng.)* **1995**, *72*, 389.

3. Labadie, J. W.; Hedrick, J. L.; Wakharkar, V.; Hofer, D. C.; Russell, T. P. *IEEE Trans. Compon., Hybrids, Manuf. Technol.,* **1992**, *15*, 925.

4. Hedrick, J. L.; Labadie, J. W.; Russell, T. P.; Hofer, D. C.; Wakharkar, V. *Polymer,* **1993**, *34*, 4717.

5. Tummala, R. R.; Rymaszewski, E. J. *Microelectronics Packaging Handbook;* Van Nostrand Reinhold: New York, 1989.

6. Tummala, R. R.; Keyes, R.W.; Grobman, W.D.; Kapur, S. "Thin Film Packaging", Tummala, R. R.; Rymaszewski, E. J. eds.. *Microelectronics Packaging Handbook;* Van Nostrand Reinhold: New York, 1989.

7. Aspnes, D. E. *Thin Solid Films,* **1989**, *89*, 249.

8. Prask, H.J.; Rowe, J.M; Rush, J.J.; Schroder, I.G.; *J. Res. Natl. Inst. Stand. Technol.,* **1993**, *98*, 1

Electric Field Induced Control of Thin Film Diblock Copolymer Domain Orientation

T. L. MORKVED, W. A. LOPES, M. LU, A. M. URBAS, H. M. JAEGER,
The James Franck Institute and Department of Physics, The University of Chicago,
Chicago, Illinois 60637.

P. MANSKY, AND T. P. RUSSELL
Conte Center for Polymer Research, University of Massachusetts, Amherst, MA 01003

ABSTRACT

Local control of domain orientation in diblock copolymer thin films is demonstrated through the use of external electric fields. Thin films of a polystyrene-polymethylmethacrylate diblock copolymers, denoted P(S-b-MMA), were spin coated onto silicon nitride membrane substrates with prefabricated in-plane electrodes, forming cylindrical PMMA microdomains. Films annealed under an applied electric field (E $\leq$ 37V/μm) at 250°C for 24h under an argon atmosphere showed an alignment of the cylindrical microdomains parallel to the electric field lines. A quantitative measure of the degree of alignment was obtained by correlating the local field strength, E, and direction with the observed cylinder orientation. The alignment was found to saturate above E$\approx$30V/μm, and to decrease rapidly as E falls below this value.

INTRODUCTION

Domain formation by block copolymer thin films is being investigated as a means towards patterning substrates. The self-assembly property of block copolymers opens new possibilities for the fabrication of mesoscopic structures and devices. In principle each of the two polymer blocks can be designed to exhibit specific chemical, electrical, or optical functionality. For example, differences in the chemical etch rates of the blocks have been used to prepare templates suitable for nanolithographic pattern transfer[1]. By using diblocks with different wetting characteristics, selective decoration of one of the blocks with nanometer-sized metal particles has been demonstrated[2]. Alternatively, organo-metallic groups can be incorporated directly into one of the polymer blocks, which can then be released inside selected domains upon further processing[3].

A diblock copolymer, the simplest of the block copolymers, consists of two chemically distinct polymer chains joined by a covalent bond[4]. Below an order-disorder temperature, these polymers form periodic patterns of nanometer size domains. In bulk, the morphology is dominantly controlled by the ratio of the volume fraction of the two polymer blocks. Morphologies ranging from alternating lamellae to arrays of cylinders or spheres of one of the blocks embedded in a matrix of the other have been found. Equilibrium morphologies consist of highly symmetic arrays of these domains. The resulting structure, however, depends significantly on preparation parameters, such as annealing conditions and applied forces. Typically, many defects are present, and only short range order exists. Often, a bulk sample is made by casting with solvent in a mold, allowing the solvent weeks to evaporate, to facilitate long range order. The alignment of the microphase separated domains into highly uniform patterns can be accomplished by applying external forces such as shearing the polymer melt[5].

In thin films, surface forces play an important role in the resulting morphology. Lamellar domains usually form parallel to the substrate, due to selective wetting of surfaces[6]. The high energy cost in forming stretched domains results in the quantization of the polymer film thickness to integer repeat spacings (with symmetric boundaries). Excess (deficient) polymer volume results in island (hole) formation. In cylindrical phase films, surface forces can also produce thickness quantization. Ordering in the plane, however, typically is still poor because

109

Mat. Res. Soc. Symp. Proc. Vol. 461 © 1997 Materials Research Society

preparation parameters play a much more important role than in lamellar phase films. Specifically, in thicker films, cylindrical domains can form both parallel and perpendicular to the substrate[7]. In either the lamellar or cylindrical case, shear techniques for ordering require bulk samples, and cannot easily be applied to thin films. Control over the morphology in the thin film limit has remained a challenge and is key to any use of diblock-copolymers for device fabrication.

Here we demonstrate highly controlled alignment of diblock-copolymer thin films by external electric fields[8]. For copolymers exhibiting cylindrical microdomain structure, it is shown that the cylinders align themselves along the direction of the electric field if the field strength is sufficiently high. Field-induced copolymer alignment has previously been reported for bulk samples by Amundson *et al.* [9,10,11] but was found inferior to shear or flow techniques in terms of speed and degree of alignment. In thin films, however, electric fields offer several advantages. The high electric field strengths of order 10^6V/m, problematic in bulk samples yet required for significant alignment, are easily achieved over micron-sized or smaller distances. Furthermore, in contrast to bulk alignment techniques, electric fields can be applied locally. Sets of independent electrodes thus offer unprecedented flexibility in controlling the local pattern alignment in different places on the same film. We report on domain orientation studies on diblock copolymer thin films prepared between planar electrodes on amorphous silicon nitride membranes, observed directly by transmission electron microscopy.

EXPERIMENTAL

Asymmetric diblock copolymers of polystyrene and poly(methylmethacrylate), denoted P(S-b-MMA), having a styrene volume fraction of 0.66 and an average molecular weight of 9×10^4 with a narrow molecular weight distribution, were used. This copolymer forms cylindrical microdomains of PMMA in a PS matrix in thin films. The molecular weight is too high for this copolymer to exhibit an order-disorder transition at usable temperatures. Thin films were prepared by spin-casting a 2% solution of the copolymer in toluene onto silicon nitride substrates with prefabricated electrodes. Under an applied, in-plane electric field the films were then annealed in an argon atmosphere up to 250°C for several hours. Finally, the resulting film morphology was examined in a Phillips CM120 transmission electron microscope (TEM) operated at 120 kV. PMMA domains are easily detected by TEM (white lines in the figures). The contrast is due to radiation-induced thinning of the PMMA cylinders [12].

The substrates consisted of a 100 nm thick silicon nitride layer deposited onto a silicon wafer. Under small rectangular areas the silicon was selectively etched away from the backside of the wafer, providing self-supporting and mechanically stable, yet TEM-transparent silicon nitride membranes with lateral dimensions of 60 μm. These membranes thus form windows which are an integral part of the substrate. Planar electrodes with gap spacings of 4 μm were fabricated on top of the silicon nitride within the window areas using optical or electron-beam lithography. The electrodes were formed by evaporation of 25 nm of chromium. Small gold wires were attached with silver epoxy and connected to a programmable voltage source. Figure 1 shows a top view of a typical electrode layout, along with the corresponding electric field lines. In a number of samples slight variations of this lay-out were used. The data discussed below refers to the configuration sketched in Fig.1.

RESULTS AND DISCUSSION

Figures 2a-c show close-ups of regions between electrodes spaced four microns apart produced under identical annealing conditions. In this case, the copolymer film thickness confines one layer of PMMA cylinders to lie within a single plane, parallel to the substrate. Without an applied electric field a highly curved, unaligned pattern forms (Fig. 2c). Note that the orientation of individual cylinders is not influenced by the presence of the electrodes. Fig. 2b shows the strikingly different results if annealed in an electric field. In this case we applied a voltage difference of 15V, corresponding to E = 37 V/μm across the central gap region. This

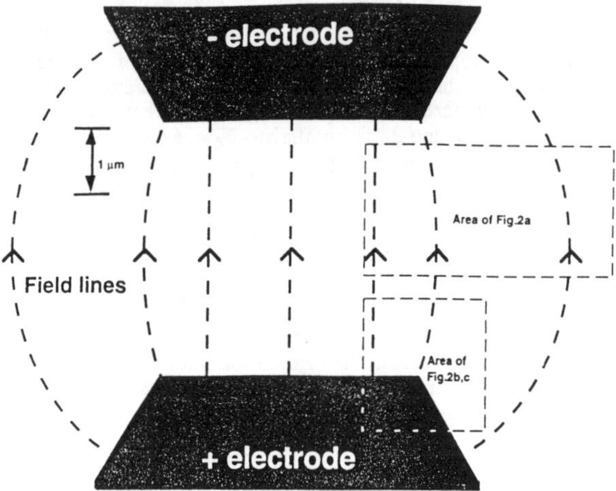

Fig.1. A sketch of the experimental electrode configuration. The electrodes are drawn with a few field lines in between as a guide to the eye. The boxes indicate regions where micrographs were taken for Fig. 2.

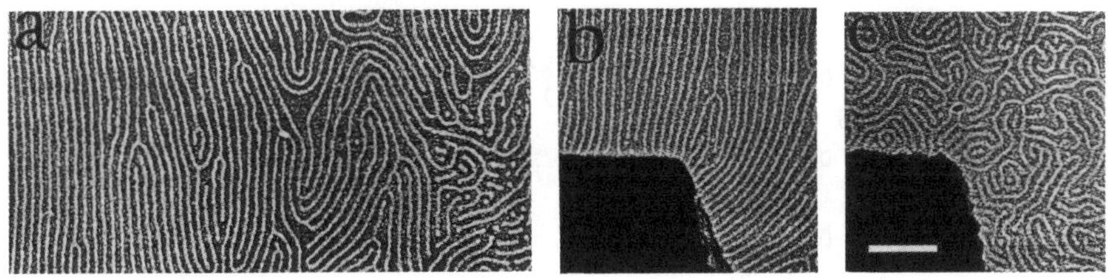

Fig.2. TEM micrographs of cylindrical phase diblock-copolymer films. The lighter lines are cylinders of PMMA surrounded by a darker PS background. The cylinder repeat spacing is 62 nm. All three samples were prepared under identical annealing conditions. The scale is the same for all three micrographs (the size bar in (c) indicates 500 nm). (a) A close-up near the central region between electrodes, indicating a transition from well aligned (left edge, $E \approx 37$ V/µm) to unaligned cylinders (right edge, $E \approx 20$ V/µm). (b) & (c) Close-ups of regions near a corner of one electrode: annealed in the presence of an applied electric field (b), and without field (c).

field produces a periodic, two-dimensional line pattern with good uniformity and a period of 60 nm, close to the cylinder repeat spacing in the absence of electric fields. Figure 2b clearly shows that the cylinder axes align parallel to the electric fields, following the curvature of the electric field lines outside the central gap region. Figure 2a indicates the gradual transition from highly aligned (left) to unaligned behavior (right) with decreasing electric field strength (cf. Fig.1).

The electrode configuration produces a different electric field strength at different positions, and thus, we can calculate the degree of alignment as a function of field strength on any individual sample by analyzing digitized TEM micrographs as in Fig. 2, assuming a two-dimensional geometry. The local field strength and direction can then be correlated with the observed orientation of cylinders. For each cylinder segment, subject to a local electric field, E, we define θ as the included angle between the direction of the cylindrical axis, $\hat{e}_c$, and the direction of the electric field, $\hat{e}_z$. We define as our two-dimensional orientational order parameter, $C(E) = 2\langle\cos^2\theta\rangle - 1$, where the brackets $\langle...\rangle$ denote an average over all regions with local field strength E within the analyzed field of view. When averaged over a sufficiently large area, $C(E)$ will be zero in a macroscopically unaligned block copolymer sample. For ideal alignment $C(E)$ approaches unity and $\hat{e}_c$ is parallel to $\hat{e}_z$ everywhere. To minimize systematic errors that arise from the finite-resolution of any digital image processing routine, we find it more convenient to consider a scaled order parameter S, defined as $S(E) = (C(E) - C_{min})/(C_{max}-C_{min})$. Here C_{max} and C_{min} are the order parameters obtained from a perfectly aligned region containing no defects, and from an unaligned sample annealed without applied electric-field, respectively.

Figure 3 shows S as a function of the square of the local field strength, E^2, for the sample in Fig.2a-b. E^2 is proportional to the energy density of the electric field (see below). We note that above a threshold value of ≈ 30 V/μm the orientation parameter assumes a field independent value (within our experimental accuracy), but decreases rapidly as E falls below this value (Fig. 3). The crossover marks the change from nearly complete to partial alignment outside the central gap region between the electrodes where the field strength drops off. The saturation at about 80% is caused by residual point defects, disclination lines and wall defects, whose removal may require different annealing conditions, or which may result from pinning to the substrate. In different samples, prepared with varying gap widths or applied voltages, a high degree of alignment was never observed for E < 30V/μm, even in the central gap region.

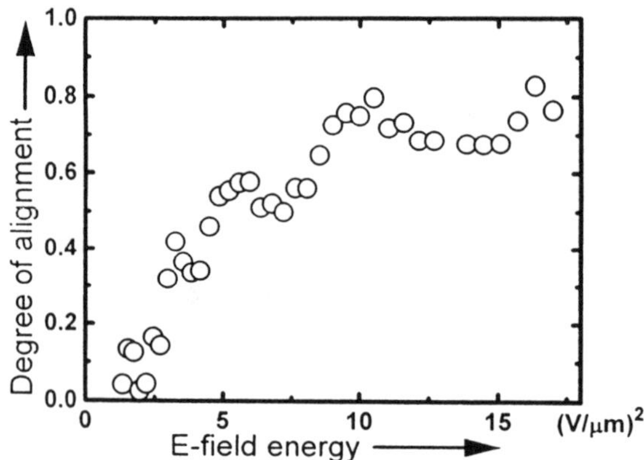

Fig.3. Normalized orientational order parameter S as function of the square of the electric field strength, E^2. S increases rapidly with electric field and saturates above $E\approx 30$ V/μm.

Without a global aligning field, the thin films do not reach their equilibrium morphology with the given preparation parameters. Instead, the cylindrical domains form with a high degree of curvature which cannot be removed by longer annealing times (Fig 2b). Here we find a large density of defects with only short-range ordering of the cylinders. Since curvature contributes an amount $K_1/(2L^2)$ to the free energy density [13], forces which would straighten domains must be prevented by energy barriers inhibiting collective motions (K_1 is the splay elastic coefficient). These barriers may be greater than those in bulk due to pinning from the substrate or the reduced dimensionality. We first estimate the pinning energy from images such as Fig.2c. Cylinder segments can straighten out up to a characteristic length L* at which the curvature term becomes comparable to the pinning energy. By analyzing images like Fig. 2c, we estimate L* $\approx$ 200 nm.

Using $K_1 \approx 10^{-7}$ dyne [13], we thus find a pinning energy of order 80 eV/μm^3 for our substrate. These same polymer films prepared on different substrate have a different characteristic length.

Amundson *et al.* [8,9] consider the electric field contribution to the free energy that arises from a difference in the dielectric constants, $\Delta\varepsilon$, of the two copolymer blocks. This approach treats bulk copolymers and does not consider the additional contributions that arise in the ultrathin film limit from interactions with the substrate. Nevertheless, it provides a starting point from which to gain insight into the mechanisms underlying the observed behavior. Consider a domain containing cylinders oriented either all parallel or all perpendicular to the applied electric field. If the dielectric constants of the two blocks are different, there will be a body force on the domain, tending to minimize the overall electrostatic energy. Just as in a capacitor filled with alternating regions of different dielectrics, the lowest energy configuration will be the one in which the cylinders are oriented parallel to E. The free energy difference per unit volume, corresponding to the difference in capacitance between the two configurations, is proportional to $(\Delta\varepsilon E)^2$, with details depending on domain geometry [8]. For PMMA and PS the dielectric constants [14] differ by $\Delta\varepsilon \approx 3.5\pm1$. The energy gain between an aligned state like Fig. 2b and a curved state like Fig. 2c but in field, would be half this much. This energy gain should be compared to the pinning energy arising from the interface with the substrate or other energy barriers opposing collective motion. At the threshold of alignment, just enough energy gain from the electric field should be available to overcome this pinning energy. Using the estimated pinning energy of 80 eV/μm^3 , we estimate the expected threshold electric field for alignment to be E=33 V/μm. This value is consistent with the measured value found in Fig. 3. In the absence of a detailed theoretical understanding, these values should only be considered a rough estimate. They clearly demonstrate, however, that body forces induced by applied electric fields provide a viable mechanism for the alignment of copolymer domains.

CONCLUSION

We have demonstrated the control of cylindrical domain orientation by electric fields in diblock copolymer thin films. These films offer the opportunity to have electric fields and interfacial interactions work in synergy. In bulk samples, or in thin films with electrodes on the top and bottom surfaces, the aligning E-field is perpendicular to the largest surfaces. These surfaces by themselves produce strong alignment [6] which counteracts the electric field forces, prohibiting perfect alignment [9]. In thin films with planar electrodes, as in this work, the surfaces and the electric field both produce forces in the plane of the film. In particular, in ultrathin films the copolymer thickness becomes quantized in integer repeat spacings. Figure 2 shows how this confinement can be exploited to produce exactly one layer of cylinders parallel to the substrate. In this work it has been shown that electric fields coupled with interfacial interactions can produce very highly aligned structures in diblock copolymers. Unlike in bulk, electric field alignment is quite effective for thin films. Thus, exquisite control over the orientation of the microdomains can be achieved providing a key to tailoring mesoscopic structures.

ACKNOWLEDGMENTS

We thank T. A. Witten for many helpful discussions, and E. E. Ehrichs for assistance with lithography. This work was supported in part by the MRSEC program of the National Science Foundation (NSF) under Award No. DMR-9400379. The silicon nitride membranes were fabricated at the National Nanofabrication Facility at Cornell, which is supported by NSF under Grant ECS-8619049. HMJ acknowledges fellowship support from the David and Lucile Packard Foundation. TPR and PM acknowledge the support of the US Department of Energy, Office of Basic Energy Sciences under contract FG03-88ER-45375.

REFERENCES

1. P. Mansky, C.K. Harrison, P. M. Chaikin, R. A. Register, N. Yao, Appl. Phys. Lett **68**, 2586 (1996); P. Mansky, P. M. Chaikin, E. L. Thomas, J. Mat. Sci. **30**, 1987 (1995). .

2. T. L. Morkved, P. Wiltzius, H. M. Jaeger, D. G. Grier, and T. A. Witten, Appl. Phys. Lett. **64**, 422 (1994); R. Saito, S. Okamura, K. Ishizu, Polymer **33**, 1099 (1992); K. Ishizu et al., Polymer **34**, 2256 (1993).

3. Y. Ng C. Chan, R. R. Schrock, and R. E. Cohen, Chem. Mater. **4**, 24 (1992); R. Tassoni, R. R. Schrock, Chem. Mater. **6**, 744(1994).

4. F.S Bates and G. H. Fredrickson, in Annual Reviews of Physical Chemistry (Annual Reviews, Palo Alto, Ca, 1990), Vol. 41, pp.525-557; ibid 1996.

5. A. Keller, E. Pedemonte, F. M. Willmouth, Nature **225**, 538 (1970); K. A. Koppi, M. Tirrell, F. S. Bates, K. Almdal, R. H. Colby, J. Phys. (Paris) **2**, 1941 (1993).

6. T. P. Russell, G. Coulon, V. R. Deline, D. C. Miller, Macromolecules **22**, 4600 (1989).

7. Y. Liu et al., Macromolecules **27**, 6559 (1994).

8. A previous account of this work has appeared in: T. L. Morkved, M. Lu, A. M. Urbas, E. E. Ehrichs, H. M. Jaeger, P. Mansky, T. P. Russell, Science **273**, 931 (1996).

9. K. Amundson, E. Helfand, D. Davis, X. Quan, S. Patel, S. D. Smith, Macromolecules **24**, 6546 (1991).

10. K. Amundson, E. Helfand, X. Quan, S. D. Smith, Macromolecules **26**, 2698 (1993).

11. K. Amundson, E. Helfand, X. Quan, S. D. Hudson, S. D. Smith, Macromolecules **27**, 6559 (1994).

12. E. L. Thomas and Y. Talmon, Polymer **19**, 225 (1978).

13. K. Amundson and E. Helfand, Macromolecules **26**, 1324 (1993).

14. For PS, $\varepsilon \approx 2.45 \pm 0.1$ at 250°C. (R. F. Boyer, in Encyclopedia of Polymer Science and Technology, (Intersciences, New York, 1970), Vol.13 pp 251-277.) For PMMA, $\varepsilon \approx 6 \pm 1$ at 250°C., (N. G. McCrum, B. E. Read, G. Williams, Anelastic and Dielectric Effects in Polymeric Solids (J. Wiley, New York, 1967) p. 264).

DETERMINING THE MORPHOLOGY AND THE INTERACTION BETWEEN TERMINALLY-ANCHORED POLYMER LAYERS

A.C. BALAZS, C. SINGH
Chemical and Petroleum Engineering Department, University of Pittsburgh, Pittsburgh, PA 15261

ABSTRACT

Using a two dimensional self-consistent field theory, we investigate the interactions between two planar surfaces that are coated with terminally-anchored homopolymers. One surface is coated with A chains and the other is covered with B homopolymers. The chains are grafted at low densities and the B polymers are chosen to be solvophobic, while the A chains are relatively solvophilic. We determine the morphology of the layers and the energy of interaction as the surfaces are compressed. Our results provide guidelines for controlling the interaction between polymer-coated colloidal particles.

INTRODUCTION

The interaction between polymers tethered to solid surfaces provides the steric stabilization that keeps colloidal particles suspended in solution and the lubrication that reduces the friction and wear between mechanical components. While considerable attention has been focused on the interaction of tethered chains in good solvents, less is known about the properties of such polymers in poor solvents. When homopolymers are terminally-anchored onto a substrate and immersed in a poor solvent, the incompatibility between the polymer and solvent drives the chains to cluster into distinct aggregates, or "pinned micelles" [1-7]. These micelles have a well-defined size and spacing [8-12]. In a recent study, we considered the interaction between two surfaces that are coated with solvent-incompatible (solvophobic) homopolymers and are compressed [13]. The free energy of interaction versus distance profile shows a distinct attractive region. The attraction arises from the merging of the pinned micelles on the respective surfaces as the layers are brought into close proximity. We noted similar behavior in the case where each surface contained <u>both</u> solvophobic and solvent-compatible (solvophilic) chains [14] or diblock copolymers that contain both solvophobic and solvophilic blocks [15].

Another means of controlling the interaction between surfaces is to coat one substrate with homopolymer A and the other with homopolymer B, where A and B have different solvent affinities [14]. When the surfaces are far apart, the grafted homopolymers that are in a good or a theta solvent form stretched brushes, while those in a poor solvent form pinned micelles. In this paper, we determine of the effects of bringing such surfaces into contact. We investigate the behavior of the system as a function of the relative polymer-solvent interactions and the polymer-polymer interactions. As we show below, as the surfaces are compressed, the layers self-assemble into unique morphologies in order to minimize the surface tension between the solvent and the solvophobic chains.

THE MODEL

To carry out our investigations on tethered polymers, we use a two dimensional self-consistent mean field (SCF) theory [13-15]. Through a 2D SCF theory, we determine both the vertical and the lateral density profiles, which are crucial for visualizing and characterizing such laterally inhomogeneous structures as pinned micelles. Our SCF method is derived from the lattice theory of Scheutjens and Fleer [16]. In this treatment, the phase behavior of polymer systems is modeled by combining Markov chain statistics with a mean field approximation for the free energy. Given the probability of finding a single monomer in a particular layer, and the fact that all the monomers in a chain are connected, the statistics for chains of arbitrary length can be obtained through a series of recursion relations. These recursion relations involve the potential of mean-force acting at a site $\mathbf{r}$. This potential, in turn, is determined from the local distribution of all the components at the point $\mathbf{r}$, as well as the Flory-Huggins interaction parameters, or χ's, between the different components. Solving this series of equations numerically and self-consistently yields

Mat. Res. Soc. Symp. Proc. Vol. 461 © 1997 Materials Research Society

the equilibrium profiles for the polymers and solvent in the system [16]. In the two dimensional SCF theory [13-15], the equations are written explicitly in terms of both the vertical (Z) and lateral (Y) directions. All the quantities are assumed to be translationally invariant in the X direction.

The grafted polymers are characterized by three sets of parameters: grafting density, chain length and the relevant χ parameters. We fix the grafting density per line along the Y direction at 0.5 for all calculations, i.e., the polymers are grafted on alternate lattice sites. The grafting density per line along the X direction is denoted by ρ and is fixed at $\rho = 0.025$ for all of the results presented below. (Thus, the average area per chain, s, has s=2/ρ.) Our calculations are thus restricted to the case of low grafting densities, where the formation of pinned micelles is most pronounced [3].

The length of the both the A and B chains is held fixed at N = 80. We let χ_{AS} and χ_{BS} represent the polymer-solvent interactions for the A and B components, respectively, and χ_{AB} is the interaction parameter for the two different monomers. The strength of interaction between the polymers and the planar, impenetrable surface is assumed to be the same as that between the polymer and the solvent. We fix the polymer-solvent interaction for the B chains at $\chi_{BS} = 2$, while gradually varying χ_{AS} from 0 to 1. We also set $\chi_{AB} = 0$ and subsequently determine the effect of increasing $\chi_{AB} > 0$.

RESULTS AND DISCUSSION

<u>Effect of Varying</u> χ_{AS}; $\chi_{BS} = 2$, $\chi_{AB} = 0$.

Figure 1 shows the effects of bringing the surfaces together when one surface is covered with $\chi_{BS} = 2$ homopolymers and the other surface is coated with $\chi_{AS} = 0$ chains. As the surfaces start interacting (Figs. 1a and b), there is an increase in the density of the soluble A component around the B micelles. This shields the solvophobic B's from the unfavorable solvent and thereby lowers the surface tension in the system. As the surfaces get closer, the shielding of the B chains by the soluble A component becomes more effective. Soon the morphology of the layers resembles a "flower-like" structure: the B's form a dense core and the A's form a corona of extended "petals" around the B domain. We note that the decrease in the surface separation causes the overall density of the soluble component to gradually increase since more and more of the solvent gets squeezed out of the system. However, the density of the B core remains relatively unchanged.

The increased shielding due to solvent expulsion leads to an enthalpic gain in the free energy, while the surface confinement results in entropic losses. This competition between the enthalpic and entropic free energies results in a net cancellation of the two effects. Therefore, the free energy of interaction plotted in Figure 2a appears to be relatively flat over a wide range of surface separations, even though the layers actually start interacting near L ~ 20 (as shown in Figure 1a). For $\chi_{AB} = 0$, however, the free energy of interaction has a small attractive region where the gain in enthalpy due to better shielding of the B component wins over the losses in the entropy due to the compression of surfaces. After this small attractive region, further compression of the surfaces leads to considerable entropic losses and the interaction profile becomes increasingly repulsive.

We note that if we compress a homopolymer surface with $\chi_{BS} = 2$ with an A layer that is not in a good solvent but in a theta solvent ($\chi_{AS} = 0.5$), the features are qualitatively similar to those described above for the good solvent case. In the theta solvent, however, the brush height is smaller and the brush density is higher because it is not as energetically favorable for the chains to stretch into the theta solvent. Moreover, since the gain in the free energy in forming a highly stretched brush is less in a theta solvent than in a good solvent, a much better shielding of the B component is obtained in the theta solvent for the same surface separation. Thus, in a theta solvent the enthalpic gains due to the shielding of the B chains outweigh the entropic losses due to

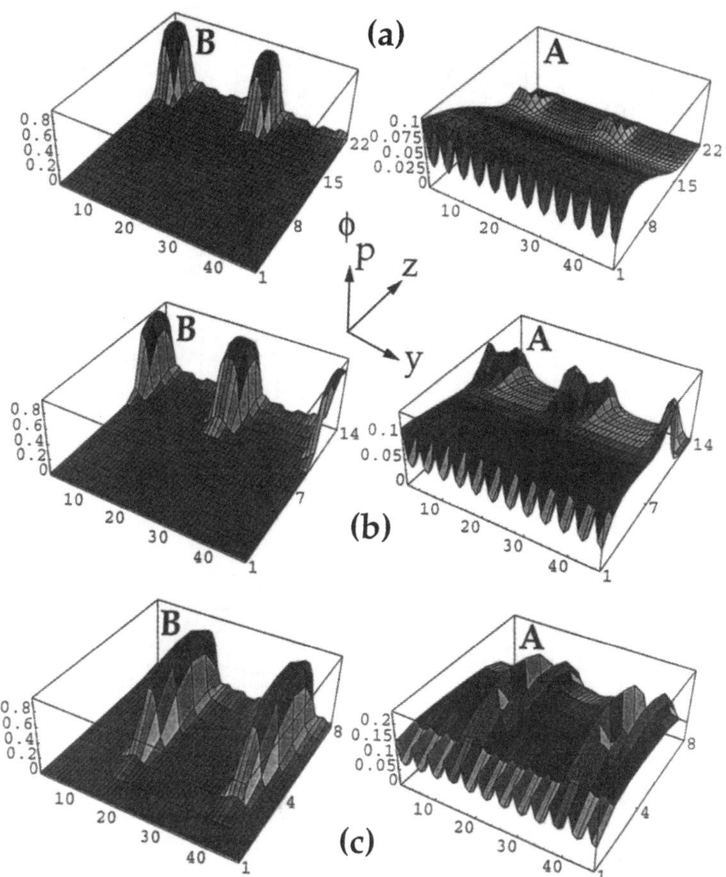

Figure 1.   SCF density profiles showing the effect of decreasing the separation, L, between the surfaces. Here, N = 80. The grafting density is $\rho = 0.025$, $\chi_{BS} = 2$, $\chi_{AS} = 0$ and $\chi_{AB} = 0$. In (a), L = 21 and in (b), L =13, while in (c), L = 7. The chains are grafted in the XY plane and the two surfaces are separated along Z. The parameter ϕ_p is the polymer density. The plots marked "B" show the polymer density of the B chains, while the plots marked "A" show the density of the A chains.

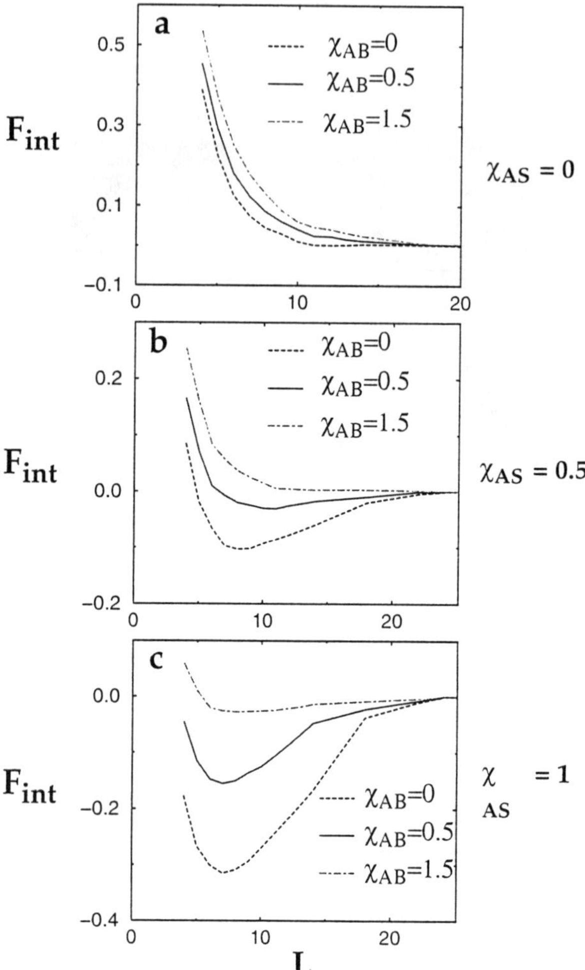

Figure 2.   Free energy of interaction, F_{int}, as a function of surface separation, L.

confinement over a wider range of surface separations than for $\chi_{AS} = 0$. This can be seen in Figure 2b, which shows a very distinct minimum for $\chi_{AB} = 0$.

Figure 3 shows the effects of bringing together two surfaces that are both covered with homopolymers in poor solvents; for one, $\chi_{BS} = 2$ and for the other, $\chi_{AS} = 1$. When the layers are far apart, the chains form pinned micelles on both surfaces. As the surfaces come together, some of the A chains stretch towards the B surface to shield the more solvophobic component, while the remaining A chains still form pinned micelles near the grafting surface, as shown in Figure 3a. With further compression of the surfaces, all of the A chains stretch out to form a thin shell around the B component. In this way, the structures now resemble "onion-like" micelles. As the surfaces are brought closer together, the density of the B core and the outer A layer remain unchanged (Fig. 3c). Eventually, when the surface separation becomes comparable to the vertical extent of the onion shell, the density of the A layer increases. The free energy of interaction in Figure 2c shows a much more pronounced attractive region in this case (compared to Figures 2a and 2b) due to a greater enthalpic gain in this system. Of course, when the surfaces are highly compressed entropic losses will eventually dominate and the interaction becomes repulsive.

<u>Effect of Increasing χ_{AB}; $\chi_{BS} = 2$.</u>

We now consider the effect of introducing a repulsion between the two homopolymers. When A is in a good or theta solvent, the behavior of the system is similar to that described above. The shielding of the B component by the A chains, which form a stretched brush, gradually decreases as χ_{AB} increases. For $\chi_{AB} = 1.5$ there is very little shielding left. As the surfaces are compressed, the density of the homopolymer brush increases monotonically (except for the presence of depletion regions where the B micelles are located).

The structural changes in the presence of $\chi_{AB} > 0$ are more pronounced when both components are in poor solvents. We describe the changes in the self-assembly of the layers when $\chi_{BS} = 2$, $\chi_{AS} = 1$ and $\chi_{AB} = 1.5$ [14]. As the surfaces are brought together, the micelles formed near both the surfaces start overlapping to some extent. With further compression of the surfaces, the micelles formed by the A chains attempt to shield the B micelles. In particular, the micelles formed by the A chains split into two and shield both sides of the B micelles. If the separation between the surfaces becomes smaller than the size of the micelles, the size of the A micelles increases laterally although their core density does not change noticeably. In the highly compressed state, two of the A micelles, which shield two adjacent B micelles, merge so that the chains form alternate A and B micelles between the surfaces.

The free energy of interaction as a function of surface separation for $\chi_{AB} = 0.5$ and 1.5 are shown in Figure 2 along with $\chi_{AB} = 0$. As can be seen, the attractive region shrinks and for the same surface separation, the interaction becomes less attractive as χ_{AB} increases.

CONCLUSIONS

We determined the morphology and interaction between surfaces that are coated with A <u>or</u> B chains. At least one of the polymers is in a poor solvent. The chains are grafted at low densities and form such self-assembled structures as "onion" and "flower" -like micelles. We found that the free energy of interaction has a large attractive region if the relatively more solvophilic A component is in a theta or poor solvent. The attraction between the surfaces is due to the formation of novel self-assembled structures as the surfaces are brought into contact.

Our findings provide guidelines for controlling the interactions between polymer-coated substrates in solution. The results reveal that specified interactions can be obtained by grafting a low volume fraction of solvophobic (B) and solvophilic (A) chains and tailoring the A-solvent or A-B energies.

ACKNOWLEDGEMENTS

This work was supported in part by ONR grant N00014-91-J-1363 to A.C.B.

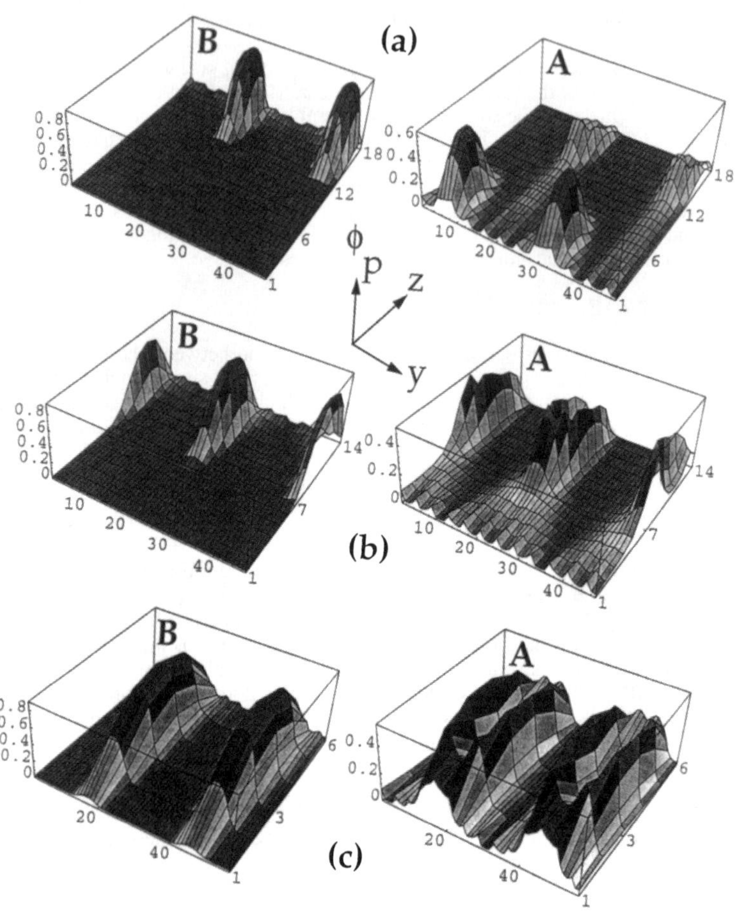

Figure 3.   SCF density profiles showing the effect of decreasing the separation, L, between the surfaces. Here, N = 80. The grafting density is $\rho = 0.025$, $\chi_{BS} = 2$, $\chi_{AS} = 1$ and $\chi_{AB} = 0$. In (a), L = 17 and in (b), L =13, while in (c), L = 5. The chains are grafted in the XY plane and the two surfaces are separated along Z. The parameter ϕ_p is the polymer density. The plots marked "B" show the polymer density of the B chains, while the plots marked "A" show the density of the A chains.

REFERENCES

1. L. I. Klushin (unpublished).
2. P. Lai and K. Binder, *J. Chem. Phys.* **97**, 586 (1992).
3. C. Yeung, A. C. Balazs and D. Jasnow, *Macromolecules* **26**, 1914 (1993)
4. K. Huang and A. C. Balazs, *Macromolecules* **26**, 4736 (1993).
5. D. R. M. Williams, *J. Physique II* **3**, 1313 (1993).
6. G. S. Grest and M. Murat, *Macromolecules* **26**, 3108 (1993).
7. K. G. Soga, H. Guo and M. J. Zuckermann, *Europhys. Letts.* **29**, 531(1995).
8. The size of the micelles is determined by a balance between the entropic losses due to the stretching of the chains to form these aggregates and the energetic gain due to the mutual shielding of the chains from the poor solvent.
9. D. F. Siqueira, K. Kohler and M. Stamm, *Langmuir* **11**, 3092(1995).
10. W. Zhao, G. Krausch, M. Rafailovich and J. Sokolov, *Macromolecules* **27**, 2933 (1994).
11. S. J. O'Shea, M. E.Welland and T. Rayment, *Langmuir* **9**, 1826 (1993).
12. A. Stamouli, E. Pelletier, V. Koutsos, E. van der Vegte, G. Hadziioannou, *Langmuir* **12**, 3221(1996).
13. C. Singh and A. C. Balazs, *J. Chem. Phys.* **105**, 706 (1996).
14. C. Singh, G. Pickett and A. C. Balazs, *Macromolecules* **29**, 7559(1996).
15. C. Singh, and A. C. Balazs, *Macromolecules* , in press.
16. G. Fleer, M. A. Cohen-Stuart, J. M. H. M. Scheutjens, T. Cosgrove, B. Vincent, *Polymers at Interfaces* (Chapman and Hall, London 1993).

The Role of Copolymer Architecture on the Interfacial Structure of a Ternary Polymer Blend

M. D. DADMUN
Chemistry Department
University of Tennessee
Knoxville, TN 37996-1600

ABSTRACT

The role of copolymer sequence distribution on the interfacial characteristics of a ternary polymer blend containing 2 homopolymers and a copolymer in the phase separated state are examined using Monte Carlo Simulation. The copolymer does migrate to the biphasic interface in the phase separated regime while the configuration and expansion of the copolymer at the interface is a function of sequence distribution within the copolymer. This effect is interpreted in terms of the efficiency of the copolymer to strengthen the biphasic interface. These results suggest that block, alternating, and block-ran structures show promise as interfacial modifiers, while the purely random and alt-ran copolymers will be less efficient as an interfacial strengthener. It is also found that a variation of the sequence distribution away from a purely random structure can dramatically effect the ability of the copolymer to modify the interface. As most polymers which are not block nor alternating are termed 'random', this differentiation may have an effect on experimental studies of 'random' copolymers as compatibilizers in polymer blends.

INTRODUCTION

Polymer blends offer an attractive route for the design of new materials with tailored properties. A mixture of two polymers that are mixed on a molecular level should offer a macroscopic sample with properties which are an average of the component polymers. Unfortunately, most polymers do not form thermodynamically miscible mixtures due to the low entropy of mixing of two long chain polymers. This immiscibility results in alloys that often exhibit properties which are inferior to those desired. Therefore, a method by which the properties of the immiscible system can be improved is needed. The addition of a copolymer which is composed of two monomers which have an affinity for the two homopolymers offers a compatibilizer which can migrate to the biphasic interface of the phase separated polymer blend and act as a surfactant or interfacial modifier. It is hoped that this interfacial modification will result in a lower surface tension between the two phases, promote adhesion, and improve the macroscopic properties of the blend. One structure that has been examined extensively towards this goal is a block copolymer.[1-5] Blends which have been modified by block copolymer addition exhibit enhanced toughness and improved minor phase dispersion. Unfortunately, block copolymers have not been widely used for this purpose as their synthesis is not trivial and often expensive. Therefore, it would be desirable to offer other polymeric structures which are more simply synthesized and yet still efficient interfacial modifiers. One example that immediately comes to mind is a copolymer with a different, more easily synthesized, sequence distribution than a block copolymer.

Mat. Res. Soc. Symp. Proc. Vol. 461 © 1997 Materials Research Society

The sequence distribution of a linear copolymer chain can vary from a block copolymer to a random copolymer to an alternating copolymer as is shown in figure 1. In this figure, A and B signify the two component monomers. The sequence distribution for the examples shown here (always 50% A and 50 % B) can be parameterized by the parameter P, the probability of finding two consecutive monomers along the chain which are the same type. In the block copolymer, except at the junction point, all A monomers are surrounded by A monomers and all B monomers are surrounded by B monomers, $\therefore$ P $\approx$ 1.   In the alternating copolymer, all A monomer are surrounded by B monomers and B by A, $\therefore$ P = 0. The purely random copolymer is one that is half way between the alternating and block copolymers, where the probability of finding neighboring similar monomers along the chain is 50%, $\therefore$ P = 0.5. An important distinction here is that our definition of a random copolymer is one where there is no correlation between the identity of a given monomer to the identity of it's immediate neighbors. A variation of this probability from 0.5 yields a copolymer with a different sequence distribution and potentially different properties. Therefore, the sequence distribution of the copolymer can be *continuously* varied from P=0 to P=1 to produce a set of copolymers whose microstructure can be thought of as varying continuously. The pertinent question then becomes "which copolymer sequence distribution offers the most benefit as an interfacial modifier in a polymer blend?"

Monte Carlo simulation will be utilized to examine the role of copolymer sequence distribution and copolymer concentration on the configuration of the copolymer at the biphasic interface of a phase separated polymer blend. The results will be interpreted to predict which copolymer structure provides the most promise as an interfacial modifier in polymer blends. Only linear chain architectures will be considered here.

Copolymer Structure	Type	P
ABABABABABABABAB	Alternating	0
ABBAABBAAABBABBA	Random	0.5
AAAAAAAABBBBBBBB	Block	ca. 1

Figure 1 - Sequence distributions of linear copolymers

MODEL

The model used has been previously described.[6] The salient features will be summarized here, but the reader is referred to the original manuscript for further details. The system in this study is a three dimensional lattice model. Periodic boundary conditions are applied in all three orthogonal dimensions to simulate an infinite set of chains. The only interaction energy is a nearest neighbor monomer-monomer interaction, ε_{A-B}. ε_{A-B} is zero if two neighboring monomers are of different type (A-B) and is negative otherwise. In this study, N_p = 742 is the number of polymer chains present, N=10 is the length of all chains, homopolymers and copolymers, and L=21 is the size of the cubic lattice. The concentration of chains on the lattice is 81.25%.

One half of the homopolymers are of type A and half are type B. The percentage and sequence distribution of copolymer present is varied, however, the composition of the copolymer is always 50% A and 50% B. These model parameters insure that the number of A monomers and B monomers in the blend will remain constant, only the way they are covalently bonded together is changed. The sequence distribution of the copolymer is parameterized by P, the probability of finding similar neighboring monomers along the copolymer chain. P can also be related to the experimentally relevant reactivity ratios of the component monomers.

Various chain configurations are created by applying a modified reptation technique to the chains. The new configuration is accepted according to the Metropolis sampling technique.[7] The miscibility and structure of the system is characterized by the heat capacity of the system, C_v, which is calculated as the fluctuation in the total energy of the system; the density profile of the two homopolymers, ρ_a and ρ_b, which is calculated as the proportion of occupied sites in a y-z plane which are populated by homopolymer A (B); and the radii of gyration of the copolymer along each axis, r_gx, r_gy, r_gz. These parameters allow the analysis of the effect of copolymer architecture on the miscibility temperature of the compatibilized polymer blend through the heat capacity, the morphology of the phase separated polymer blend through the density profiles, and the configuration of the copolymer at the biphasic interface through the radii of gyration.

RESULTS

The results from the previous study[6] can be summarized as follows. The addition of 1% copolymer does not significantly lower the miscibility temperature of the blend and is independent of copolymer structure. It was also found that the copolymer migrates to the biphasic interface regardless of copolymer structure for the model parameters. It was found, however, that the configuration of the copolymer at the biphasic interface is a function of copolymer structure for the 1% copolymer system. These results were interpreted in terms of the ability of the copolymer to entangle with the homopolymers to strengthen the interface and are summarized in Table I. Two new copolymer structures are referenced in table I, alt-ran and block-ran. As the names imply, alt-ran is a copolymer that is halfway between alternating and random copolymers (P=0.25) and block-ran is a copolymer that is blocky in character but also random (P=0.75). The results of the previous study led to the conclusion that alternating and block copolymers show promise as interfacial modifiers as they exhibit expanded structures which should promote entanglements between the copolymer and homopolymers. The purely random copolymer exhibits behavior that suggests collapse of the copolymer at the biphasic interface. This was interpreted to signify that the random copolymer would be less efficient as an interfacial

Table I

Structure	P	Configuration	Comments
Alternating	0	Disk	Expand across and along interface
Alt-Ran	0.25	Compact disk	Collapsed Disk
Random	0.5	Dense Disk	semi-collapsed disk
Block-Ran	0.75	Disk	Expanded disk
Block	1	cylinder	expands across interface

compatibilizer as the collapsed structure would not promote interaction between the copolymer and the homopolymers. Alt-ran and block-ran structures were also analyzed. These results led to the prediction that the alt-ran copolymer would be the poorest interfacial modifier as it was the most collapsed structure while the block-ran copolymer shows promise as an compatibilizer, exhibiting substantial expansion along and across the interface. The results to be presented in the current paper will expand the previous results to higher copolymer concentrations and analyze the results in terms of the response of the copolymer volume to a quench into the phase separated regime to gain a more intuitive understanding of the effect of copolymer sequence distribution under these conditions.

In the current study, rather than merely interpreting the effect of sequence distribution on the configuration of the copolymer at the biphasic interface, the expansion of the copolymer that is confined to the biphasic interface will be evaluated. This will be completed by transforming the experimentally determined radii of gyration of the copolymer to a volume. The volume that the copolymer occupies as a function of reduced temperature will provide information on the response of the copolymer to confinement at the interface. Figures 2-4 show the change in the copolymer volume as a function of reduced temperature τ for various copolymer sequence distributions at 1%, 5%, and 10% copolymer concentrations, respectively.

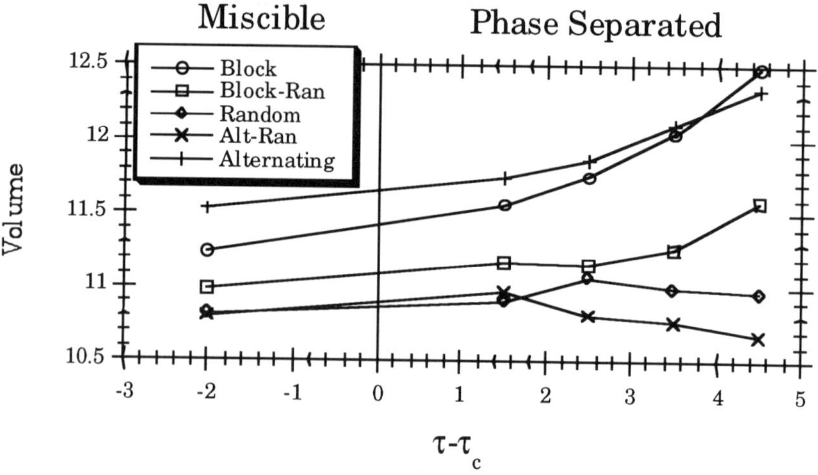

Figure 2 - Change in copolymer volume in the miscible and immiscible regime for the system with 1% copolymer.

DISCUSSION

The volume results offer intuitive insight into the importance of copolymer sequence distribution on the compatibilizing ability of copolymers. At high temperature, the homopolymers and copolymers are miscible and isotropically dispersed throughout the system. As the system is brought below the miscibility temperature, the two homopolymers phase separate and the copolymer migrates to the biphasic interface. To strengthen the interface, the

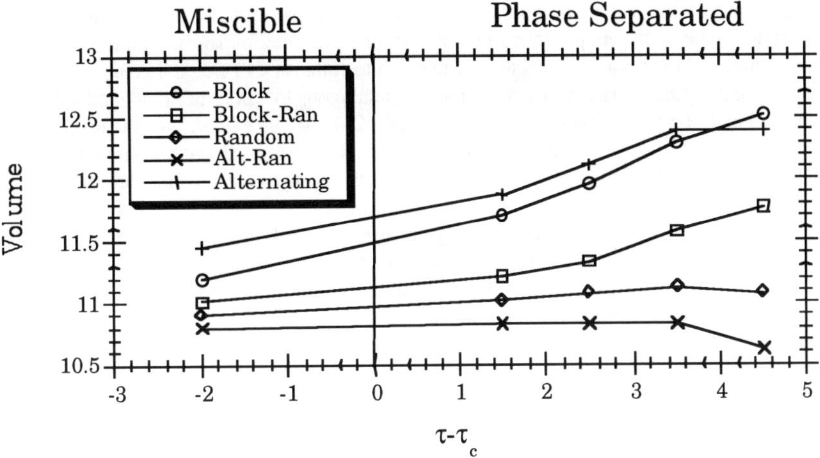

Figure 3 - Change in copolymer volume in the miscible and immiscible regime for the system with 5% copolymer.

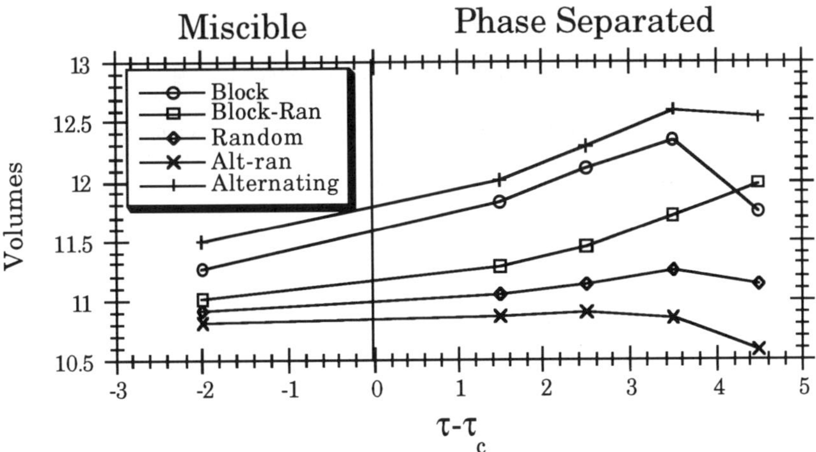

Figure 4 - Change in copolymer volume in the miscible and immiscible regime for the system with 10% copolymer.

copolymer must promote adhesion between the two phases by entangling with the two homopolymer phases. The copolymer, however has two options. It can expand into the homopolymer phases or collapse into a globule. If the copolymer expands, interaction between the copolymer and the homopolymers is feasible, while collapse will create a system with three non-interacting regimes. These two conditions are analogous to a polymer in a good solvent (expansion) and in a poor solvent (collapse). Figures 2-4 provide evidence of the response of each copolymer structure to containment at the biphasic interface of a phase separated blend. For each concentration, the block, alternating, and block-ran copolymer each show an expansion in their volume as the system is quenched into the biphasic region. This is reminiscent of a polymer in a good solvent, suggesting that these copolymer structures will prove to be effective interfacial modifiers. There are some subtle changes with concentration, however. Most notably, for the 10% copolymer system, (Fig. 4) the alternating and block copolymers exhibit a collapse in volume at the lowest temperature. This may be a result of the copolymers reaching a limit in their minor axis under these conditions. Further analysis will be completed to examine the reasons for this surprising result. The random copolymer exhibits a stronger dependence on the copolymer concentration. At 1% copolymer (Fig. 2), the random copolymer seems to collapse as the thermodynamic driving force for phase separation increases. At 5% copolymer (Fig. 3), the random copolymer shows very little change in expansion, neither expansion nor collapse, while at 10% (Fig. 4) the copolymer expands moderately upon quenching into the phase separated regime. The alt-ran copolymer, however, collapses for all copolymer concentrations as the system becomes more phase separated. This response is similar to a polymer in a poor solvent and further suggests that this copolymer sequence distribution will be very inefficient at strengthening the blend. From these results, predictions of the compatibilizing ability of these copolymers can be can be made. The block, alternating, and block-ran copolymer expand most readily as the thermodynamic driving force for phase separation increases, suggesting that these copolymers should promote adhesion between the two homopolymer phases and prove to be efficient modifiers. It is interesting to note that block-ran copolymers will eventually be formed at the interface between two reactive polymers, indicating that reactive engineering may be a useful method to promote interfacial adhesion. The alt-ran copolymer collapses as it is confined at the biphasic interface, denoting that this structure will not promote interaction between the two homopolymer phases. Finally, the random copolymer exhibits concentration dependent behavior, suggesting that dilute random compatibilizers will not be efficient strengtheners, but increasing the concentration may improve their behavior.

REFERENCES

1.) H.R. Brown, Macromolecules, **22**, 2859 (1989).

2.) H.R. Brown, K. Char, V.R. Deline, and P.F. Green, Macromolecules, **26**, 4155 (1993).

3.) C.F. Creton, E.J. Kramer, C.-Y. Hui, and H.R. Brown, Macromolecules, **25**, 3075 (1992).

4.) K. Char, H.R. Brown, and V.R. Deline, Macromolecules, **26**, 4164 (1993).

5.) C.F. Creton, H.R. Brown, and V.R. Deline, Macromolecules, **27**, 1774 (1994).

6.) M. D. Dadmun, Macromolecules, **29**, 3868 (1996).

7.) N. Metropolis, A.N. Rosenbluth, M.N. Rosenbluth, A.H. Teller, and E. Teller, J. Chem. Phys, **21**, 1087 (1953).

Simulations of Polymer Blends and Interfaces

MARTIN-D. LACASSE and GARY S. GREST

Corporate Research Science Laboratories, Exxon Research and Engineering Company, Annandale, New Jersey 08801.

Abstract

An efficient continuum-space model for simulating polymer blends and copolymers is presented. In this model, the interactions are short-range and purely repulsive, thus allowing for excellent computational performances. The driving force for phase separation is a difference in the repulsive interaction strength between like and unlike mers. To demonstrate the effectiveness of the model, we study the phase behavior of a symmetric binary blend of polymers as well as the interface between the two immiscible phases. As predicted by theory, we find the critical interaction parameter to scale with the inverse of the chain length of the polymers. The structure of the interface between two immiscible phases is investigated as a function of chain length and immiscibility. Capillary waves are observed and their measurement allows us to determine the surface tension accurately. Finally, the surface tension is related to the interface width.

Introduction

The mixing of two different homopolymer species in the melt state often results in a system of two immiscible phases. This behavior is often a severe technical problem especially when mixing homopolymers is used to create materials that would combine the properties of each component. Although the Flory-Huggins theory provides a good picture of the thermodynamics of mixing, it is still a coarse approximation that is often improved through an *ad hoc* functional dependence of a fitting parameter χ. In the immiscible regime, where the two phases are segregated, these systems exhibit many features that are only partially predicted by existing theories, especially when one considers the structure of the interface. It is the purpose of the present paper to introduce a simple continuous-space (CS) model to study binary blends of homopolymers. Thus, our approach differs significantly from previous studies which used lattice models almost exclusively.[1] Although off-lattice methods tend to be slower, the difference in performance is well-compensated for by other advantages such as the capability of simulating various ensembles (e.g. isobaric), or that of representing more complicated molecules (e.g. branched polymers). For systems with interfaces, CS models have one more advantage: besides their spatial isotropy, they offer a simple way to measure the surface tension from the measured pressure tensor.[2] In order to keep the possibility of studying dynamical phenomena (e.g. shear flow), we use molecular-dynamics (MD) as the simulation algorithm. For equilibrium calculations however, the MD moves are complemented by Monte-Carlo (MC) type-exchange moves in order to improve the sampling of phase space. We shall now briefly introduce the model and then describe a subset of our results obtained for binary blends and interfaces.

Model and Method

The magnitude of the relevant length and time scales responsible for the macroscopic properties of polymeric systems allows one to use simplified models for representing the polymer molecules. For a CS model, a convenient representation of a polymer molecule consists in attaching N soft or hard spherical beads of mass m together to form a chain. The resulting object, representing one chain, has $3N$ spatial coordinates. A system is built by constructing a large number M of such chains and enclosing them in a virtual box having the desired boundary conditions. The time evolution of the coordinates of all chains is resolved

129

through Newton's equation of motion (EOM), integrated with a velocity-Verlet algorithm.[3] The softness and stickiness of the beads, or mers, is set by the interaction potential.

The motion of the chains is coupled to a heat bath (that can be thought of as the solvent), acting through a weak stochastic force $\mathbf{W}(t)$ and a corresponding viscous damping term with friction coefficient Γ. Besides improving the diffusion of the system in phase space, this coupling has the practical advantage of stabilizing the numerical calculation. Including these terms, the EOM of mer i reads

$$m\frac{d^2\mathbf{r}_i}{dt^2} = -\nabla_i U - m\Gamma\frac{d\mathbf{r}_i}{dt} + \mathbf{W}_i(t). \tag{1}$$

The last term $\mathbf{W}_i(t)$ is a white noise having an average strength determined by temperature and the friction coefficient through the fluctuation-dissipation theorem. In the viscous drag term, Γ has to be carefully chosen to avoid overdamping so that the motion of the mers be dominated by their inertia. Finally, the conservative force term derives from a potential energy U having two contributions: the interation potential U_{IJ} acting between all the mers, responsible for excluded volume effects, and an attractive potential holding adjacent mers along the chain U^{ch}.

For the first contribution in a mixture of two homopolymer species, say A and B, the interaction potential $U_{IJ}(r_{ij})$ between two beads i and j of types $I, J = \{A, B\}$ separated by a distance r_{ij} is taken as the repulsive core of a central-force Lennard-Jones (LJ) 6:12 potential,

$$U_{IJ}(r_{ij}) = 4\epsilon_{IJ}\left[\left(\frac{\sigma}{r_{ij}}\right)^{12} - \left(\frac{\sigma}{r_{ij}}\right)^{6} - \left(\frac{\sigma}{r_c}\right)^{12} + \left(\frac{\sigma}{r_c}\right)^{6}\right], \tag{2}$$

for $r_{ij} < r_c = 2^{1/6}\sigma$ and zero otherwise. Instead of using the Lorentz-Berthelot rule, we choose $\epsilon_{IJ} = (1 + \delta_{IJ}\delta)\epsilon$, where $\delta \geq 0$ is a small immiscibility parameter and δ_{IJ} is the Kronecker delta. Our choice in energy parameters can be seen as a special (symmetric) case of

$$\delta = \epsilon_{AB}/\epsilon - \frac{1}{2}(\epsilon_{AA} + \epsilon_{BB})/\epsilon, \tag{3}$$

which represents the kind of interactions used in simple lattice systems such as in the Flory-Huggins theory. Here ϵ and σ are, respectively, parameters fixing the energy and length scales. Accordingly, all of our results are reported in terms of these natural units, using $\tau = \sigma(m/\epsilon)^{1/2}$ for the time scale.

Simply by varying δ, while leaving the temperature T constant, one can induce a phase transition from a homogeneous blend to a system with two segregated coexisting phases. This effect is reminescent to what is done in recent experiments reported by Gehlsen et al.[4] in which the phase separation of binary mixtures is studied as a function of the difference in deuterium between two otherwise identical polymer species.

For the second contribution to the interaction potential, we use an attractive anharmonic potential (so-called FENE) only acting between neighboring beads along the same chain:

$$U^{ch}(r) = -\frac{1}{2}kR_o^2\ln\left[1 - \left(\frac{r}{R_o}\right)^2\right], \tag{4}$$

for $r < R_o$ and ∞ otherwise. The parameters are set to $k = 30.0\epsilon/\sigma^2$ and $R_o = 1.5\sigma$. This particular choice prevents the chains from passing through each other as is described in more detail elsewhere.[5] For all the systems described here we use $\Gamma = 0.5\tau^{-1}$ at an overall number

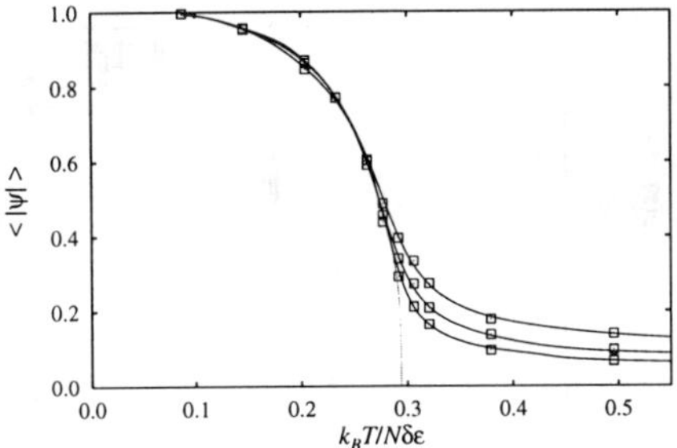

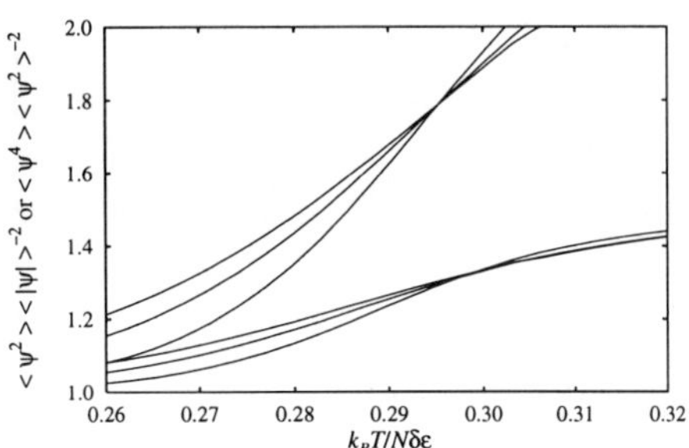

FIG. 1. The coexistence curve for a binary AB blend of 20-mers interacting with an immiscibility δ. The order parameter is defined as $\psi = (\rho_A - \rho_B)/\rho$. Curves are for different system sizes, top to bottom: the number of chains $M = 100, 200, 400$. The dotted line is the binodal.

FIG. 2. The crossing of dimensionless ratios $\langle\psi^{2n}\rangle\langle|\psi|^n\rangle^{-2}$ of the order parameter for 20-mers. Series are for $n = 2$ and 1, top to bottom. Curves of each series are for systems having a number of chains $M = 100, 200, 400$. Steeper curves have larger M.

density $\rho = 0.85\sigma^{-3}$ and a temperature $T = 1.0\epsilon/k_B$. We use an isochoric ensemble although this model could be used at constant pressure.[8] All systems are cubic and monodisperse.

In principle, the model and simulation method we described so far would be adequate to study the kinetics of phase separation in binary blends. However, the time scale involved in the unmixing transition —which normally occurs through the diffusive motion of the polymer chains— is extremely long, even more so near T_c. To properly sample the phase space in a reasonable calculation time, we supplement the MD moves with an MC procedure relabeling the type of the homopolymer chains ($A \leftrightarrow B$) with a Metropolis transition rule. This procedure was first introduced in lattice simulations and has been very successful.[6] Our exchange attempt rate is near one M per τ. For more detail see Ref. [8].

Results

Our results divide into two categories: systems with full periodic boundary conditions, used to study the bulk properties of our model, and systems having an antiperiodic* boundary condition in one direction, used to study the interfacial structure. We shall start by describing the bulk results of our model.

In Fig. 1, the equilibrium order parameter $\psi = (\rho_A - \rho_B)/\rho$, where ρ_I is the overall density of species I, is shown for systems of 20-mers of different sizes as a function of the immiscibility factor δ. A finite-size scaling analysis of our results using the multiple-histogram technique[7] is shown in Fig. 2. This allows us to extract the critical immiscibility value δ_c associated with a system of infinite size (value at the crossing of the curves). Critical values of binary systems of homopolymers of different chain length N were extracted in a similar way. Figure 3 shows the functional dependence of $\delta_c(N)$ in terms of the chain length, clearly evidencing a $1/N$ behavior of the critical immiscibility, as predicted by the Flory-Huggins theory.

With the help of finite-size scaling the properties of infinite systems can be extracted from rather small systems. It is thus possible to study the critical properties of our model,[8] but rather than doing this here, we shall look at the interfacial properties in detail. However, the finite-size effects on the interfacial properties of multiphase systems are not as well-known as

---

*An antiperiodic boundary condition acts like a periodic one except than a mer of type A leaving one side of the simulation box reenters as B on the other side, thus forcing the existence of an odd number (most likely one) of interfaces.

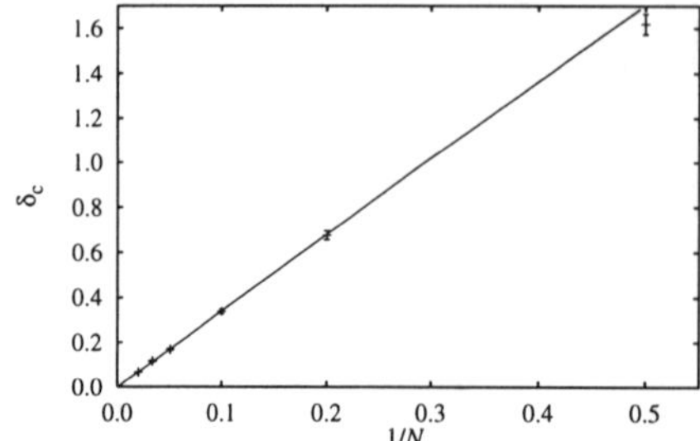

FIG. 3. The critical incompatibility δ_c required to induce a phase separation in a infinite system of a binary blend of homopolymers of length N. For $N > 2$, our data is well described by $N\delta_c = 3.40(5)k_B T/\epsilon$. $N = 2, 5, 10, 20, 30, 50$.

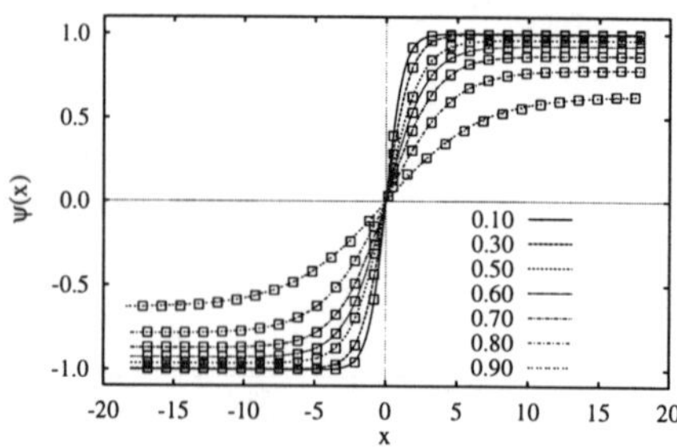

FIG. 4. The interfacial profiles of the yz-averaged order parameter for a system of 4000 10-mers with antiperiodic boundary conditions. The values of δ_c/δ are as indicated. The lines are from a fit $\psi(x) = \psi_o \tanh(2x/w)$.

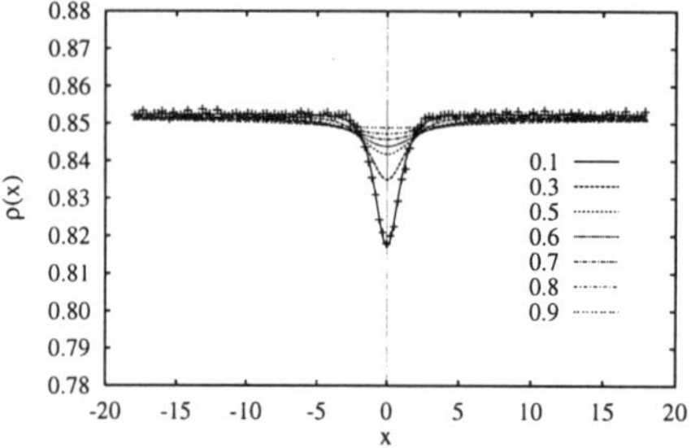

FIG. 5. The number density profile $\rho(x)$ evidencing the density hole at the interface. Systems are as in Fig. 4. For clarity, data points are only shown for $\delta_c/\delta = 0.1$. The curves are fits to the functional form $\rho(x) = \rho_o[1 - \Delta/\cosh(x/2w_\rho)]$.

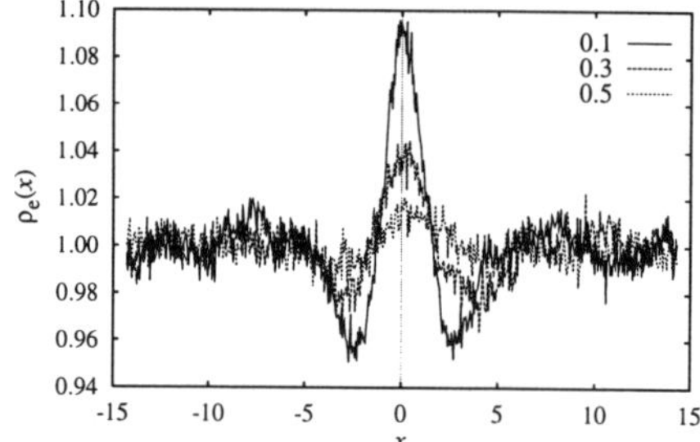

FIG. 6. The interfacial profile of the number density of chain ends $\rho_e(x)$ normalized to 1. The system contains 2000 20-mers at the immiscibility values δ_c/δ indicated in the plot.

for the bulk. Apart from the capillary-wave effects, the finite-size corrections to properties such as the surface tension are not as easily obtainable, and therefore we shall concentrate on binary systems either strongly or moderately segregated.

Since we know the critical immiscibility of our model, we can systematically build antiperiodic immiscible systems (i.e., with $\delta_c/\delta < 1$) and study the interfacial properties of the model. Figure 4 shows the interfacial profile of the order parameter $\psi(x)$ obtained by spatially averaging $\psi(\boldsymbol{r})$ in yz slices, perpendicular to the antiperiodic direction x. Each order parameter profile fits very well to a $\tanh(2x/w)$ functional form, thus associating a width w to the interface. Similar measurements for the number density profile $\rho(x)$ across the interface reveal that the interface between strongly immiscible liquids has a dip in the number density. This is shown in Fig. 5 where $\rho(x)$ has been fit to a form $\rho_o[1 - \Delta/\cosh(2x/w_\rho)]$, where Δ represents the amplitude of the density dip, and w_ρ is another measure of the interfacial width. For strong immiscibility, we also observe a strong ordering of the interfacial layer, with the chain perpendicular to it. This is shown in Fig. 6, where the normalized number density of chain ends is averaged in slabs parallel to the interface. For such values

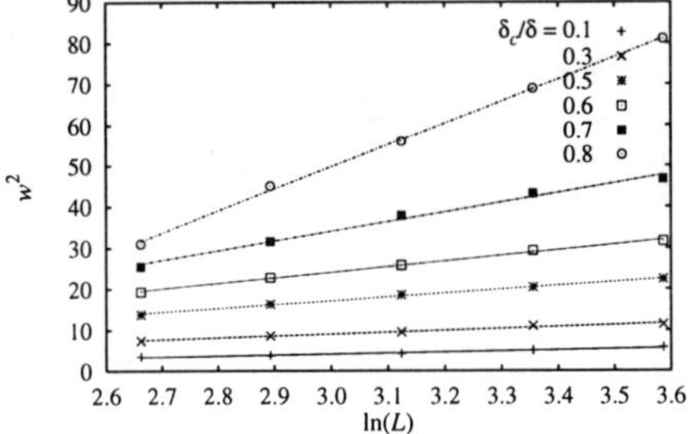
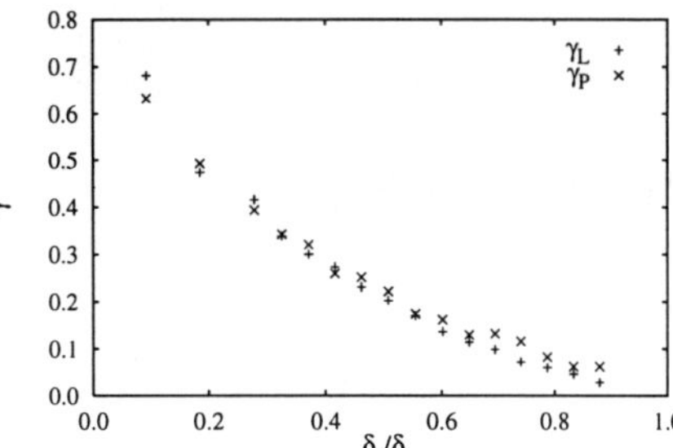

FIG. 7. Capillary waves and immiscibility factor dependence of the interfacial width for various systems of 10-mers. The values of δ_c/δ are as indicated. The lines are fits to $w^2 = a_\delta + b_\delta \ln(L)$.

FIG. 8. Two different estimates of the surface tension for a monomeric systems. γ_P is obtained from the difference in pressures parallel and perpendicular to the interface and the surface tension $\gamma_L = 1.1/b_\delta$, where b_δ is obtained as in Fig. 7.

of δ, the interface creates an ordering of the chains near the interface, at a length scale much larger than w.

It is interesting to measure the interfacial width as a function of the immiscibility δ. Our measure of w however, should include capillary-wave effects and a systematic study of w for systems of different sizes should reveal their presence. This is shown in Fig. 7, where the dependence of w on both the linear size L of the system and δ is shown, clearly evidencing the presence of capillary waves. To our knowledge, this is the first time that capillary waves are observed in computer simulations of polymers. We fit our data to the theoretical[9] expression $w^2 = a_\delta + b_\delta \ln(L)$, where $b_\delta \sim 1/\gamma_L$, the inverse of the surface tension (the subscript L indicates that γ has been derived from capillary wave effects). Very similar results are obtained from fitting w_ρ with the same functional form. By measuring the surface tension γ_P independently from the difference in the pressures (obtained from the virial,[3] hence the subscript P) perpendicular and parallel to the interface,[2] one can get another estimate of the surface tension, and compare it with γ_L. Figure 8 shows these two different estimates of γ for a monomeric (Lennard-Jones fluid) system. Our data show that with $\gamma_L = 1.1/b_\delta$, both measures are consistent. The values obtained from capillary waves (γ_L) are less noisy than those obtained from the pressure γ_P.

In Fig. 9, the effect of the chain length on the surface tension is studied for chains of length $N = 5$, 10, and 20. The data are plotted so that the immiscibility δ_c/δ is equivalent for all three chain lengths. Our data could be collapsed on a single curve with the scaling form $\gamma = N^\alpha f(\delta_c/\delta)$. This scaling form contains two distinct effects: the critical approach to $\gamma = 0$ and the effect of the chain length at fixed relative immiscibility δ_c/δ. For the first effect, it is well-known that the value of N changes the size of the critical region, with a width decreasing as N^{-1}. This effect should be taken care of by measuring the "temperature" scale with δ_c/δ. For fixed N, our data is consistent with a crossover to $f(\delta_c/\delta) \sim (\delta_c/\delta - 1)^\mu$, with the critical exponent approaching the Ising value $\mu = 1.23$ as $\delta \to \delta_c$. Concerning the second effect, i.e. the effect of the chain length at fixed δ_c/δ, our data could be collapsed with an exponent $\alpha \approx 1/4$ to $1/3$. Systems of longer chains are presently under investigation and we should provide a better value in a near future.

Lastly, in Fig. 10, we present the scaling law between the surface tension and the interfacial width. Our data for both monomers and 20-mers are consistent with the predicted[10]

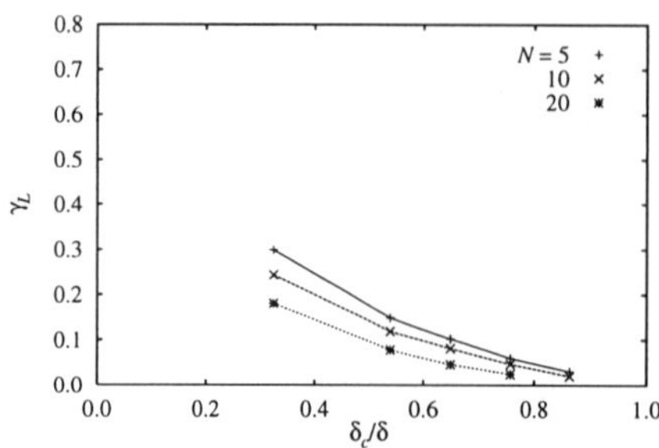

FIG. 9. The immiscibility dependence of the surface tension for systems of different chain lengths. Our data collapse when plotted as $N^\alpha \gamma$, with $\alpha \approx 1/4$–$1/3$.

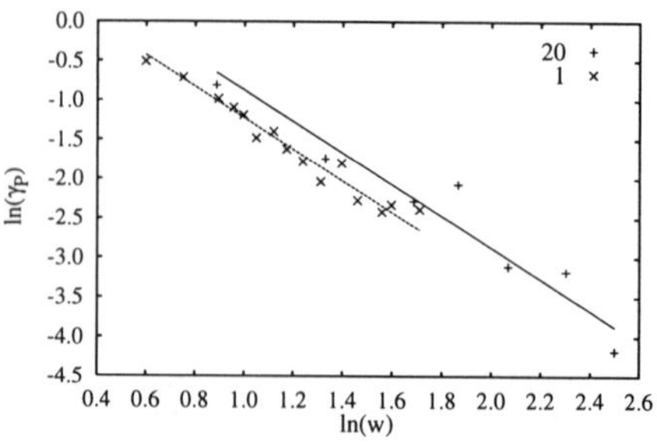

FIG. 10. A logarithmic plot of the surface tension γ_P as a function of the interfacial width w. The straight line is a fit to $\gamma_P \sim w^{-2}$. These data are from systems of 2000 20-mers and 8000 monomers.

scaling law $\gamma_P \sim w^{-2}$ for more than a decade in w.

Conclusion

We have shown that a continuous-space model with a rather simple potential allows us to study in detail the interface between immiscible binary blends of homopolymers. The natural isotropy of an off-lattice model coupled with a weak mismatch between the polymer chains made it possible for us to observe strong capillary wave effects at the interface of moderately small systems. This model has also been used for studying the phase behavior of diblock copolymers.[8] In a near future, the present work will be extended to larger binary systems and to binary systems containing nonanionic surfactants such as diblock copolymers at the interface.

Acknowledgments

We would like to thank Scott Milner for stimulating discussions and le *Fonds FCAR du Québec* for financial support.

References

[1] For a collection of recent reviews see *Monte Carlo and Molecular Dynamics Simulations in Polymer Science*, edited by K. Binder (Oxford University Press, New York, 1995).

[2] T.L. Hill, *An Introduction to Stastistical Thermodynamics* (Dover, New York, 1986) p. 316.

[3] M.P. Allen and D.J. Tildesley, *Computer Simulation of Liquids* (Clarendon, Oxford, 1987).

[4] M.D. Gehlsen *et al.*, Phys. Rev. Lett. **68**, 2452 (1992).

[5] K. Kremer and G.S. Grest, J. Chem. Phys. **92**, 5057 (1990); *Ibid.* **94**, 4103 (1991) (Erratum).

[6] A. Sariban and K. Binder, J. Chem. Phys. **86**, 5859 (1987); Macromolecules **21**, 711 (1988).

[7] H.-P. Deutsch, J. Stat. Phys. **67**, 1039 (1992).

[8] G.S. Grest, M.-D. Lacasse, K. Kremer, and A. Gupta, *J. Chem. Phys.* **105**, 10583 (1996).

[9] A.N. Semenov, *Macromolecules* **27**, 2732 (1994).

[10] M. Stamm and D.W. Schubert, Ann. Rev. Mater. Sci. **25**, 325 (1995).

STRENGTHENING EPOXY/HIGH IMPACT POLYSTYRENE INTERFACES USING END GRAFTING POLYMER CHAINS

Y. Sha*%, C.Y. Hui*%, E.J. Kramer**%, S. F. Hahn***, C. A. Berglund***
* Theoretical & Applied Mechanics, Cornell University, Ithaca, NY 14853
** Materials Science and Engineering, Cornell University, Ithaca, NY 14853
% The Materials Science Center, Cornell University, Ithaca, NY 14853
***Dow Chemical Co., Central Research and Development, Midland, MI 48674

Abstract

The fracture toughness G_c of epoxy/high impact polystyrene (HIPS) interfaces was measured as a function of grafting chain density Σ of dPS-COOH chains of various degrees of polymerization N.   For short chains, no effective entanglements can be formed between the dPS chains and the PS matrix of the HIPS, thus no enhancement in G_c over that of a bare interface. For very long chains, even though each chain is well entangled with the HIPS, the maximum grafting density achievable is very low and such an interface fails by scission of the grafting chains so that the interface fracture toughness is also low.   Large values of G_c are observed at intermediate chain lengths where both effective entanglements can be formed and a large Σ can be achieved.   Under these conditions the interface fails initially due to the formation of crazes and the subsequent break down of one of these crazes at the interface.   The transition areal chain density from chain scission to crazing, Σ_c, is independent of N.

1. Introduction

It has been well known that the interfaces of two immiscible polymers can be strengthened by adding block polymers to the interfaces[1-5].   In this work, we studied the reinforcing of the epoxy and high-impact polystyrene (HIPS) interfaces using deuterated polystyrene chains with carboxylic acid end functional groups (dPS-COOH).   The dPS chains penetrate into the bulk high-impact polystyrene whereas the -COOH end reacts with the epoxy to form a chemical bond.   The interface adhesion is quantified by determining the interface fracture toughness G_c, which is measured by the asymmetric double cantilever beam (ADCB) technique [1,4].   After fracture, forward recoil spectrometry (FRES) was used to measure the amount of deuterium on each side of interface. Since deuterium is used to label the grafted chains, the sum of the deuterium on the two surfaces could be used to determine the areal chain density, Σ, of the grafted polymer at the interface.   By analyzing the ratio of deuterium on the epoxy side after fracture to the total deuterium, Σ_{ep}/Σ, we also gain important information about the interface failure mechanism.   We studied the effect of the degree of polymerization, N, of the dPS chains and the grafting areal density, Σ, on G_c.   These results are compared with those for epoxy/PS interfaces reinforced with the same dPS-COOH chains [5].

2. Experiment

2.1 Materials

The epoxy resin, consisting of the diglycidyl ether of bisphenol A, had a molar mass of 350 g/mol.   Triethylenetetramine (TETA) is used as crosslinker. The HIPS, containing 8.48% by weight of polybutadiene rubber, *ca.* 1.2 mm in diameter, was provided by the Dow Chemical Co.   The HIPS contained 0.14% by weight of an antioxidant, but did not contain any plasticizers,

135

Mat. Res. Soc. Symp. Proc. Vol. 461 © 1997 Materials Research Society

mold release agents or other additives that could diffuse to and weaken an interface. The carboxylic acid terminated deuterated polystyrene (dPS-COOH) was prepared using anionic polymerization. Details of the polymerization and termination procedure can be found in reference [5]. The degree of polymerization of the dPS chains N, the weight average molecular weights M_w, and the polydispersity indices M_w/M_n for the dPS-COOH samples are listed in the table below:

N	160	410	540	690	840	1290	1480	1790
M_w	16,600	42,900	55,700	71,500	87,100	134,900	153,700	186,000
M_w/M_n	1.03	1.02	1.04	1.07	1.1	1.1	1.06	1.2

<u>2.2 Sample Preparation</u>

The epoxy resin was heated up to form a melt and mixed with TETA for 2 to 17 minutes. The mixture was then injected into a glass sandwich mold and transferred to an oven to carry out a B-stage cure. After the epoxy was removed from the mold, a 2% solution of dPS-COOH in toluene was spun cast onto the epoxy surface. The dPS coated epoxy slab was then annealed at $160\,^{\circ}$C in vacuum for 6 hours to allow the dPS-COOH chains to react with epoxy network. The unreacted chains were removed by washing the epoxy slab in tetrahydrofuran (THF) in an ultrasonic bath. The mixing time of the epoxy resin and TETA will determine the grafting density Σ, e.g., a mixing time of 2 minutes will give a low Σ while a mixing time of 15 minutes (near gelling) will give a high Σ, for each chain length N. The HIPS slab was made by compression molding at 160°C. The epoxy plate was then joined to the HIPS slab at 160°C under contact pressure to allow the tails of the grafted dPS chains to entangle with the HIPS. The sandwich was then cooled down to room temperature in air and cut into strips 50mm long and 9mm wide. The thickness of epoxy and HIPS strips were 0.8 mm and 1.6 mm respectively. After fracture, the areal density of chains Σ on both epoxy and HIPS surface were measured using forward-recoil spectrometry (FRES).

<u>2.3 Fracture toughness measurement</u>

The fracture toughness was measured by the asymmetric double cantilever beam (ADCB) technique [1-5]. The measurement was performed by driving a single edged razor blade, of thickness Δ, along the interface at a constant rate of 3×10^{-6} m/sec using a servo-motor driver. The fracture toughness of the interface G_c is given by the critical energy release rate of the interfacial crack. The energy release rate G was computed using[4]:

$$G = \frac{3\Delta^2 E_{ep}h_{ep}{}^3 E_{hp}h_{hp}{}^3}{8a^4}[\frac{E_{ep}h_{ep}{}^3 C_{hp}{}^2 + E_{hp}h_{hp}{}^3 C_{ep}{}^2}{(E_{ep}h_{ep}{}^3 C_{hp}{}^3 + E_{hp}h_{hp}{}^3 C_{ep}{}^3)^2}] \tag{1}$$

In eq 1, h_{hp} and E_{hp} denote the thickness and the Young's modulus of the HIPS slab and $C_{hp} = 1 + 0.64 h_{hp}/a$. Quantities C_{ep}, E_{ep} and h_{ep} are similarly defined for epoxy and are labeled by the subscript ep. The crack length, a, is measured from the crack tip to the edge of the razor blade. To allow many measurements of a, and thus G_c to be made, the entire history of the crack growth was recorded using a video tape for subsequent play back and measurement.

3. Results

Plots of fracture toughness G_c versus interface areal chain density Σ for grafted dPS-COOH chains with polymerization indices N=160, 540, 1290, 1480 and 1790 are shown in figures 1a-5a. The corresponding plots of the fraction of deuterium found on the epoxy side versus areal chain density for the grafted dPS-COOH chains are shown in figures 1b-5b. The fracture toughness of the bare interface is found to be 4.8 ± 0.5 J/m^2 and is shown as a star in

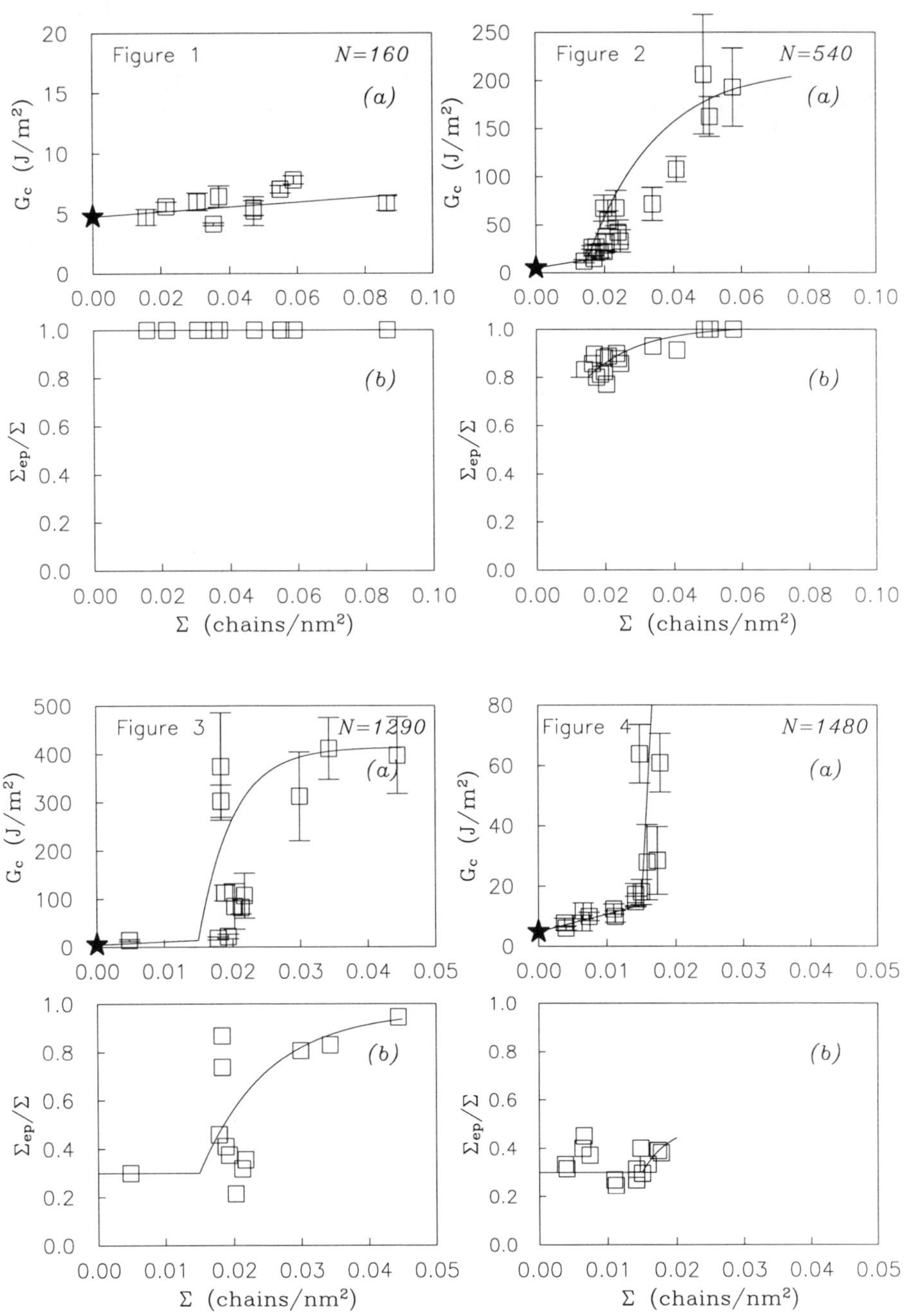

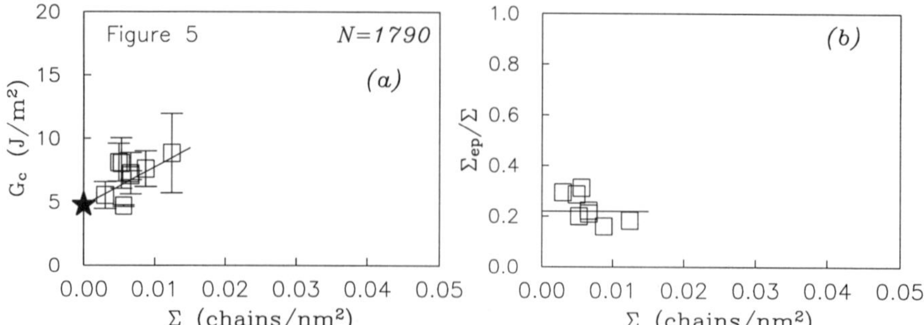

Figures 1-5 Epoxy/HIPS interface: (a) Fracture toughness vs. areal chain density (b) Fraction of deuterium on epoxy side after fracture.

figures 1a-5a.   Each data point of the fracture toughness curve presented in Figures 1a-5a is obtained by averaging 16 measurements.   The error bars in these figures corresponds to the standard deviation.

<u>Short chains (N=160)</u>

The entanglement length N_e for PS is 173[7].   Thus for N=160 practically no effective entanglements can be formed between the dPS chains and the PS matrix of HIPS.   In figure 1a the observed fracture toughness for the interface reinforced using N=160 dPS-COOH chains is only slightly higher than that of the bare interface, and is nearly independent of Σ.   Post fracture FRES analysis figure 1b shows that all the dPS is found on the epoxy side of the fracture surface, which clearly demonstrates that the interface fails by pull-out of the dPS chains from the HIPS side of the interface.   This result agrees well with the expectation that when chains are not sufficiently entangled to provide stress transfer across the interface, there will be little increase in G_c over that of the bare interface, despite the fact that large grafting densities Σ can be reached.

<u>Long chains (N = 1790)</u>

For long chains, the maximum areal chain density achievable at the interface is low and thus we expect the fracture toughness to be low, even though the anchored chains are well entangled with HIPS.   For N=1790, as shown in Figure 5a, the interface fracture toughness is close to that of the bare interface because the maximum areal chain density that can be reached for this case is only 0.014 chains/nm². The interface fails by chain scission of the dPS-COOH chains near the -COOH anchoring group prior to crazing.   The deuterium distribution in Figure 5b shows that about 20% of the deuterium is left on epoxy surface after fracture.   This amount of deuterium can be explained by the penetration of dPS chains into the epoxy so that the -COOH end-functional group is grafted some distance below the epoxy surface[5].   Since the crack propagates along the epoxy/HIPS interface, this result implies that a length of about 300-400 dPS monomers on average are left on the epoxy.

<u>Chain scission to crazing transition(N=1480)</u>

Theory about polymer/polymer interface failure mechanism[1-3] suggests that at low Σ the interface fails by scission of the grafted dPS chain.   At higher Σ there should be a critical areal chain density Σ_c where the transition occurs from chain scission to crazing which is independent of the polymerization index N, i. e.,

$$\Sigma_c = \sigma_{craze} / f_b \tag{2}$$

where f_b is force needed to break one C-C bond. This prediction is well supported by figure 4a for N=1480, which shows a sharp increase in G_c at $\Sigma_c \approx 0.015$ chains/nm^2. Using $f_b = 2 \times 10^{-9}$ N / bond [4] in eq 2, this result implies that the crazing stress of HIPS is about 30MPa, significantly lower than the crazing stress of PS (~ 55 MPa) under the same condition[5].

<u>Chains of intermediate length (N=540,690,840,1290)</u>

For short chain lengths, the maximum areal chain density that can be grafted on the interface is high(~ 0.09 chains/nm^2) but the fracture toughness is low due to poor entanglement between the grafted chains and HIPS. In contrast, for very long grafted chains, the chains are well entangled with the PS matrix of HIPS but the maximum grafting density that can be achieved is low due to the slow kinetics of grafting[6]. Therefore we anticipate that the fracture toughness for intermediate chain lengths will be the highest. Figures 2a and 3a (for N=540 and 1290) show that the interface fracture toughness G_c for the intermediate chain lengths are low until a value of $\Sigma > \Sigma_c \approx 0.015$ chains/nm^2 is exceeded, upon which the G_c increases rapidly with further increases in Σ, eventually achieving a value as high as *ca.* 500 J/m^2, which is almost 100 times of that of the bare interface, for N=1290. Experimental measurements for N=690 and 840 show similar trends[3]. This result supports the prediction of the fracture theory(eq 2) that the transition areal density from chain scission to crazing, Σ_c, is independent of N[1-3,5].

4. Discussion and Conclusions

The Young's modulus of polybutadiene rubber particles in HIPS is three orders of magnitude lower than that of the glassy PS matrix. This modulus mismatch leads to a stress concentration at the equator of the particle during mechanical deformation. The presence of the stress concentration can lead to crazes that propagate from particle to particle and hence to crazing throughout a large volume of material, rather than just at the crack tip. Hence the polymer absorbs a large amount of energy during deformation and is toughened. Also the yield stress of HIPS will be much lower than that of PS. This is indeed what we observed if we compare our results with the PS/epoxy interface reinforced with the same dPS-COOH polymer chains. Figure 6a shows the G_c vs. Σ for the N=1290 dPS chains at the PS/Epoxy interface. The G_c is low until an areal chain density $\Sigma \approx 0.03$ chains / nm^2 is reached, then G_c increases rapidly with increasing Σ. The deuterium fraction on epoxy side after fracture is shown in figure 6b. The interface fails by chain scission at low Σ, but failure occurs due to craze formation followed by breakdown of the craze at high Σ. This result shows that the chain scission to crazing transition for PS is 0.03 chains/nm^2, in agreement with the value found for PS/epoxy interfaces[5], and is consistent with the measured crazing stress of PS of about 55 MPa[4]. The crazing stress we obtained for HIPS is ~ 30 MPa. This shift in crazing stress and transition areal chain density from PS to HIPS is thus as expected.

Figure 7a shows that the maximum grafting density Σ_{max} of dPS-COOH on the epoxy surface decreases almost linearly with degree of polymerization N. Similar trends were found by Norton et al.[5] for the PS/epoxy system. The Σ_{max} values obtained by Norton et al.[5] are slightly higher than ours. This difference may be due to small differences in sample preparation procedure between the two sets of experiment. A plot of the maximum fracture toughness G_c^{max} (corresponding roughly to the value at Σ_{max}) vs. N is shown in Figure 7b. In the low N region, i.e. N<200, the interface fails by pull-out of the dPS without crazing and G_c^{max} is very low. As N is increased still further to 1290, chain scission is the dominant craze failure

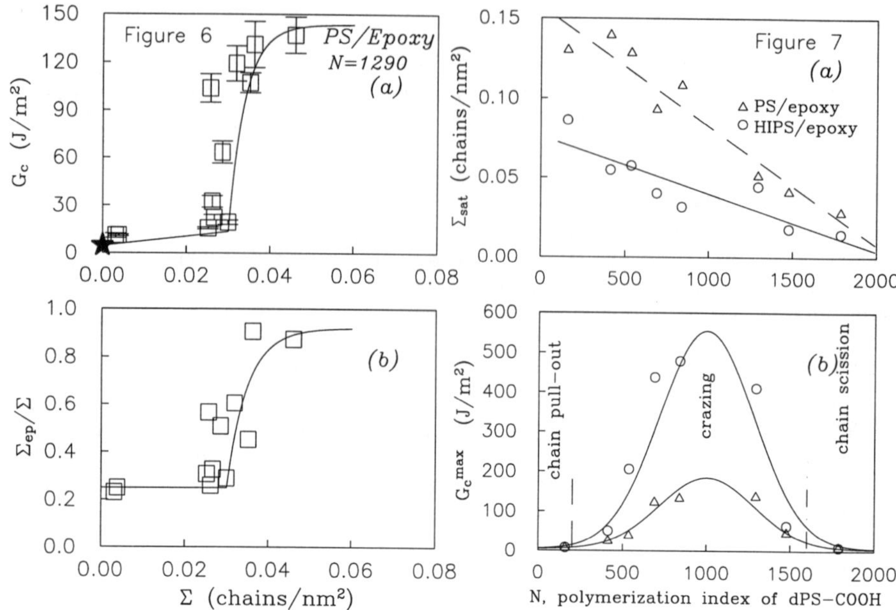

Figure 6. PS/Epoxy interface for N=1290:
(a) Fracture toughness vs. areal chain density
(b) Fraction of deuterium stayed on epoxy side
after fracture

Figure 7. (a) Maximum grafting density as a
function of polymerization index N
(b) Maximum fracture toughness as a function of
polymerization index N

mechanism, but chain disentanglement still plays a role in craze failure. The interface fails by chain scission at low Σ and craze formation followed by craze break down at high Σ. The G_c^{max} can reach as high as 500 J/m² for the HIPS/epoxy interface. At N=1480, the craze fails by chain scission but Σ_{max} barely reaches $\Sigma_c \approx 0.015$ chains/nm² so the G_c^{max} decreases significantly. Finally at N= 1790, Σ_c can not be reached and the interface fails by chain scission prior to crazing. To optimize the interface adhesion, a value of N around 1000 appears to be needed, where not much disentanglement will be observed but where relatively large values of the grafting density are still possible.

References

[1] J. Washiyama, E. J. Kramer, C. Y. Hui, <u>Macromolecules</u>, **1993**, 26, 2928.
[2] Dai, K. H.; Kramer, E. J.; Fréchet, J. M. J.; Wilson, P. G.; Long, T. E.; <u>Macromolecules</u>, **1994**, 27, 5187
[3] Y. Sha, C.Y. Hui, E.J. Kramer, S. Hahn and C. Berglund, <u>Macromolecules</u>, **1996**, 29,4728
[4] Creton, C.; Kramer, E. J.; Hui, C. Y.; Brown, H. R., <u>Macromolecules,</u> **1992**, 25,3075
[5] L. J. Norton, V. Smigolova, M. U. Pralle, A. Hubenko, K. H. Dai, E.J. Kramer, S. Hahn, C. Berglund and B. DeKoven, <u>Macromolecules</u>, **1995**, 28,1999.
[6] E. J. Kramer, <u>Israel J. Chem.</u> **1995**, 35, 49
[7] Onogi, S.; Masuda, T.; Kitagawa, K., <u>Macromolecules,</u> **1970**, 3 ,109.

Acknowledgments

The financial support for this project at Cornell was provided by the Dow Chemical Company. Partial support by the Materials Science Center which is funded by NSF is acknowledged.

THE DISTRIBUTION OF SIS COMPATIBILIZER IN MELT-MIXED POLYSTYRENE-POLYETHYLENE HOMOPOLYMER BLENDS

G. Kim,[*] M. Libera,[*] R. Potluri,[+] C.G. Gogos[*+]
[*] Stevens Institute of Technology, [+] Polymer Processing Institute, Hoboken, NJ 07030

ABSTRACT

This paper presents the results of transmission electron microscopy (TEM) studies on the distribution of polystyrene/polyisoprene/polystyrene (SIS) triblock copolymer compatibilizer in polystyrene-polyethylene homopolymer blends. The principal independent variable studied is the major-phase rheological properties (PS_{major}/PE_{minor} and PE_{major}/PS_{minor} blends). The polyisoprene block was preferentially stained by OsO_4 and could be clearly identified in cryo-ultramicrotomed TEM sections. The microstructural data show that when the rheologically strong PS phase is the major constituent, the compatibilizer follows the classical model localizing itself at the PS/PE interface. In the PE_{major}/PS_{minor} blends, however, the SIS compatibilizer is not localized at the PE/PS interface. Instead microdispersions of PS in SIS form which in turn are dispersed in the PE matrix. While the various microstructures are affected by the mixing protocol and polymer rheology, quiescent annealing after mixing indicates that the compatibilizer distribution is in part driven by thermodynamic effects.

INTRODUCTION

Melt-mixed homopolymer blends have been a subject of intense study for many years. [1-3] Blends are designed to produce a hybrid material which combines the desirable mechanical properties of each component. Significant processing and microstructure problems arise, however, because most homopolymers display little or no miscibility in each other. [4-8] These problems can be in part overcome when two immiscible polymers are made compatible by the addition of a third polymer agent called a compatibilizer. Compatibilizers act both as polymer surfactants to enhance the dispersion of one homopolymer phase in another and as stabilizers against coarsening of the dispersed phase. [9] The compatibilizer usually increases the interfacial adhesion between the two polymers, resulting in enhanced mechanical properties relative to the uncompatibilized blend.

Compatibilizers in blends have classically been believed to localize themselves at the interface between the matrix and dispersed phase. There have been many theoretical and experimental efforts to show that this distribution results in minimum total surface tension of the polymer blend system. [10-14] Debier et al. [10] showed the compatibilizer distribution at the interphase boundary using electron microscopy and found rheological improvement in MBS-modified in PC/SAN blends. Cheng et al. [11] and Hobbs et al. [12] showed theoretically that surface tension is minimized when the compatibilizer localizes itself at the interface in polycarbonate-based blends and confirmed their predictions using TEM. The interfacial location of compatibilizer was also found by Tremblay using electron energy-loss spectroscopy on unstained specimens. [13] In contrast, Fayt et al. [14] observed HPB-PIP-PS(10% blend concentration) compatibilizer in PS/LDPE blends not only at the interface between the PS and the PE phases, but also inside of PE matrix. They argued that the strong affinity to LDPE of the HPB block in the compatibilizer is responsible for the non-classical movement of compatibilizer toward the LDPE phase in this PS/LDPE blend. The present research uses transmission electron microscopy (TEM) with differential staining to determine the distribution

Mat. Res. Soc. Symp. Proc. Vol. 461 © 1997 Materials Research Society

of polystyrene-polyisoprene-polystyrene (SIS) triblock copolymer compatibilizer in a series of polystyrene(PS)/polyethylene(LDPE) homopolymer blends as a function of major-phase rheological properties and mixing protocol. The microstructural data indicate that the compatibilizer behaves classically localizing itself at the interphase boundary when the matrix is rheologically strong (polystyrene) and the dispersed phase is relatively weak (polyethylene). In contrast, in the inverse case where the major phase (polyethylene) is rheologically weak and the dispersed phase is rheologically strong (polystyrene), the compatibilizer is distributed throughout the dispersed PS phase.   This morphology is largely independent of various possible permutations of mixing the polyethylene major phase, the polystyrene minor phase, and the SIS compatibilizer.   The results are attributed to thermodynamic and rheological properties of the various blend components.

EXPERIMENTAL

LDPE (LDPE1409, Mn=15,000, Chevron) and the PS (PS685, Mn=28,000, Dow) were used as the blend components.   SIS (18% styrene content, Mw=125,000, Dexco) was used as a compatibilizer.   The solubilities and viscosities of these materials are presented in Table 1. The homopolymers were mixed in an intensive batch mixer under a nitrogen atmosphere.   The principal mixing protocol involved precompounding of the minor homopolymer component (9.3% blend concentration) with SIS (7% blend concentration) followed by subsequent mixing with the major component (83.7% blend concentration). Some specimens were annealed after mixing at 200 °C for 2 hrs. under a vacuum of 10^{-6} torr .

Electron-transparent thin films of as-mixed or annealed specimens were cut using a Reichert Jung Ultramicrotome with a cryogenic attachment at -110 °C. Specimens were studied in a Philips CM30 Supertwin TEM using a single-tilt holder at room temperature after exposure to vapor from OsO_4 crystals at 50 °C.

Table 1. Solubilities and viscosities of blend components

	LDPE	PS	PI	SIS
Solubility Parameters	8.0	9.2	8.1	
Shear viscosity (@200 °C) (Pa.s)	130	850		650

RESULTS AND DISCUSSION

For the microstructural study of multiphase polymers by traditional bright-field TEM techniques, the choice of staining agent to induce image contrast is very important. In general, heavy-element stains induce dark contrast by electron scattering outside an objective aperture from stained specimen regions. [15] It is well known that OsO_4 reacts with unsaturated double carbon bonds. [16] RuO_4 oxidizes aromatic groups and, to a somewhat less extent, unsaturated double carbon bonds. [17] Fig. 1 illustrates the effect on image contrast of these stains on specimens of SIS-compatibilized LDPE(major)/PS(minor) blends. When stained with OsO_4 the compatibilizer has dark contrast, because OsO_4 preferentially stains the isoprene block in SIS. However, when this same blend is stained by RuO_4, the compatibilizer is not clearly distinguished from the other phases (fig. 1b). These phases all show varying levels of contrast which are insufficient to unequivocally distinguish the distribution of compatibilizer.

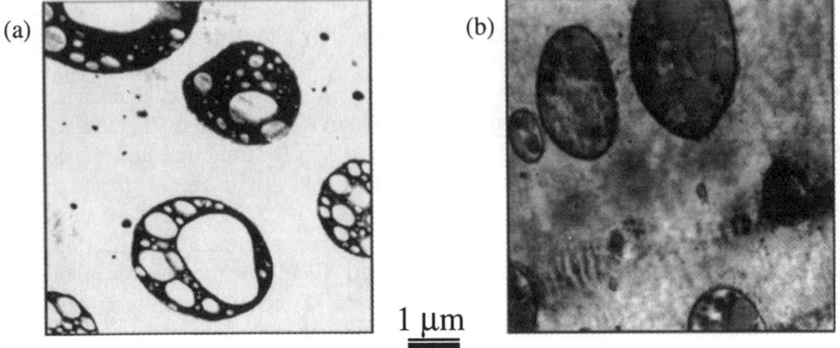

Fig. 1. Bright-field TEM images of SIS compatiblized LDPE(major)/PS(minor) blends:
a) OsO$_4$ stained; b) RuO$_4$ stained.

In the SIS-compatibilized PS(major)/PE(minor) blend, the compatibilizer localizes itself at the interphase boundary between the PS matrix and the PE dispersed phase (fig. 2a), as one would expect from the classical model of compatibilizer behavior. The PE dispersed phase is largely present as individual droplets separated from the PS matrix by a thin layer of SIS compatibilizer. This morphology is the result of a phase-inversion process which occurs when the premixed PE and SIS master batch is subsequently mixed with the PS major phase. The morphology of the premixed PE/SIS master batch is illustrated by fig. 2b. The rheologically weaker PE phase forms a continuous matrix around a dispersion of SIS (dark contrast) particles. This morphology must invert itself when mixed with PS homopolymer in order for the SIS to become the continuous phase encapsulating droplets of PE (fig. 2). This suggests that even the small fraction of PS in the SIS copolymer has sufficient thermodynamic affinity to the PS homopolymer matrix to enable radical rearrangement of the master batch morphology within the short duration time of mixing (5 min.). The speed with which this phase inversion occurs and the SIS-encapsulated PE dispersion forms is also tremendously enhanced by the rheological properties of the blend. The rheologically strong PS major phase can effectively transfer stress to both the PE and SIS phases to deform them and change the overall blend morphology.

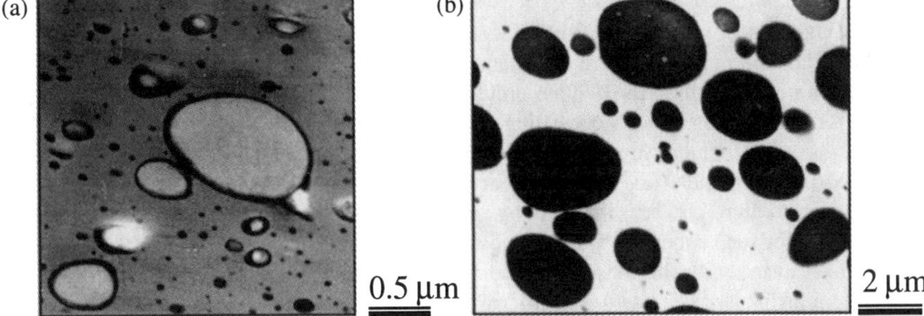

Fig. 2. Bright-field TEM images: (a) the morphology of PS major PS/SIS/LDPE blend stained by OsO$_4$. The SIS compatibilizer has dark contrast and is localized at the PS(83.7%)/LDPE(9.3%) interface; (b) the morphology of LDPE/SIS binary blend (master batch) stained by OsO$_4$; SIS is dispersed in LDPE matrix.

A dramatically different compatibilizer distribution is generated when the combination of homopolymers is inverted and the rheologically strong PS phase is dispersed in the rheologically weaker PE matrix. Rather than forming an interfacial layer between the matrix and the dispersed phase, fig. 3 shows that the PS homopolymer forms a microemulsion in an SIS matrix. The SIS/PS microemulsion in turn forms a dispersion in the continuous PE matrix. The origin of this dispersed SIS/PS microemulsion morphology lies in the structure and rheological properties of the PS and SIS master batch prior to subsequent mixing with the PE major phase. Fig. 3b shows this master-batch morphology. The lower-viscosity SIS phase (dark contrast) forms a continuous matrix around a dispersion of PS homopolymer particles. The morphology of this master batch is quite similar to that of the dispersed SIS/PS microemulsion in the PE major phase (fig. 3a). Because the viscosities of both SIS and PS are substantially higher than that of PE, one would expect the master batch to behave as a very hard phase in a rheologically soft matrix. Dispersion of the master batch material would then proceed largely by brittle fracture processes rather than by traditional elongation and deformation processes.

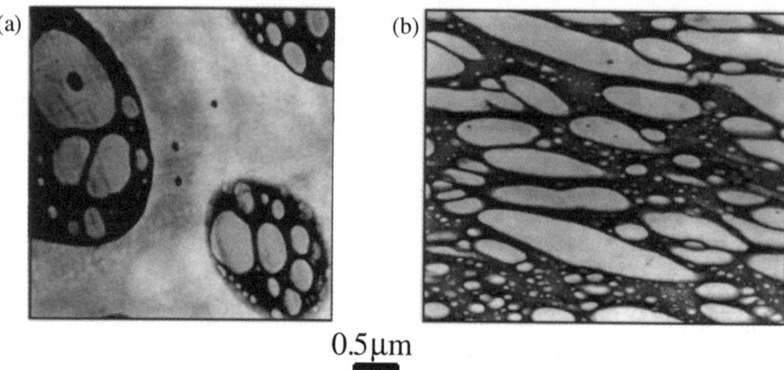

Fig. 3. Bright-field TEM images : (a) the morphology of LDPE major LDPE/SIS/PS blend stained by OsO_4; the SIS(7%) is distributed throughout the dispersed PS(9.3%) phase in the LDPE(83.7%) matrix; (b) the morphology of LDPE/SIS binary blend stained by OsO_4; SIS is dispersed in LDPE matrix.

Unclear at this point is what form the SIS-compatibilized PE(major)/PS(minor) blend would take if the mixing process were continued beyond the duration of 5 minutes studied here. One can question whether the SIS/PS microemulsion would itself become further dispersed and ultimately form a microstructure consisting of SIS compatiblizer localized at the interface between dispersed PS minor phases and the PS major phase as found in the inverse case of PS(major)/PE(minor) blends (fig. 2). The effects of thermodynamic and rheological forces on the blend microstructure can be separated by annealing experiments after mixing. Used as a starting material for such annealing was a blend formed by a mixing protocol where the PE major phase was first mixed with the SIS compatibilizer. This binary blend was subsequently mixed with the PS minor phase. The as-mixed microstructure is shown in fig. 4a. In contrast to the microstructure formed by dispersing the SIS/PS master batch in the PE major phase (fig. 3), this mixing protocol produces a significant population of dispersed PS particles coated by SIS. These all disappear upon annealing (200 °C; 2 hours; 10^{-6} torr vacuum), and only SIS/PS microemulsified dispersed phases can be found (fig. 4b). This finding indicates that the SIS has a thermodynamic affinity for the PS homopolymer. A similar phenomenon was found by

Koizumi et al. [18] who studied the diffusion of polystyrene-polyisoprene diblock copolymer into a matrix of PS homopolymer. The relative influence of this thermodynamic affinity and the substantial differences in PE, PS, and SIS rheological properties in developing the final SIS-compatibilized PE(major)/PS(minor) blend morphology is not yet known. However, given the fact that an SIS/PS microemulsion can form in response only to annealing in the absence of shear, suggests that this particular blend is unlikely to ever achieve a morphology consistent with classical compatibilizer behavior.

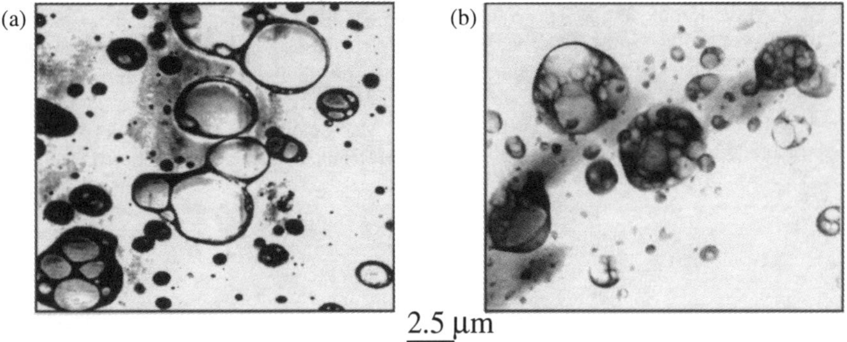

Fig. 4. (a) Mixing PS (minor) with precompounded PE(major)/SIS (compatibilizer) generates a morphology with many PS particles coated by SIS compatibilizer;
(b) Post-mixing quiescent annealing generates the SIS/PS microemulsion structure.

CONCLUSIONS

OsO$_4$ is an effective stain to preferentially decorate the polyisoprene block in SIS triblock copolymer and can be used to establish the location of this compatibilizer in PS/PE homopolymer blends. When the rheologically strong PS homopolymer is the major phase, the compatibilizer localizes itself as a film at interfaces between the PS matrix and PE dispersion as expected from classical model of compatibilized blends. This morphology is generated by a phase inversion when the PS major component was mixed with premixed PE/SIS. When the rheologically weak LDPE is the major phase, the SIS is not localized at PE(major)/PS(minor) interface but is instead distributed throughout the dispersed PS minor phase. This is a substantial departure from the classical picture of compatibilizer behavior. Quiescent annealing of PE(major)/PS(minor) blends after mixing shows that the SIS compatibilizer forms a microemulsion with the dispersed PS minor phase indicating that there is a thermodynamic component to the observed nonclassical compatibilizer behavior.

REFERENCES

1. M. Folks and P. Hopes, <u>Polymer Blends and Alloys</u>, (Blackie A & P, New York ,1993).
2. B. D. Favis, Canadian J. Chem. Eng. **69**, 619 (1991).
3. D. R. Paul and S. Newman, <u>Polymer Blends</u> , (Academic Press Inc., New York, 1978).
4. S. Wu, Polymer Eng. and Sci. **27**, 335 (1987).
5. P. C. Heimenz, <u>Polymer Chemistry</u>, (Dekker, New York, 1984).
6. W. M. Barentsen and D. Heikens, Polymer **14**, 579 (1973).

7.   E. Helfand and Y. Tagami, J. Chem. Phys. **56**, 3592 (1971).
8.   J. A. Brydson, <u>Plastics Materials</u>, 5th ed. (Butterworth-Heineman, New York, 1969).
9.   C. C. Chen and J. L. White, Polymer Eng. and Sci. **33**, 923 (1993).
10. D. Debier, J. Devaux and R. Lergas, Polymer Eng. and Sci. **34**, 613 (1994).
11. T. W. Cheng, H. Keskkula and D. R. Paul, Polymer **33**, 1606 (1992).
12. S. Y. Hobbs, M. E. J. Dekkers and V. H. Watkins, Polymer **29**, 1598 (1988).
13. A. Tremblay, S. Tremblay, B. D. Favis, A. Selmani and G. L. Esperance, Macromolecules **28**, 4771 (1995).
14. R. Fayt, R. Jerome and P. Teyssie, J. Poly. Sci. (Poly. Lett.) **24**, 25 (1986).
15. L. Sawyer and D. Grubb, <u>Polymer Microscopy</u>, (Chapman and Hall, New York, 1987).
16. W. P. Griffith., <u>The Chemistry of the Rare Platinum Metals</u>, (Wiley-Interscience, New York, 1967).
17. J. S. Trent, J. Scheinbeim and P. R. Couchman, Macromolecules **16**, 589 (1983).
18. S. Koizumi, H. Hasegawa and T. Hashimoto, Macromolecules **23**, 2955 (1990).

NANOMETER-SCALE METAL DISPERSIONS IN POLYMERIC MATRICES

K. R. Shull*, D. H. Cole* **, L.E. Rehn** and P. Baldo**

* Department of Materials Science and Engineering, Northwestern University, Evanston, IL
    60208

** Materials Science Division, Argonne National Laboratory, Argonne, IL 60439

ABSTRACT

We have investigated the diffusive properties of model metal nanoparticle dispersions in
polymeric matrices of several different molecular weights. Rutherford Backscattering
Spectrometry was used to measure the depth distribution of gold nanoparticles within thin layers
of poly(t-butyl acrylate) (PTBA). The gold nanoparticles were created by evaporation of a
discontinuous gold layer onto a thin film of PTBA. A second PTBA film was placed onto these
samples to create "sandwiches" in which the gold existed between two PTBA films. Gold particle
diffusion coefficients were obtained from measured gold particle depth distributions in annealed
samples for which the molecular weights of the two PTBA layers were identical. The experiments
revealed that particle mobility was decreased by two to three orders of magnitude compared with
the predictions of the Stokes-Einstein model of particle diffusion. These results are attributed to
bridging interactions between particles arising from slow exchange kinetics of polymer segments
at the polymer/metal interface. Experiments for which the molecular weights of the two polymer
films are different are sensitive to the ability of polymer molecules to pass through the gold
particle layer. Experiments done with thermally evaporated particles are consistent with a picture
in which polymer molecules are able to freely pass through the gold particle layer. Results
obtained with gold deposited by electron-beam evaporation are strikingly different. The gold in
this case is not able to diffuse, and polymer molecules are not able to penetrate the gold layer.
These results, in addition to preliminary results from optical absorption experiments, indicate that
much smaller particles are obtained by electron-beam evaporation than by thermal evaporation.

INTRODUCTION

Metal particle dispersions have a variety of interesting optical, magnetic and electronic
properties which enable them to be utilized in a variety of technologies. Many of these properties
require that the particles be dispersed in an electrically insulating, optically transparent medium,
such as a polymer. Appropriate choice of the polymer allows one to control the morphology of
the metal particle dispersion, and hence control the properties of interest. Several groups have
used the self-organizing ability of block copolymers, for example, in an attempt to control both
the size and spatial distribution of nanometer-scale metal particles.[1-5] The idea in this case is to
use a highly organized block copolymer microstructure, with characteristic domain sizes of tens of
nanometers, as "templates" for controlling the distribution of the metal particles. Block
copolymer systems are chosen so that the metal particles preferentially segregate to one type of
polymer domain. If one particle exists in each domain, and the block copolymer morphology is
maintained during the incorporation of the metal particles, then one should have excellent
morphological control over the metal particle dispersion. A significant complication arises,
however, in that there is almost always more than one metal particle in an individual domain, even

Mat. Res. Soc. Symp. Proc. Vol. 461 © 1997 Materials Research Society

for domains as small as 10 nm. Coalescence of these metal particles is generally very slow, and one therefore loses control of the size of the metal particles. In this example, one must first understand how particles are able to diffuse and coalesce within a single domain of uniform composition. These processes can be more readily understood by studying the properties of metal particle dispersions in homopolymers of different compositions. These simpler types of dispersions are the subject of the work described here.

Previously, it was shown that gold nanoparticles, produced either by thermal evaporation or by reduction of $HAuCl_4$, behave very differently in polymer matrices that in many ways are quite similar.[6, 7] Consider, for example, the three polymer structures shown in figure 1:

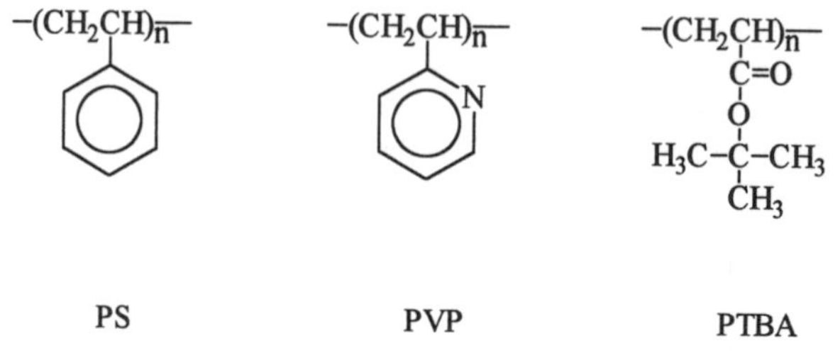

Figure 1: Chemical structures of polystyrene (PS), poly(2-vinylpyridine) (PVP) and poly(t-butyl acrylate) (PTBA).

Poly(2-vinylpyridine) and polystyrene are both amorphous polymers with a glass transition temperature near 100 °C. In addition, PS and PVP polymers of similar molecular weight have a nearly identical viscosity vs. temperature relationship.[8] Nevertheless, the development of the gold particle morphology above the glass transition temperature is remarkably different for the two polymers. In particular, the coarsening rate of the particles in a PVP matrix is much lower than the coarsening rate in a PS matrix[6]. We have characterized the PVP/Au system as a "strongly" interacting system, in that the exchange of polymer segments in intimate contact with the metal particles is very slow. The reduced rate of exchange allows for polymer molecules in contact with different particles to act as bridges between them. For sufficiently high particle concentrations, a network of bridges is formed, which retards particle diffusion, and reduces the rate at which particles can come into contact with one another in order to coalesce into a larger particle. At low particle concentrations, the average particle separation exceeds the characteristic dimensions of the polymer molecules. In this case, particle bridging is not a factor. Particle diffusion at low concentrations still takes place for strongly interacting systems, although the hydrodynamic radius of the particles may exceed the actual particle radius by an amount which is related to the polymer chain dimensions.[7]

Polystyrene and poly(2-vinylpyridine) have several disadvantages with respect to their use as model polymer matrices for studying the properties of polymer/metal nanocomposites. The PVP/Au system is so strongly interacting that particle coalescence, and particle diffusion at high particle concentrations, are essentially eliminated altogether.[6] The strength of the PS/Au interaction is not as high, but photo-oxidation of the PS is potentially a problem. Even very low levels of oxidation can affect the PS/Au interactions significantly. We believe these effects are responsible for the difficulties we have had obtaining reproducible results from the PS/Au system. These difficulties are avoided with the model system we are investigating here, consisting of gold nanoparticles in a poly (t-butyl acrylate) (PTBA) matrix. As with the previous studies, gold was chosen for its high atomic number and relative inertness to chemical reaction. In addition to its excellent oxidative stability, PTBA is easily spun-cast into uniform films. This polymer can also be converted to other acrylic systems, allowing for future work to probe the effects of different functional groups on the polymer-metal interactions.[9] PTBA also has a glass transition temperature (T_g) of about 45°C, which is conveniently accessible. Finally, the PTBA/Au interactions appear to be uniquely weak. The relative lack of interaction is thought to be derived from the presence of the bulky t-butyl group, which shields the gold particles from the more interactive ester group and its delocalized π-bonded electrons. Molecular orbital calculations performed by Ho *et. al.* have, for example, shown that delocalized electrons are responsible for strong interactions between an individual gold atom and a benzene ring,[10] Despite the relatively weak interactions between gold and PTBA, significant effects due to bridging interactions between particles are observed, as described in the following sections.

EXPERIMENT

Information about the mobility of individual gold particles and the polymer molecules themselves was determined from measurements of the depth distribution of the gold particles after different annealing treatments above the glass transition temperature of PTBA. The polymers were synthesized in our lab using anionic polymerization to achieve polydispersities (the ratio of weight average to number average molecular weight) of less than 1.15. Our synthetic procedure is similar to the procedure which has been described by Fayt, *et al.*[11] Diffusion experiments were conducted by creating sandwich samples of PTBA containing a buried layer of gold nanoparticles. The samples were made by spin casting 0.3 - 0.4 μm PTBA films of varying molecular weights onto highly polished single crystal silicon substrates. The substrates were then placed into an evaporation system, and gold was deposited on the samples. Two different evaporation techniques, thermal evaporation and electron-beam evaporation, were utilized. These techniques are illustrated in Figure 2. In thermal evaporation, the gold is resistively heated, whereas the second method uses an electron-beam to heat the sample. The morphology and diffusive properties of the gold particles obtained by these two evaporation methods were dramatically different, as discussed in more detail below. The thickness of gold on the sample for each deposition method was monitored with a quartz crystal thickness monitor. For the diffusion experiments, the equivalent thickness of gold deposited was typically between 0.4-0.5 nm. This thickness is insufficient to achieve continuous coverage, and the film actually consists of islands, approximately 3 nm in diameter, of roughly spherical shape.[4]

Following the evaporation of the gold, another polymer film was spun-cast onto a glass slide, and the film was floated off the slide onto a water bath. This film was then placed on the silicon substrate, on top of the evaporated layer, creating a sandwich. The excess water was evaporated in vacuum at room temperature, producing a uniform sample with a gold layer buried in the center of the film. Diffusion experiments were then performed by annealing the sandwich in vacuum, at temperatures between 50-165°C, depending on the polymer molecular weight. Higher temperatures were avoided, since PTBA will degrade at ~180°C, and partial conversion to poly (acrylic acid) may occur at slightly lower temperatures[12].

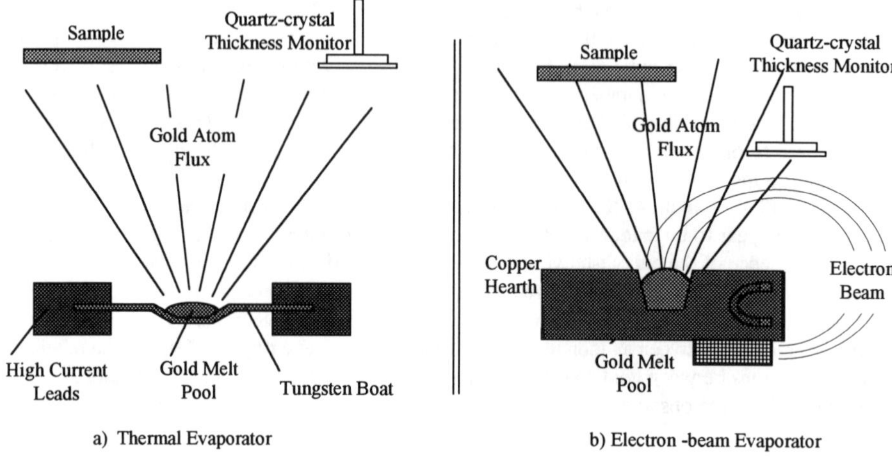

Figure 2: Schematic diagrams illustrating thermal evaporation (part a) and electron-beam evaporation (part b) of gold.

Upon annealing, the particles diffuse through the polymer by Brownian motion. This motion was quantified by utilizing Rutherford Backscattering Spectrometry (RBS) as a depth profiling technique. The polymer sandwich samples are well suited for RBS, as the sensitivity to marker atoms increases as the square of the atomic number.[13] This sensitivity allows extremely small fractions of a high mass element, such as gold, to be detected reliably while embedded in a low atomic mass matrix, such as the PTBA film. The separation in recoil energy of incident ions scattered from gold versus those scattered from the lower atomic mass constituents allows the gold concentration profile to be isolated from the signal from the substrate to a depth of ~ 1 μm. The samples showed extensive mass loss due to the ion beam striking the sample, but further experimentation revealed that despite the degradation of the polymer, the distribution of particles in the sample remained unaltered, allowing quantitative analysis despite the beam damage. To extract diffusion coefficients from the measured depth profiles, simulated concentration profiles were fit to the actual data. These simulated profiles corresponded to the appropriate solutions to the diffusion equation, convoluted with a Gaussian broadening function to account for the RBS depth resolution.

Finally, in order to assess the potential interplay between particle motion and polymer motion, diffusion samples were made in which the molecular weights of the polymers on either side of the gold layer were different. We refer to these experiments as "marker motion" experiments, because in addition to any diffusive broadening of the gold particle distribution, there is a net translation of the center of the distribution toward the lower molecular weight polymer. The gold particles in this case can be viewed as markers which are sensitive to the diffusive motions of the polymer molecules themselves.

RESULTS AND DISCUSSION

<u>Gold particle Diffusion</u>

Figure 3 shows the results of a gold particle diffusion experiment in which the polymer on either side of the gold layer was PTBA with a molecular weight of 7,000 g/mol. The sample was heated for one hour at 65 °C. The gold particle diffusion coefficient, D, in this case is 1.58×10^{-14} cm^2 s^{-1}. A convenient way to analyze the data is by comparison to the Stokes-Einstein equation for diffusion of isolated spherical particles in a viscous medium:

$$D = \frac{k_B T}{6 \pi \eta R} \tag{1}$$

Here k_B is Boltzmann's constant, T is the absolute temperature, η is the viscosity of the medium, and R is the particle hydrodynamic radius.[14] This theory contains the implicit assumption that the viscosity of the medium is not affected by the particles themselves. It is this assumption that leads to our definition that a system which shows particle diffusion in close accord with the Stokes-Einstein prediction should be classified as 'non-interacting,' a system in which diffusion is present but substantially slowed compared to Stokes-Einstein predictions is 'weakly interacting,' and a system where diffusion is eliminated entirely (as is the case for polyvinylpyridine)[6] is 'strongly interacting.' In the context of our introductory remarks concerning the segregation of metal particles to individual domains in ordered morphologies, the same interactions which cause preferential segregation into one of the domains are also responsible for hindering the motion of particles in that domain.

In Figure 4, the measured diffusion coefficients are plotted versus temperature and compared to predictions from the Stokes-Einstein equation, using effective values of the hydrodynamic radius to force agreement with the predictions. The viscosity was determined separately for each polymer molecular weight and temperature (in the absence of gold particles) by oscillatory shear rheometry. Our analysis here in terms of the Stokes Einstein equation is instructive. Note that the values of the effective hydrodynamic radius determined in this way are quite large, varying from 70 nm for PTBA with a molecular weight of 7000 g/mol, up to 1000 nm for PTBA with a molecular weight of 100,000 g/mol. The highest of these values is nearly three orders of magnitude larger than the actual particle radii. These radii are approximately 1 nm, and even with the presence of some particle coarsening should not exceed 10nm.[6] These large effective hydrodynamic radii are compelling evidence of the formation of bridging interactions between particles.

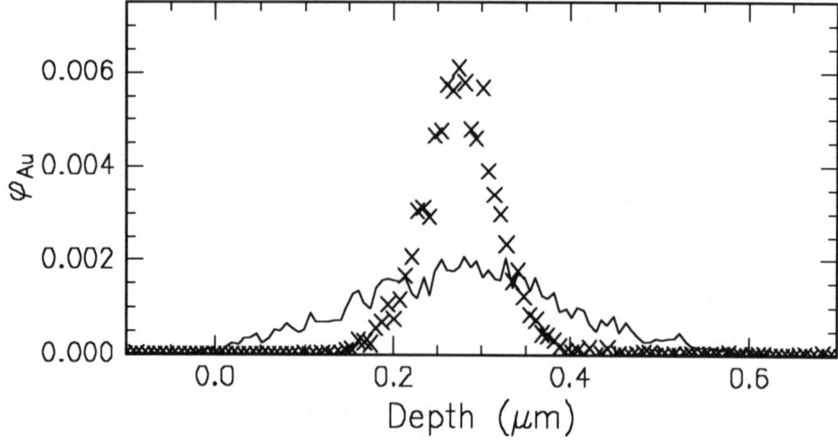

Figure 3: Typical gold particle concentration profiles before annealing (crosses)and after annealing for 1 hour at 65°C (solid line). The width of the distribution prior to annealing is a measure of the RBS depth resolution.

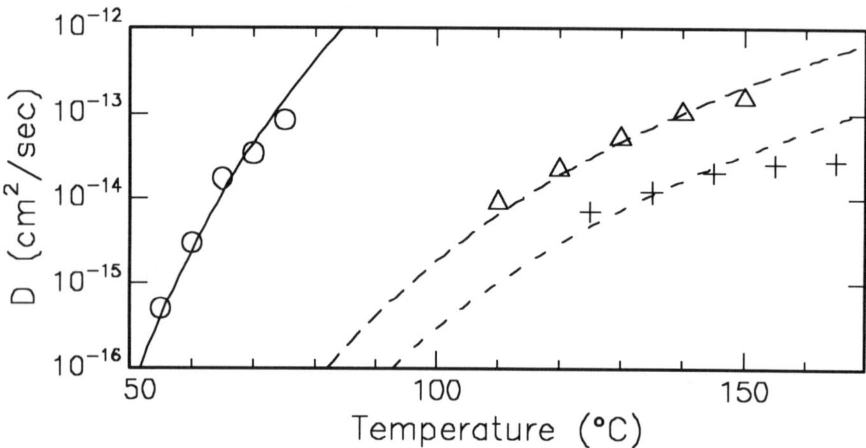

Figure 4: Gold particle diffusion coefficient vs. temperature for three polymer molecular weights. M = 7,000 g/mol (O), M = 77,000 g/mol (Δ), and M = 100,000 g/mol (+). The Stokes-Einstein predictions are indicated for each polymer. The effective hydrodynamic radii are 70 nm (solid line), 400nm (upper dashed line), and 1000nm (lower dashed line).

A comparison of the length scales of a polymer strand and the interparticle spacing shows that even for relatively short chains such as the 7,000 g/mol PTBA, bridging interactions are likely. The statistical segment length, **a**, defined as the effective length of a monomer in a random walk model of polymer chain dimensions, is roughly 0.7 nm for PTBA.[15] The root-mean-square end-to-end distance, R_O, for a polymer molecule with a molecular weight of M is $\mathbf{a}(M/M_0)^{1/2}$, where M_0 is the repeat unit molecular weight. For PTBA, M_0 = 128 g/mol, from which we obtain R_0 =5.2 nm for PTBA with M = 7000 g/mol, as compared with R_0 = 17 nm for M = 77,000 g/mol, and R_0 = 20 nm for M = 100,000 g/mol. The average particle separation, L, will depend on the size and distribution of the particles. For particles which reside in a two-dimensional layer, simple geometric arguments yield $L = (2r)^{3/2}h^{-1/2}$, where r is the particle radius and h is the equivalent thickness of the metal particle layer. For our samples r~1 nm and h~0.4 nm, yielding L=4.4 nm. If the particles are distributed through the thickness of a polymer film of thickness H, the scaling changes to $L = 2r(H/h)^{1/3}$. For a 100 nm film L=12.6 nm, which exceeds the end-to-end distance for only the lowest molecular weight polymer in our experiments. Clearly, at a fixed value of L, the importance of bridging interactions will increase as the polymer molecular weight increases. This effect is evident in our experiments, where the effective hydrodynamic radii of the gold particles increase dramatically as the polymer molecular weight increases. Also, we have found in preliminary experiments that the effective hydrodynamic radius decreases as the particles diffuse, and does not necessarily remain constant throughout a given diffusion experiment. The decrease in effective particle radius can be attributed to increases in L as the particles diffuse throughout the film. Increases in L due to particle coarsening may also play a role here. Other experiments using colloidal gold suggest that reducing the concentration of particles to the point where L far exceeds typical polymer chain dimensions leads to fewer bridging interactions and better agreement with the Stokes-Einstein predictions based on actual particle dimensions.[7].

<u>Marker motion experiments</u>

Figure 5 shows a representative sample geometry for the marker motion experiments. In these experiments, gold particles were sandwiched between PTBA polymers with very different molecular weights (420,000 g/mol vs. 7,000 g/mol). Two types of samples were prepared. For the first type of sample, the higher molecular weight polymer was spun-cast on the silicon substrate. Gold was then evaporated on this sample, and the low molecular weight film was floated on top of the higher M film. The second sample had the reverse of this geometry, with the lower molecular weight film directly against the substrate. Both samples were annealed for 30 minutes at 85°C, and in both cases the interface moved toward the low molecular weight layer.

Several factors determine the rate of gold particle motion in the marker motion experiments. The simplest approaches assume that the gold particles are "inert" markers which do not alter the diffusive behavior of the polymers. In this case the rate of marker motion is determined by the relative fluxes of molecules in the two directions across the interface. Suppose that the fast-diffusing, low molecular weight molecules exist at higher concentrations to the right of the markers, so that the flux of material to the left will be higher than the flux of material to the right. The net flux of atoms to the left leads to the buildup of compressive, osmotic stresses to the left of the markers. These stresses reduce the net flux of molecules to the left, unless these stresses are able to relax on the time scale of the experiment. From a theoretical standpoint, two limiting cases have been considered.[16] In the "slow" theory of mutual diffusion, it is assumed that these stresses cannot relax over the timescale of the experiment, so that that the net fluxes to the right and to the left of the markers are equal to one another. In this case there is no net motion of the markers. In the "fast" theory of mutual diffusion, the opposite assumption is made, i.e., that the osmotic stesses relax to zero very quickly, so that these stresses do not affect the marker motion. In this case the marker motion is determined primarily by the faster diffusing species. The displacement of the markers, Δx_m, varies with the square root of the diffusion time according to the following equation:

$$\Delta x_m = C\sqrt{D^* t}.$$

Here C is a constant related to the ratio of diffusion coefficients of the two polymer layers, D^* is the tracer-diffusion coefficient of the lower molecular weight polymer, and t is the annealing time. In our case, the diffusion constants of the two polymers are very different, and C can be assumed to be equal to the limiting value of 0.48 for a divergent ratio of diffusion coefficients. In this manner we obtain $D^* = 8 \times 10^{-12}$ cm^2 s^{-1}. for the sample with the higher molecular weight polymer as the bottom layer, and $D^* = 9 \times 10^{-12}$ cm^2 s^{-1} for the sample with the high molecular weight polymer as the top layer.   This comparison indicates that there is no preference for the markers to move toward the substrate or toward the free surface. Evaluating the marker motion as a function of time reveals that the marker displacement scales as $t^{1/2}$. Figure 6 shows the data from a series of marker motion experiments carried out at 60°C. A best fit to the data produces $D^* = 2.52 \times 10^{-14}$ cm^2 s^{-1}.

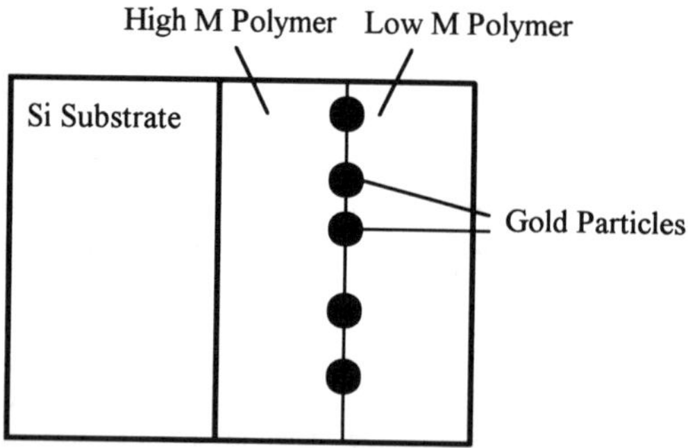

Figure 5: Schematic representation of the marker motion experiments.

The central quantity relating diffusive and rheological behavior of polymer melts is the monomeric friction factor, ζ_0. The tracer diffusion coefficient of a relatively low molecular weight polymer molecule is given by the following expression:[17]

$$D^* = \frac{k_B T M_0}{M \zeta_0} \tag{2}$$

where it has been assumed that M is low enough so that the diffusing molecule does not form entanglements with molecules in the polymer matrix. This assumption is certainly valid for PTBA with M=7,000 g/mol, since the entanglement molecular weight of this polymer is near 20,000 g/mol. The monomeric friction factor can also be related to the zero shear viscosity of an unentangled polymer melt through the Rouse theory of polymer dynamics.[18] In this context ζ_0 is given by the following expression:

$$\zeta_0 = \frac{36\eta_0 M_0^2}{M N_{av} \rho a^2} \tag{3}$$

where ρ is the bulk density of PTBA and $N_{av} = 6.02 \times 10^{23}$. Previous comparisons of polymer self diffusion coefficients and melt viscosities in the unentangled regime have confirmed that friction factors determined from these two types of experiments are quite consistent with one another.[16, 17, 19] Deviations in the friction factors in our case can therefore be attributed either to the presence of the gold particles, or to deviations from the prediction of the fast theory of mutual diffusion which we have used here. The friction factor obtained from rheological data between 60 °C and 160 °C is given by log ζ_0 = A+B/(T-T_{ref}) with A= -12.4, B=706 and T_{ref} = -10.3°C. At 60 °C, we obtain $\zeta_0 = 4.7 \times 10^{-3}$ g s^{-1} from the bulk viscosity of the polymer. The calculated friction factor at 60 °C from the marker motion experiments is $\zeta_0 = 3.34 \times 10^{-2}$ g s^{-1}.

The comparison made above indicates that the friction factor obtained from the marker motion experiments is seven times larger than the value obtained from the viscosity measurements. One possible interpretation of this discrepancy is that the gold particles are not necessarily acting as "inert" markers, but that they somehow hinder the diffusion of the low molecular weight PTBA molecules. Recall, however, that our analysis is based on the "fast" theory of polymer interdiffusion, where osmotic pressure gradients in the sample are assumed to relax very quickly. In fact, the relaxation time for the high molecular weight polymer (obtained as the ratio of viscosity to plateau modulus) is approximately 2 days at 60 °C, which is several times longer than the longest annealing times in our experiments. Some slowdown of the marker motion compared to the predictions of the fast theory is obviously expected, as this experiment was almost certainly conducted in the crossover regime between the validity of the fast and slow theories of mutual diffusion. An increase of the "effective" friction factor by a factor of seven seems plausible. At this point, we can only conclude that the observed marker motion is potentially consistent with the ability of polymer molecules to pass through a relatively concentrated dispersion of metal particles, provided that these particles are produced by thermal evaporation.

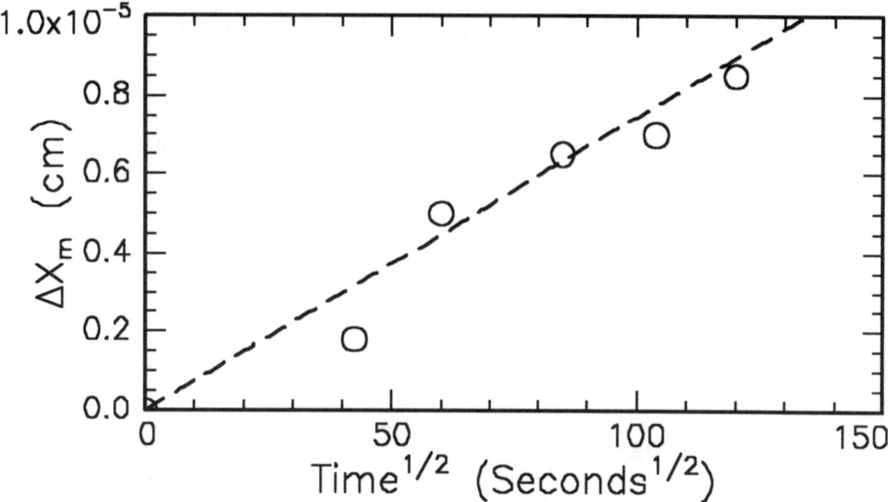

Figure 6: Time dependence at 60 °C for thermally evaporated gold particles sandwiched between PTBA layers with molecular weights of 7,000 g/mol and 420,000 g/mol. The equivalent thickness of the gold layer is 0.4 nm.

Very different results were obtained when the gold particles were produced by electron-beam evaporation. In this case, no diffusion or marker motion was observed for a gold particle layer with an equivalent thickness of 0.4 nm. In this case, not only are the bridging interactions between particles strong enough to eliminate any particle diffusion, but motion of polymer molecules through the gold layer is inhibited as well. One potential explanation for this is that the gold does not exist as a layer of discrete particles, but exists as a continuous gold film with a thickness of only 0.4 nm. This explanation does not seem likely to us, given that continuous gold films on inorganic glasses are only formed for thicknesses of several nanometers.[20] Nevertheless, we have not been able to rule out this possibility explicitly. Preliminary measurements of the optical adsorption spectra of the gold films indicate that the particle size is, at the very least, considerably smaller for the gold films produced by electron-beam evaporation than for the films produced by thermal evaporation. Our conclusions here are based on the nature of the plasmon resonance peak near 550 nm which is characteristic of nanometer-scale metal particles.[21] This adsorption maximum is reduced substantially in intensity for the films produced by electron-beam evaporation, which is the expected result for a very small particle size, perhaps approaching the continuous film limit. Clearly the structure and properties of the gold films depend not only on the polymer, but on the deposition method as well.

CONCLUSIONS

The gold/PTBA system is an ideal system for studying the behavior of concentrated metal particle dispersions, where the distance between particles is less than the dimensions of the polymer molecules. Interparticle bridging interactions arising from the slow exchange kinetics of polymer molecules in contact with different particles significantly retard particle diffusion in these systems. Polymer/metal interactions in the gold/PTBA system are "weak, in that measurable diffusion coefficients are obtained for the gold particles. Nevertheless, these diffusion coefficients are two to three orders of magnitude less than predictions based on the Stokes-Einstein equation for particle diffusion in a viscous medium. For gold particles produced by thermal evaporation, the temperature dependence of particle diffusion follows the temperature dependence of the bulk viscosity, but the effective hydrodynamic radius of the gold particles is up to 50 times larger than the end-to-end distance of the polymer molecules, which is in turn larger than the gold particles themselves. The long relaxation time of the high molecular weight polymer used in our marker motion experiments makes it difficult for us to assess the degree to which diffusion of polymer molecules through the metal particle dispersions is actually hindered. At any rate, any potential slowdown of polymer diffusion is certainly much less than the slowdown of metal particle diffusion, at least for the thermally evaporated particles. For gold particles produced by e-beam evaporation, particle diffusion is eliminated altogether, and polymer molecules are not able to diffuse throughout the gold layer. Our overall conclusion is that particle mobility at high concentrations in polymer melts is dramatically reduced, even in systems where the polymer/metal interactions are quite weak.

REFERENCES

1. Y.N.C. Chan, R.R. Schrock and R.E. Cohen, *J. Am. Chem. Soc.* **114**, 7295 (1992).
2. Y.N.C. Chan, R.R. Schrock and R.E. Cohen, *Chem. Mater.* **4**, 24 (1992).
3. Y.N.C. Chan, G.S.W. Craig, R.R. Schrock and R.E. Cohen, *Chem. Mater.* **1992**, 885 (1992).
4. T.L. Morkved, P. Wiltzius, H.M. Jaeger, D.G. Grier and T.A. Witten, *Applied Physics Letters* **64**, 422 (1994).
5. J.P. Spatz, A. Roescher and M. Möller, *Adv. Mater.* **8**, 337 (1996).
6. M.S. Kunz, K.R. Shull and A.J. Kellock, *J. Appl. Phys.* **72**, 4458 (1992).
7. M.S. Kunz, K.R. Shull and A.J. Kellock, *J. Coll. Int. Sci.* **156**, 240 (1993).
8. K.R. Shull and A.J. Kellock, *J. Poly. Sci., Polym. Phys.* **33**, 1417 (1995).
9. S.K. Varshney, C. Jacobs, J.-P. Hautekeer, P. Bayard, R. Jérôme, R. Fayt and P. Teyssié, *Macromolecules* **24**, 4997 (1991).
10. P.S.B. Ho, R.A. Haight, R.C. White, P.N. Sanda and A.R. Rossi, *IBM Journal of Research and Development* **32**, 658 (1988).
11. R. Fayt, R. Forte, C. Jacobs, R. Jérôme, T. Ouhadi, P. Teyssié and S.K. Varshney, *Macromolecules* **20**, 1442 (1987).
12. V.O. Cherkezyan and A.D. Litmanovich, *European Polymer Journal* **21**, 623 (1991).
13. W. Chu, J.W. Mayer and M. Nicolet, Backscattering Spectrometry, (Academic Press, New York, 1978).
14. G.K. Batchelor, *Journal of Fluid Mechanics* **74**, 1 (1976).
15. *Polymer Handbook (3rd. Edition)* J. Brandrup and E.H. Immergut, Ed. Wiley: New York, 1989.
16. P.F. Green, C.J. Palmstrøm, J.W. Mayer and E.J. Kramer, *Macromolecules* **18**, 501 (1985).
17. W.W. Graessley, *Adv. Polym. Sci.* **47**, 67 (1982).
18. J.D. Ferry, Viscoelastic Properties of Polymers, (J. Wiley and Sons, New York, 1980).
19. D.S. Pearson, L.J. Fetters, W.W. Graessley, G. Ver Strate and E. Von Meerwall, *Macromolecules* **27**, 711 (1994).
20. T. Andersson and C.G. Granqvist, *J. Appl. Phys.* **48**, 1673 (1977).
21. U. Kreibig and L. Cenzel, *Surf. Sci.* **156**, 678 (1985).

EFFECT OF DRYING ON PERMEABILITY OF MULTILAYER POLYMERIC FILMS

P.V. NAGARKAR and C.S. KO
Film Imaging Research and Development Division, Polaroid Corporation,
1265 Main St., Waltham, MA 02254

ABSTRACT

Photographic applications rely greatly on the permeability and swelling behavior of multilayer thin films. This behavior is, at least, a function of (I) material properties such as particle size, percent solids and solubility, and (II) processing conditions that affect film formation, such as rate of drying, maximum web temperature and nature of underlying layers.

In order to test permeability of multilayer structures, an alkaline medium is allowed to permeate sample. In the structure under test, the bottom most layer is acidic in nature. This acid neutralizes the alkali as it permeates through the layers. Neutralization causes a change in ionic concentration in the sample and this is measured using a high frequency cavity sensor (commercially available as the conductivity Puntex technique) that changes its resonant frequency in response to change in conductivity of the sysytem. This frequency change can be monitored to to give a time transient for permeability. In the event that frequency changes even before neutralization has occured, it can be correlated to capacitance changes within the layers produced by swelling and water uptake.

For a porous layer within the sample, the permeation proceeds along two paths. One through the porosity and the other through the bulk. In this case the bulk signal is superimposed over the signal from porous component. For such a system, a transient of the time derivative of resonant frequency proved to be a useful tool to study interfacial interactions between the acid layer and the layer above it. Impact of drying was also inferred from similar measurements. Data indicate that the higher temperature of drying of constituent layers resulted in a microstructurally tighter film, that took longer time for the alkali to permeate.

INTRODUCTION

Properties of thin film multilayer structures are very sensitive to the method of deposition and to their thermal history. The extent of interlayer mixing and inter-facial interaction can depend greatly on particle size, percent solids, solubility, and processing conditions. Factors such as rate of drying, maximum web temperature and nature of underlying layers also affect coalescence or film formation. For such a system, attempting to determine its permeability and/or swelling behavior is best done under actual use conditions. Conventional techniques for free standing films with external electrodes are not applicable under such conditions.

The technique of Puntex measurements allows one to measure permeation under actual use conditions. The method is based upon measuring the change in

Mat. Res. Soc. Symp. Proc. Vol. 461 © 1997 Materials Research Society

resonant frequency of a microwave cavity due to change in electrical conductivity within the sample. **Figure 1** shows a schematic of the sample on the sensor. The sample under investigation is placed in good contact with the top wall of the cavity sensor. All other walls of the cavity are enclosed in a box. Changes in conductivity of the sample alter the resonant frequency of this sensor and this is measured externally with a frequency counter. Many modifications of this technique have been used in the past for moisture measurements in oils and papers [1-3] in the late sixties to early seventies. The current technique for permeability measurements is from the same time period and has been in use at Polaroid for over two decades.

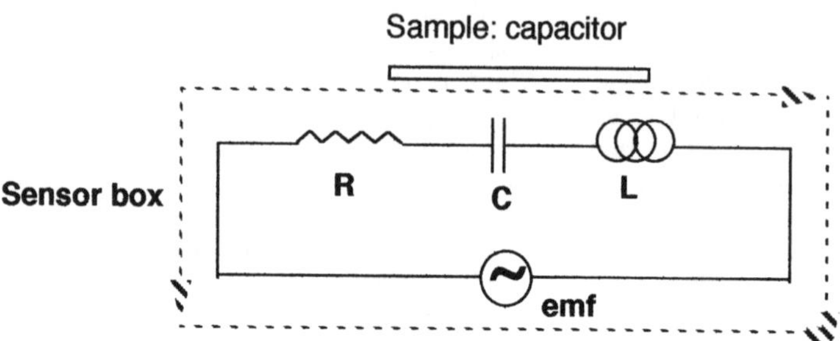

Figure 1: A schematic diagram for permeation measurements using a high frequency resonant cavity.

EXPERIMENT

Background:

In the present work the samples being tested are made up of four layers: the bottom most layer is an acidic layer (L, 20 μm) comprising of polymeric acid. Above this is the polymeric dispersion layer of polyurethane blend and is called the timing layer (TM, 3μm). The layer above this is the image receiving layer (D, 3μm) made up of copolymers. Above this is an hydrophobic layer called top coat/strip coat (SC, <0.2μm). All of these are coated on a photographic paper substrate.

Changes in permeability of polymeric layers can also be detected by allowing a highly alkaline solution to permeate through the layers and by monitoring its neutralization as it reaches the lower-most acidic layer and vice-versa with an acidic solution and alkaline layer. The time taken from the onset of permeation to completion of neutralization can be used to determine permeability and swelling of the complete system. A routine test, therefore, to determine the time required for the system to reach a certain pH involves using an indicator dye in the permeant solution and observing the point of change in color.

A point to note is that indicator tests give surface pH changes. These changes are dependent upon neutralization that occurs <u>within</u> the sample and are not easily measured by indicators. The method of Puntex measurement probes the interface between the acid layer (L) and the timing layer (TM). Besides it requires no indicator to be added to the permeant and avoids any interferences. In the present sytem, the

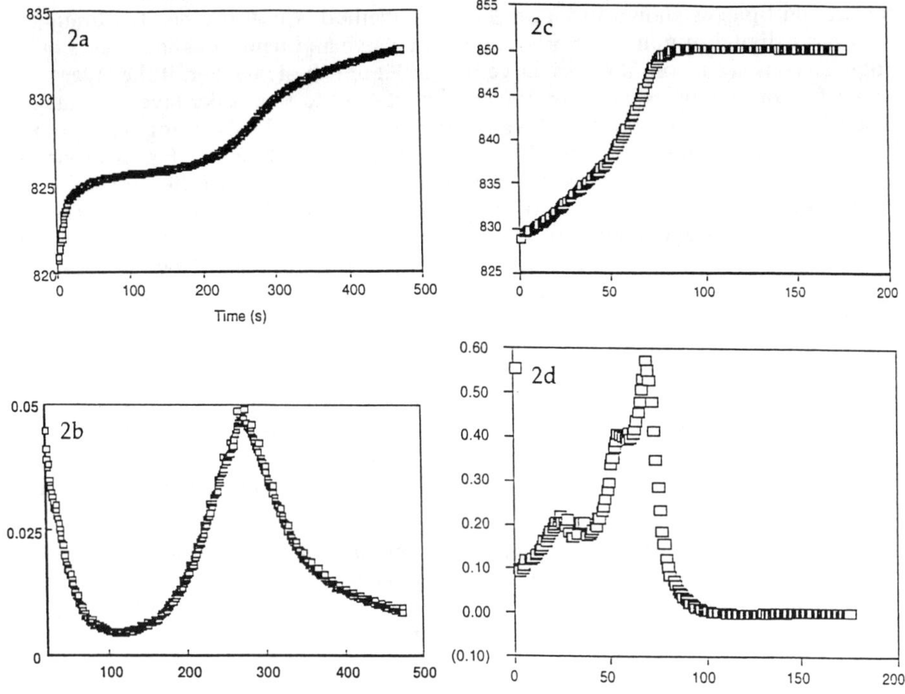

Figure 2a: A plot of change in resonant frequency with time (fo vs t), used to estimate permeation behavior of alkali through a sample with good hold-release character.

Figure 2b: Aplot showing the transient of a time derivative of resonant frequency (dfo/dt vs t) for the sample in figure 2a.

Figure 2c: A plot of change in resonant frequency with time (fo vs t), cannot be used to estimate permeation behavior of alkali through a sample with poor hold-release character.

Figure 2d: A plot showing the transient of a time derivative of resonant frequency (dfo/dt vs t) for the sample in figure 2c, which can be used to estimate permeation time.

161

puntex technique is used in the frequency mode to give a plot of resonant frequency as a function of time. The mid point of the transient is used to determine permeation time, as shown in **Figure 2a**. This method is useful when the timing layer has a digital response ; or acts as a binary valve and neutralization occurs in a single step, as seen from the derivative plot in **Figure 2b**. However, if the layer allows for continuous permeation, it could be referred to as a leaky layer, the major signal from neutralization would be superimposed over a slowly rising background. In this case, the frequency mode of Puntex measurement, shown in **Figure 2c** yields a gradually rising curve that levels off eventually and makes accurate time measurement of neutralization difficult, if not impossible. In such cases, a time derivative of the resonant frequency was found to be very effective in increasing the signal to noise ratio. This is depicted in **Figure 2d**. The presence of two peaks in the derivative spectrum indicate one peak that corresponded to mixing of the two layers, in this case the acid layer and the timing layer, to form an interphase of sorts. This mixing was seen in cross-sectional SEM and was seen to be a function of drying of the corresponding layers. The second peak (I), corresponds well with neutralization of the alkali in the acid layer. This peak position was found to correlate well with surface pH measurements determined using indicator dyes, mentioned earlier. It provides a simple, in-situ method to probe buried layers in a multilayer structure. This method development formed an important part of this work.

The derivative mode (df/dt) of Puntex measurements was used to examine the effect of drying and film formation on permeation behavior of the four layer structure described earlier. A full factorial design experiment with two levels of drying temperatures for three layers L, D and SC, was carried out.

RESULTS AND DISCUSSION

Time to neutralization could be measured as various layers were coated after the first two layers, acid layer L and timing layer TM. **Table 1** summarizes the drying matrix and the permeation times measured for structures up to D coat, and through full structure, i.e. through SC.

Table 1: Permeation times measured using peak I of Puntex derivative.

Sec.	L	D	SC	D:	SC:	Sc-D: Diff.
1	control	control	control	120	231	111
2	control	control	HTC	116	285	169
3	control	HTC	control	141	254	113
4	control	HTC	HTC	134	291	157
5	HTC	control	control	111	249	138
6	HTC	control	HTC	109	291	182
7	HTC	HTC	control	163	276	113
8	HTC	HTC	HTC	150	294	144

Higher D coat drying gave a tighter sheet at control SC drying, indicating a slower permeation rate through the structure. A higher SC coat drying gave a higher D to SC Puntex delta as seen from the right most column in Table 1. This indicates that depending on the film formation and drying of the final layer, permeation through the underlying layers can be severely impacted. Also, Puntex data appeared to be more sensitive compared to Thymol data in determining the effect of drying as seen from the statistical analysis below.

Statistical Results for changing drying from control to higher than control (HTC):

Factors	Puntex Time	Thymol time
L drying	+12	---
D drying	+15	+43
SC drying	+38	+34
LxSC interactions	yes	---
DxSC interactions	yes	yes

The sensitivity of the Puntex technique to changes due to drying was greater than the conventional indicator dye measurements [4] on the surface, and are tabulated above as Thymol time, the number of seconds increase in permeation time seen due to variation of each factor. The most influential factors affecting permeation time in descending order are SC drying, D coat drying, D and SC drying interactions and L coat drying. Of these, SC drying also correlated strongly to product performance such as dye transfer to the sheet. In conclusion, the technique of Puntex measurement can be used as a simple and effective method to monitor permeation and swelling of multilayer films of polymeric origin inproduct development of permeation sytems.

ACKNOWLEDGMENTS

The authors would like to thank D. Radkowski, F. Pierce and K. McCarthy for their valuable comments and suggestions.

REFERENCES

1.	R. G. Bosisio, Canadian Patent No. 797,999 (22 June 1971).

2.	R.G. Bosisio and M. Giroux, "Paper Sheet Moisture Measurements by Microwave Perturbation Techniques," TAPPI, Vol. 53, No. 3, 1970.

3.	R.G. Bosisio, L. Nappert and H.Q. Huy, IEEE Trans. on Industrial Electronics and Control Instrumentation vol.IECI-21, no.4 p. 257-61 (1974).

4.	P.V. Nagarkar, " Indicator Dye Testing For Receiving Sheet" (private Communication).

THERMAL CHARACTERIZATION OF A MICROLAYERED POLYCARBONATE/ POLYMETHYL METHACRYLATE COMPOSITE

Alex J. Hsieh and Alex W. Gutierrez, US Army Research Laboratory, Weapons and Materials Research Directorate, AMSRL-WM-MA, Aberdeen Proving Ground, MD 21005-5069

ABSTRACT

Thermal behavior of a coextruded microlayer composite with 388 alternating layers of polycarbonate and polymethyl methacrylate (PMMA) was investigated using differential scanning calorimetry (DSC) and dynamic mechanical analysis (DMA). Two distinct glass transition temperatures were observed with DSC for the coextruded composite, however both were shifted very slightly towards each other, compared to the glass transition temperatures of the pure components, indicating limited miscibility. Adhesion between the alternating microlayers appeared to be very good; delamination did not occur after the microlayer composite was subjected to a high speed impact test. Adhesion is attributed to limited miscibility since little mixing resulted from the laminar flow which was required in the coextrusion process. DMA results revealed an additional damping peak, which was not observed with DSC, at a temperature between the glass transitions of the two components. This intermediate transition peak is more sensitive to change in frequency compared to the response for the individual pure components.

INTRODUCTION

An innovative coextrusion process for layering dissimilar polymeric materials was originally patented by Dow Chemical Company (1) and has been further developed by Case Western Reserve University. The coextrusion process involves a series of dies which repeatedly split, spread, and recombine the polymer melts. To date, coextruded composites of several thousand continuous layers have been achieved with layer thickness on the order of 30-50 nanometers (2).

One of the material systems of particular interest to the Army is a microlayered composite of polycarbonate and PMMA; both materials have been widely used in many military and civilian transparent glazing applications. PMMA is brittle upon impact; polycarbonate, on the other hand, has excellent ballistic performance, however it can scratch easily. Earlier attempts used laminates of thick polycarbonate and thick PMMA sheets with proper adhesives; results of the ballistic impact tests showed that in many cases spallation occurred in the polycarbonate when it was used as a backup for PMMA in these thick laminated composites (3).

Preliminary evaluation of ballistic impact for the microlayered polycarbonate/PMMA composite revealed that delamination did not occur between the alternating microlayers and that PMMA did not fracture in a brittle manner (4). Since both polycarbonate and PMMA are amorphous polymers, a better understanding of adhesion between the microlayers has great importance for producing coextruded composites with high durability. The goal of this work was to investigate the interfaces between the microlayers by examining the thermal behavior of the coextruded polycarbonate/PMMA composite with DSC and DMA.

Mat. Res. Soc. Symp. Proc. Vol. 461 © 1997 Materials Research Society

EXPERIMENTAL

A microlayer composite with 388 alternating layers of polycarbonate and polymethyl methacrylate was fabricated by the Dow Chemical Co. (received through Case Western Reserve University). The polycarbonate was Merlon[TM] (Mobay Chemical Company) M-40 and the PMMA was Plexiglas[TM] (Rohm and Hass Company) V052. The coextruded sheet was approximately 1 mm thick and the volume ratio of polycarbonate/PMMA was 57.5/ 42.5. Differential scanning calorimetry (DSC) measurements were carried out using a Perkin-Elmer System 7. The temperature was calibrated with both indium and zinc standards. Samples of 5-10 mg were prepared in an aluminum pan and scanned in a nitrogen atmosphere. Thermal properties were determined typically at a heating rate of 20°C/min.

Dynamic mechanical measurements were performed with a dynamic mechanical analyzer DuPont 983 DMA. Microlayered polycarbonate/PMMA specimens of 30 mm long, 13 mm wide and 1 mm thick were mounted in the vertical clamps at a clamping distance of 8 mm. The samples were analyzed in a fixed frequency mode at 1 Hz with a peak to peak amplitude oscillation of 0.3 mm and at a typical heating rate of 3°C/min. Multifrequency scans were obtained using the frequency values of 0.05, 0.1, 0.2, 0.5, and 1 Hz.

RESULTS AND DISCUSSION

Figure 1 shows a typical DSC scan with two distinct glass transition temperatures, T_g's, for the coextruded polycarbonate and PMMA; one is the glass transition temperature for PMMA and the other is for polycarbonate. The two T_g's are shifted slightly towards each other compared to the T_g's of the pure components, indicating limited miscibility between polycarbonate and PMMA.

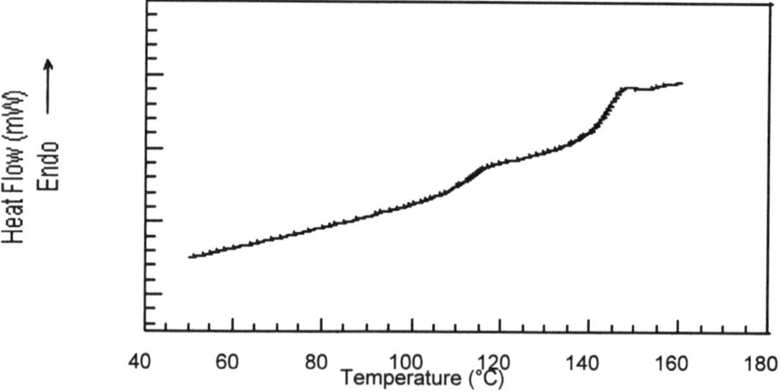

Figure 1. A typical DSC curve for the microlayered polycarbonate/PMMA composite.

Figure 2 is a typical DMA scan of storage modulus, E', and loss modulus, E", as a function of temperature for the same coextruded polycarbonate/PMMA. The peak temperature for the E" curve correlated best with the glass transition temperature measured by DSC, therefore the E" versus temperature data were used for the comparisons of T_g's obtained from DSC. The peak temperature in the E" curve is also a function of the frequency used in the DMA measurements. Similarly, T_g's obtained from DSC depend on the heating rates.

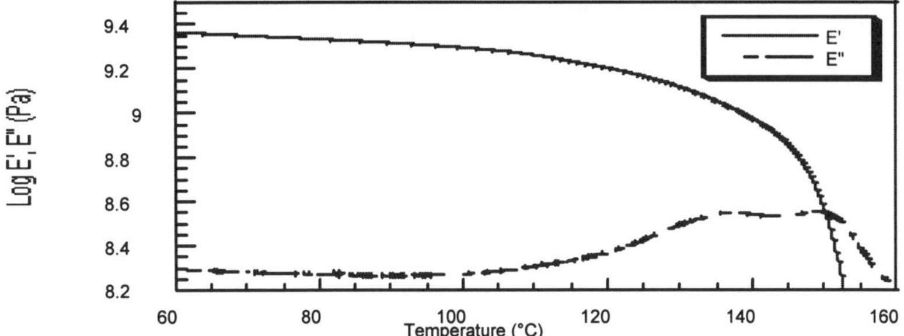

Figure 2. Plots of the dynamic storage modulus E' and loss modulus E" versus temperature at 1 Hz for the microlayered polycarbonate/PMMA composite.

In the loss modulus E" curve, three distinct transitions were observed; two were associated with sharp peaks and one with a broadened transition. Compared to the DSC data, the broadened transition is the glass transition of PMMA.

The physical state of the specimen associated with the two sharp relaxation processes was further investigated; this was done by heating the specimen to various temperatures followed by visual examination of its physical integrity. There was not any visible change noticed after the specimen was heated to temperatures slightly above the lower relaxation peak, however the specimen deformed significantly when it was heated to well above the upper peak temperature. This confirmed that the specimen remained rigid or in the glassy state when it was heated to above the lower peak temperature but still below the upper transition temperature. The deformation was the result of the transition of the microlayered composite from the glassy state to the rubbery state, which was also evidenced from the change of more than three orders of magnitude in storage modulus, E', during the upper relaxation process. Therefore, the upper peak temperature was the glass transition temperature of polycarbonate.

The remaining question is what was the origin of the intermediate relaxation peak observed by DMA at a temperature between the T_g's of PMMA and polycarbonate. This peak was of an intensity equal to that observed for the glass transition of polycarbonate. However, this additional transition was not observed with DSC. Annealing of the polycarbonate/PMMA composite was carried out using DSC at temperatures from 90°C (below T_g of PMMA) to 155°C (above T_g of polycarbonate) for various periods of time. This annealing was performed to examine whether the intermediate transition was as yet present in DSC scans. Figures 3 and 4 show the typical DSC curves of the reheating scans after annealing at 90°C and 125°C for various periods of time.

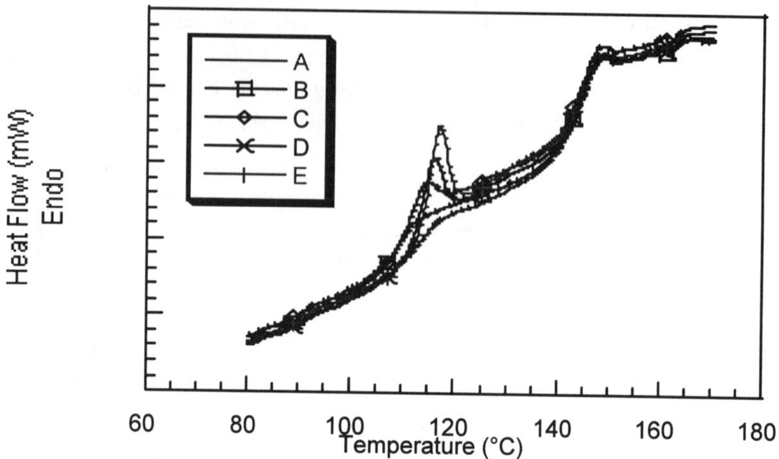

Figure 3. DSC curves for coextruded polycarbonate/PMMA annealed at 90°C for various periods of time (curve A for unannealed, B for 1/2 hour, C for 2 hours, D for 6 hours, and E for 12 hours).

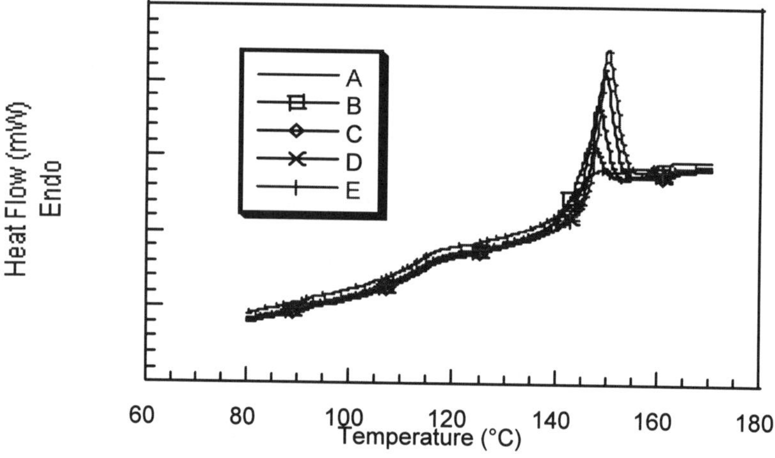

Figure 4. DSC curves for coextruded polycarbonate/PMMA annealed at 125°C for various periods of time (curve A for unannealed, B for 1/2 hour, C for 2 hours, D for 6 hours, and E for 12 hours).

These endothermic peaks are the result of enthalpy relaxation which is a function of annealing temperature and annealing time. The endotherms increased with an increase in the annealing time. For example, when annealed at 90°C there were only relaxation endotherms associated with PMMA present; endotherms were not observed in the polycarbonate. This was due to the fact that the annealing temperature at 90°C is more than 50°C lower than the glass transition temperature of polycarbonate, therefore chain mobility of polycarbonate is negligible. On the other hand, when annealed at 125°C endotherms were not observed in PMMA; this was because PMMA in this annealed condition was above the glass transition temperature and in the rubbery state where enthalpy relaxation does not occur.

Results of the annealing studies confirmed that only two distinct glass transitions were present in DSC scans for the microlayered polycarbonate/PMMA composite.

The viscoelastic behavior of the intermediate relaxation peak was further examined with dynamic mechanical measurements carried out in a multiple frequency mode. The selected frequencies were 0.05, 0.1, 0.2, 0.5, and 1 Hz; the resultant loss modulus E" versus temperature data are shown in Figure 5. All three relaxation processes were clearly seen and each individual peak temperature increased as the frequency was increased. The dependence of the peak temperature upon the shift in frequency appeared to be more significant for the intermediate relaxation than that for either glass transition of PMMA or polycarbonate.

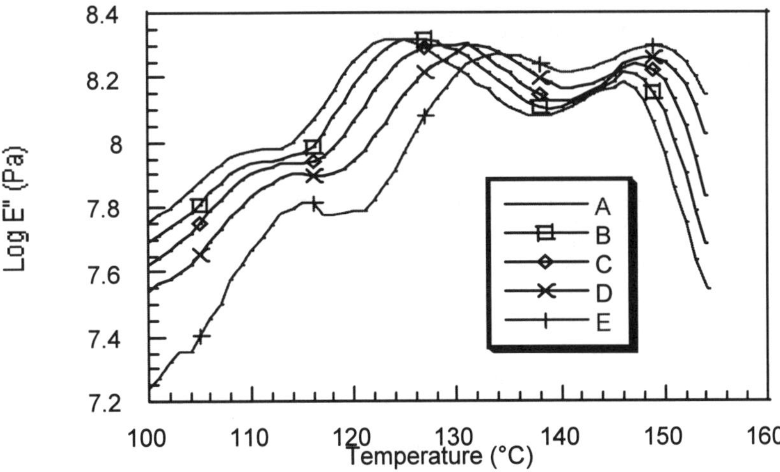

Figure 5. Plots of loss modulus versus temperature obtained from DMA using a multiple frequency mode (curves A, B, C, D, and E for 0.05, 0.1, 0.2, 0.5, and 1Hz, respectively).

Considering the effect of temperature on molecular relaxation, these data allow the determination of activation energy for these relaxation processes. Assuming that the relaxation process can be modeled by an Arrhenius temperature dependence (5), the shift in frequency of the loss peak maximum is plotted in Figure 6 as a function of the reciprocal of the temperature relevant to the respective peak. The data in Figure 6 were fit to a straight line whose slope was the apparent activation energy for the relaxation process. The apparent activation energy for the corresponding intermediate relaxation peak, determined from the available data, was 172 kJ/mol, which was much lower than 498 kJ/mol and 310 kJ/mol determined for the polycarbonate and PMMA, respectively. The lower apparent activation energy may be corresponding to the presence of an intense intermediate relaxation peak even though the miscibility between the polycarbonate and PMMA is limited. The apparent activation energy values for PMMA and polycarbonate are comparable with literature data.

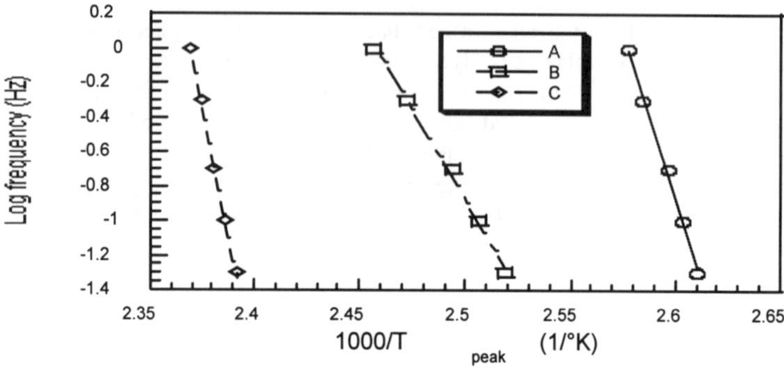

Figure 6. Experimental data of log (frequency) vs. reciprocal peak temperature for glass transition of PMMA (curve A), the intermediate relaxation peak (curve B), and glass transition of polycarbonate (curve C).

CONCLUSIONS

DMA studies revealed the presence of an additional relaxation peak at a temperature between the T_g's of polycarbonate and PMMA. The dependence of the maximal temperature on the shift in frequency was more significant for this intermediate relaxation peak and the apparent activation energy associated with this intermediate peak was smaller than that for the either components. Adhesion between the microlayers was very good when the coextruded composite was subjected to a high speed impact; this result was attributed partially to the result of interdiffusion from the layering process. The extent of interdiffusion may be limited due to the limited miscibility between polycarbonate and PMMA, as evidenced by the presence of two distinct T_g's observed in DSC. Further investigation to determine the origin and the viscoelastic response of the intermediate relaxation peak is being conducted using microlayer composites with various compositions of polycarbonate and PMMA.

ACKNOWLEDGEMENT

The authors thank Professors Baer and Hiltner at Case Western Reserve University for providing the polycarbonate/PMMA microlayer composite samples which were produced by Dow Chemical Company, and acknowledge Mr. James Kidd at ARL for proofreading this manuscript.

REFERENCES

1. J.A. Radford, T. Alfrey Jr., W.J. Schrenk, Polym. Eng. Sci., **13**, 216 (1973).

2. E. Baer, A. Hiltner, personal communication, Case Western Reserve University.

3. A.J. Hsieh, Technical Report, in preparation, Army Research Laboratory.

4. M.E. Roylance, R.W. Lewis, AMMRC TR 72-23, Army Research Laboratory (1972).

5. N.G.McCrum, B.E. Read, G. Williams, <u>Anelastic and Dielectric Effects in Polymeric Solids</u>, (John Wiley and Sons Publishers, New York, 1967).

Void Morphology in Polyethylene/Carbon Black Composites

D.W.M. Marr[*]
Chemical Engineering and Petroleum Refining Department
Colorado School of Mines, Golden, CO 80401

M. Wartenberg, K.B. Schwartz
Raychem Corporation, Menlo Park, CA 94025

M.M. Agamalian and G.D. Wignall
Solid State Division, Oak Ridge National Laboratories, Oak Ridge, TN 37830

Abstract

A combination of small angle neutron scattering (SANS) and contrast matching techniques is used to determine the size and quantity of voids incorporated during fabrication of polyethylene/carbon black composites. The analysis used to extract void morphology from SANS data is based on the three-phase model of microcrack determination via small angle x-ray scattering (SAXS) developed by W.Wu[12] and applied to particulate reinforced composites.

Introduction

Previous SANS and SAXS experiments[1] have suggested the presence of a third phase (voids) in composites of polyethylene (PE) and carbon black (CB). In this paper, we focus on a set of experiments designed to quantitatively determine the volume fraction and dimensions of voids incorporated during such composite preparation. Analysis of these systems via small angle x-ray scattering techniques is difficult because of their three-phase nature, being composed of polymer matrix, filler, and any incorporated voids. Through a combination of SANS on deuterated and protonated composites, however, the scattering contrast can be varied to reflect either a two or three phase morphology and the size and quantity of voids determined.

X-rays are scattered by fluctuations in electron density (ED) within a given sample, and thus carbon black/polyethylene composites have substantial SAXS cross sections due to the large ED difference between the components. The problem however, is that there is strong ED-contrast between carbon black, polyethylene, and voids, making interpretation of scattering data more complex. SANS uses neutrons of wavelength (5-10Å) comparable to SAXS and therefore probes similar length scales. Neutrons, however, are scattered by nuclei (as opposed to electrons), and thus the SANS contrast can be changed via isotopic substitution, which has little effect on the chemistry of the blend[2,3] .

	d [g/cm^3]	amorph. [10^{10} cm^{-2}]	cryst. [10^{10} cm^{-2}]	ave [10^{10} cm^{-2}]
hPE	0.953±.001	-0.30 (ρ_{hPE}^{a})	-0.36 (ρ_{hPE}^{x})	-0.34 (ρ_{hPE}^{t})
dPE	1.080±.002	7.30 (ρ_{dPE}^{a})	8.58 (ρ_{dPE}^{x})	8.13 (ρ_{dPE}^{t})
CB	1.917±.002			6.40 (ρ_{CB})
voids				0.0

Table 1: Scattering Length Densities

If one examines a normal protonated composite via SANS, the sample is essentially two-phase because the scattering length densities of polyethylene and voids are virtually identical (see Table 1). If one blends carbon black with deuterated polyethylene however, the presence of voids is highlighted within the carbon black/deuterated polyethylene matrix. Through a combination of these two experiments, one can now extract quantitative information about void size and quantity.

Mat. Res. Soc. Symp. Proc. Vol. 461 ©1997 Materials Research Society

Experiment

Four composites were prepared at carbon black volume fractions of 0.360 and 0.429. Two different polymer matrices were used, protonated polyethylene at a density of $0.953 \pm .001$ g/cm^3 and an analogous fully deuterated polyethylene at a density of $1.080 \pm .001$ g/cm^3. Composite samples were mixed in a DACA minicompounder[4], a small scale twin-screw mixer/extruder (for compounding and extrusion of samples up to total volumes of approximately 5 cm^3). Carbon black and deuterated polyethylene were added in the appropriate ratios to a total weight of 1.5 g and processed at a temperature of 200°C for 5 minutes in the twin screw extruder.

The experiments were performed on the W. C. Koehler 30m SANS facility at Oak Ridge National Laboratory[5]. The neutron wavelength was 4.75Å ($\Delta\lambda/\lambda \sim 5\%$) and the source (3.5 cm diameter) and sample (1.0 cm diameter) slits (irises) were separated by a distance of 7.5 m. The sample-detector distances were 4.0 m and 17.8 m and the data were corrected for instrumental backgrounds and detector efficiency on a cell-by-cell basis, prior to radial averaging to give a q-range of $0.04 < q = 4\pi\lambda^{-1}\sin\theta < 1.5$ nm^{-1}, where 2θ is the angle of scatter. The net intensities were converted to an absolute ($\pm 4\%$) differential cross section [dΣ/d$\Omega(q)$] per unit sample volume (in units of cm^{-1}) by comparison with pre-calibrated secondary standards, based on the measurement of beam flux, vanadium incoherent cross section, the scattering from water and other reference materials[6]. Procedures for transmission measurements and for subtracting the incoherent background, arising largely from the protons in the sample, have been described previously[7].

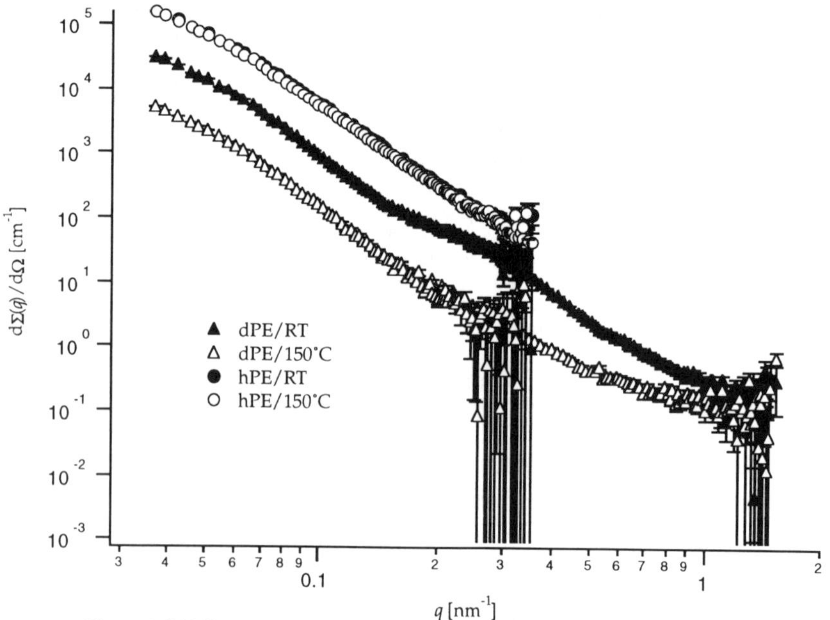

Figure 1: SANS scattering from 36 vol% CB samples at room and high temperatures

Analysis

Figure 1 shows the SANS differential cross section $d\Sigma/d\Omega(q)$ for two composite samples with a carbon black volume fraction of 36.0%. It may be seen that substituting dPE for hPE lowers the scattered intensity as expected. In addition, due to the difference in SLD between crystalline and amorphous regions in deuterated polyethylene (see Table 1) there is a "peak" associated with lamellar spacing in crystalline polyethylene around $q=.25$ nm^{-1} in the room temperature data for deuterated composites. Also of note is that the cross sections of the protonated samples at room temperature and in the melt are very similar. The SANS contrast in this system is developed primarily from the difference in scattering length density between hPE and carbon black, neither of which change significantly with an increase in temperature.

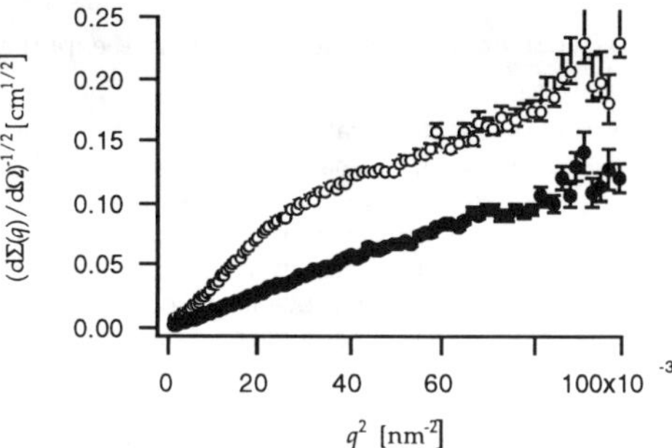

Figure 2: Debye-Bueche plot of a deuterated (°) and protonated (•) composite showing the linear behavior of the protonated sample.

This "two-phase" behavior of the protonated composites is clearly shown in the Debye-Bueche (DB) plot of Figure 2. The linear slope suggests that a single-parameter DB analysis is appropriate in this system and shows that we must use a more complex model for analysis of the deuterated systems. Interpreting the data for the protonated composites in the DB model[8] with

$$\gamma(r) = e^{-r/a_c} \text{ and } \frac{d\Sigma(q)}{d\Omega} = \frac{d\Sigma(0)}{d\Omega}\left(1 + q^2 a_c^2\right)^{-2},$$

chord lengths (or domain sizes) of the individual phases can be determined via

$$l_i = \frac{1}{1-\phi_i}\int_0^\infty \gamma(r)dr \text{ or } l_1 = \frac{a_c}{1-\phi_1} \text{ and } l_2 = \frac{a_c}{\phi_1}.$$

matrix	Temp	ϕ_{CB}	$d\Sigma/d\Omega(0)$ [cm^{-1}]	a_c [nm]	$Q^{exp} \times 10^{-22}$ [cm^{-4}]	$Q^{pred} \times 10^{-22}$ [cm^{-4}]	l_1 [nm]	l_2 [nm]
hPE	RT	0.360	$1.72\pm.02 \times 10^6$	$40.7\pm.1$	$2.00\pm.04$	2.06	$63.6\pm.2$	$113.1\pm.3$
hPE	150°C	0.335	$2.14\pm.03 \times 10^6$	$43.7\pm.1$	$2.01\pm.04$	1.97	$65.7\pm.2$	$130.4\pm.3$
hPE	RT	0.429	$1.43\pm.01 \times 10^6$	$36.4\pm.1$	$2.34\pm.03$	2.19	$63.7\pm.2$	$84.8\pm.3$
hPE	150°C	0.402	$1.67\pm.02 \times 10^6$	$38.5\pm.1$	$2.29\pm.05$	2.13	$64.4\pm.2$	$95.9\pm.3$

Table 2: Fit parameters (the volume fractions have changed at high temperature because of the expansion of the PE phase).

The scattering invariant Q

$$Q = \int_0^\infty q^2 \frac{d\Sigma(q)}{d\Omega} dq$$ can be used to verify our choice of model because $Q = 2\pi^2 \phi_1\phi_2(\rho_1 - \rho_2)^2$ for two-phase systems. To account for the lack of data at low and high q, experimental invariants were generated by integrating the model fits to the data. Our results for these fits are shown in Table 2. The domain size l_1 corresponds to the phase at volume fraction ϕ_1 (carbon black) as confirmed by the fact that it does not vary as the sample is brought through the melt transition. Note also the domain size of the polyethylene phase behaves as expected. As the samples are melted the domain size increases, indicating the polyethylene expansion; also, the domain size decreases as the polymer volume fraction is lowered. With a knowledge of the SLDs and volume fractions of the components we have generated predicted invariants for comparison to those determined experimentally. As seen in Table 2 the agreement is quite good at lower concentration, indicating that we have a good handle on the analysis for protonated polyethylene/carbon black composites and that this system effectively behaves as two-phases. To obtain experimental values for Q at higher concentrations, the data were fit to the two-correlation length Debye-Bueche model[9,10,11],

$$\frac{d\Sigma(q)}{d\Omega} = \frac{A_1}{\left(1+q^2a_1^2\right)^2} + A_2 \exp\left(-\frac{q^2a_2^2}{4}\right) \quad \text{where} \quad \frac{d\Sigma(0)}{d\Omega} = A_1 + A_2 \text{ and } f = \frac{A_1}{A_1 + \frac{8A_2}{\sqrt{\pi}}\left(\frac{a_1}{a_2}\right)^3}, \text{ and improving}$$

the agreement between the predicted and experimental invariants to within 2%.

In examining the deuterated composites, we have seen in Figure 2 that they are not well described by the single-parameter Debye-Bueche model. For two-phase systems, and as seen in the protonated composite case, this model should fit well. Such a deviation indicates that something else is present in these systems that has no SANS contrast in the protonated polyethylene but shows up in the deuterated polymer case at low q, i.e. voids[1]. This inference is supported by the difference in the calculated and predicted scattering invariants shown in Table 3. Notice that in each of the cases, the experimental invariant is higher than one would predict for a simple two-phase system, indicating once again that there is another phase at low q contributing to the total scattering.

	Temp	ϕ_{CB}	a_1 [nm]	a_2 [nm]	A_1 x 10^{-4} [cm^{-1}]	A_2 x 10^{-3} [cm^{-1}]	f	Q^{exp} x 10^{-20} [cm^{-4}]	Q^{pred} x 10^{-20} [cm^{-4}]
dPE	RT	0.360	40.3±.1	45.9±.3	22.9±1.4	17.5±.6	0.811	33.9±3.4	13.1
dPE	150˚C	0.336	38.7±.1	44.9±.1	3.20±.33	2.58±.15	0.812	5.36±.95	3.57
dPE	RT	0.429	37.0±.1	43.8±.3	19.0±.9	17.3±.5	0.801	36.8±2.9	14.0
dPE	150˚C	0.403	36.0±.1	41.1±.3	2.91±.23	2.82±.12	0.796	6.16±.83	3.85

Table 3: Results from 2-parameter DB model fits (to avoid the complications added by the crystalline lamellae in dPE, the data here were fit over a range of 0.037 nm^{-1} < q < 0.126 nm^{-1}, corresponding to length scales greater than 50 nm).

Assuming that the structure of the three-component composite (polyethylene, carbon black, and voids) is identical in the deuterated and protonated cases, one can extract quantitative information about void size and quantity. SAXS can be used to validate this hypothesis because of the sensitivity of x-rays to the scattering contrast between both polyethylenes (hPE/dPE) as well as carbon black. As displayed in Figure 3, the SAXS scattering patterns from the deuterated and protonated polyethylene composites overlap showing the microstructures to be virtually identical over the length scales investigated. Thus, it is reasonable to use the scattering data from the protonated samples as a template for the expected SANS scattering curves of the deuterated materials if composed of only two phases. We then attribute the excess scattering to the third phase.

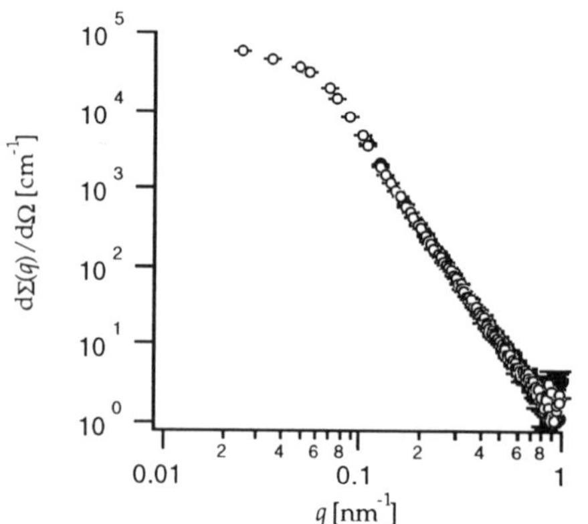

Figure 3: SAXS scattering from 36 vol% carbon black samples where the scattering from both deuterated and protonated composites overlap.

We model our analysis on the work of Wu[12] who studied microvoid formation in composite materials. By comparing the SAXS scattering from undamaged composites to damaged composites he could extract the size and quantity of the void phase. This approach is analogous to our comparisons of systems that are effectively two phase with those of similar morphology but with the addition of a small void component (deuterated composites). One begins by expressing the correlation function for the entire system as a summation of the correlations for the individual phases $\chi(\mathbf{r})$ as if the other phases composed a single phase:

$$\gamma(\mathbf{r}) \propto \phi_1(1-\phi_1)(\rho_2 - \rho_1)(\rho_3 - \rho_1)\gamma_1(\mathbf{r}) + \phi_2(1-\phi_2)(\rho_3-\rho_2)(\rho_1-\rho_2)\gamma_2(\mathbf{r}) + \phi_3(1-\phi_3)(\rho_1-\rho_3)(\rho_2-\rho_3)\gamma_3(\mathbf{r})$$

For three-phase isotropic systems (phase 1 = filler, phase 2 = matrix, phase 3 = voids) the invariant can then be determined via

$$Q = \int_0^\infty q^2 \frac{d\Sigma(q)}{d\Omega} dq = 2\pi^2 \left[\phi_1\phi_2(\rho_1-\rho_2)^2 + \phi_2\phi_3(\rho_2-\rho_3)^2 + \phi_1\phi_3(\rho_1-\rho_3)^2\right].$$

If we denote Q^0 as the invariant for a system without voids then

$$Q^0 = 2\pi^2 \phi_1^0 \left(1-\phi_1^0\right)\left(\rho_1 - \rho_2^0\right)^2 \quad \text{and}$$

$$Q - Q^0\left(\frac{\rho_1-\rho_2}{\rho_1-\rho_2^0}\right)^2 = 2\pi^2\left[(\rho_1-\rho_2)^2\left(\phi_1(1-\phi_1) - \phi_1^0\left(1-\phi_1^0\right) - \phi_1\phi_3\right) + \phi_2\phi_3(\rho_2-\rho_3)^2 + \phi_1\phi_3(\rho_1-\rho_3)^2\right].$$

If one examines isothermal transitions from deuterated to protonated composites, the filler correlations and volume fractions will not vary (i.e. $\gamma_1(r) = \gamma_1^0(r)$, $\phi_1 = \phi_1^0$) and the volume fraction of the void phase can be calculated via

$$\phi_3 = \frac{Q - Q^0\left(\dfrac{\rho_1-\rho_2}{\rho_1-\rho_2^0}\right)^2}{2\pi^2\left\{\phi_1\left[(\rho_1-\rho_3)^2 - (\rho_1-\rho_2)^2 - (\rho_2-\rho_3)^2\right] + (1-\phi_3)(\rho_2-\rho_3)^2\right\}}.$$

In addition, one can calculate the chord length of this low volume fraction domain in a similar fashion. For three-phase systems

$$Z = \int_0^\infty q\frac{d\Sigma(q)}{d\Omega} dq = 4\pi \left[\begin{array}{l}(\rho_2-\rho_1)(\rho_3-\rho_1)\phi_1(1-\phi_1)\int_0^\infty \gamma_1(r)dr + \\[2ex] (\rho_3-\rho_2)(\rho_1-\rho_2)\phi_2(1-\phi_2)\int_0^\infty \gamma_2(r)dr + (\rho_1-\rho_3)(\rho_2-\rho_3)\phi_3(1-\phi_3)\int_0^\infty \gamma_3(r)dr\end{array}\right].$$

Assuming the volume fraction of voids is small and that the chord lengths of phases 1 and 2 remain constant, the chord length of the void domain can be estimated via

$$l_3 = \frac{Z - Z^0\left(\dfrac{\rho_1-\rho_2}{\rho_1-\rho_2^0}\right)^2}{4\pi\phi_3(1-\phi_3)(\rho_1-\rho_3)(\rho_2-\rho_3)}.$$

Results and Discussion

These investigations include four experiments where the composite has gone from two to three phases under conditions of constant filler volume fraction. Using the equations developed in the previous section we can now determine the void volume fractions and associated chord lengths, the results of which are shown in Table 4. For the roo: i temperature composites, we find a void size of 44 ± 24 nm and a void content of 1.9 ± 0.3 vol% for the 36.0 vol% sample. For the 42.9 vol% sample we obtain values of 42 ± 18 nm and 2.1 ± 0.3 vol%. Once the composites are heated above the melting point, there is a significant decrease in the void content, to a value of the same order as the experimental error. This decrease might be expected and suggests that the polyethylene domains grow at the expense of the voids as the temperature is brought above the melt. Also of note, and despite the limited data, is the apparent scaling of the void content with filler concentration both at room and high temperatures.

Temp	ϕ_1 (CB)	ϕ_3 (voids)	l_1 [nm] (CB)	l_2 [nm] (PE)	l_3 [nm] (voids)
RT	0.360	0.0188±.0033	61.7±1.0	109.7±1.8	44.0±24.0
150°C	0.335	0.0018±.0011	67.1±1.2	133.1±2.5	29.7±152
RT	0.429	0.0212±.0032	57.9±1.9	77.0±2.6	42.0±18.0
150°C	0.402	0.0025±.0010	57.0±1.9	84.7±2.8	40.0±74.4

Table 4: Results

Though we currently do not have a technique that would allow us to corroborate our void size determinations, we have conducted a number of pycnometry measurements at room temperature on similarly prepared materials in order to compare to the void volume results. The void content was determined by comparing the measured densities with those predicted assuming a lack of voids via $void\ content = (1 - \rho_{meas}) / \rho_{pred}$ (see Figure 4). Despite the small range over which these pycnometry experiments were conducted, the results indicate that the void contents determined via the scattering studies are quite reasonable.

These results do not allow us to determine the mechanism of void incorporation but do provide some interesting clues. We have seen in these investigations that the void volume fraction decreases to near zero above the melt. This strongly suggests that the system is nearly completely wetted during processing and that most of the void incorporation occurs post processing during composite cool. We have also seen however, that the void content scales with filler concentration, suggesting that entrainment may be involved and that carbon black structure will play a significant role in void quantity and

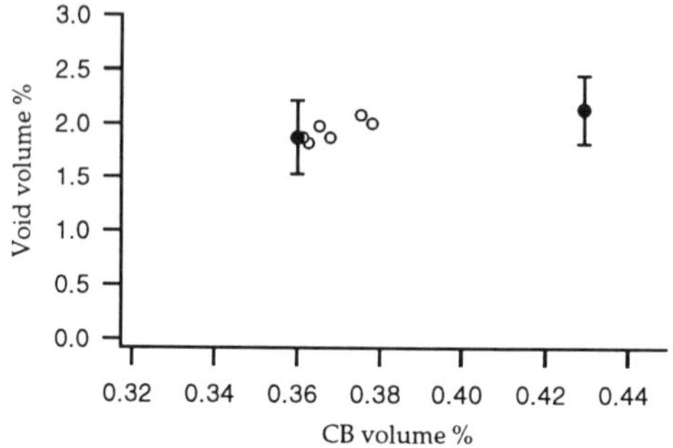

Figure 4: Void content determined via SANS (•) and pycnometry (°) at room temperature.

morphology. This apparent contradiction will serve as motivation for future studies into the influence of these factors on the system morphology.

Acknowledgments

We thank Gordon Spellman at Lawrence Livermore National Laboratories for supplying the deuterated polyethylene used in this study. The research at Oak Ridge was supported by the Division of Materials Sciences, U. S. Department of Energy under contract No. DE-AC05-96OR22464 with Lockheed Martin Energy Research Corporation.

References

[1] Wignall, G.D.; Farrar, N.R.; Morris, S. *J. Mat. Sci.* **1990**, *25*, 69.

[2] Bates, F. S.; Wignall, G. D. *Phys. Rev. Lett.*, **1986**, *57*, 1429.

[3] Wignall, G. D. Polymer Properties Handbook, American Institute of Physics, **1996**, 299.

[4] DACA Instruments, Santa Barbara, CA.

[5] Koehler, W. C. *Physica (Utrecht)*, **1986**, 137B, 320.

[6] Bates, F. S.; Wignall, G. D. *J. Appl. Cryst.*, **1986**, *20*, 28.

[7] Dubner, W. S.; Schultz, J. M.; Wignall, G. D. *J. Appl. Cryst.*, **1990**, *23*, 469.

[8] Debye, P.; Bueche, A.M. *J. Appl. Phys.* **1949**, *20*, 518.

[9] Debye, P.; Anderson, H.R.; Brumberger, H. *J. Appl. Phys.* **1957**, *28*, 679.

[10] Cheung, Y.W., Stein, R.S., Wignall, G.D, Yang, H.E. *Macromolecules* **1993**, *26*, 5365.

[11] Marr, D.W.M. *Macromolecules* **1995**, *28*, 8470.

[12] Wu, W.-L *Polymer* **1982**, *23*, 1907.

INVESTIGATION OF DIBLOCK COPOLYMER THIN FILM MORPHOLOGY FOR NANOLITHOGRAPHY

Miri Park, Christopher Harrison, and Paul M. Chaikin
Department of Physics, Princeton University, Princeton, NJ 08544

Richard A. Register
Department of Chemical Engineering, Princeton University, Princeton, NJ 08544

Douglas Adamson and Nan Yao
Princeton Materials Institute, Princeton University, Princeton, NJ 08544

ABSTRACT

The microphase separated morphology of diblock copolymers can be used to generate well-ordered nanometer scale patterns over a large area. To achieve this goal, it is important to understand and control the behavior of diblock copolymer thin films on substrates, which can differ from the bulk behavior. We have investigated the morphologies and ordering in thin polystyrene-polybutadiene (PS-PB) diblock copolymer films on bare silicon and silicon nitride substrates, and also on polymethylmethacrylate (PMMA) coated substrates. The PS-PB copolymers are synthesized to form, in bulk, PB cylinders or spheres in a PS matrix. In thin films (10-60 nm thick), prepared by spin-coating, we observe that the morphology and ordering of the microdomains are affected by strong wetting constraints and a reduced chain mobility on the substrate. The thinnest self-assembled layer of the copolymer films shows no in-plane microphase separation on both types of substrates. The PS blocks wet the PMMA substrates whereas the PB blocks wet the bare substrates as well as the air interface. Hence, different film thicknesses are necessary on the two types of substrates to obtain a uniform film of the first self-assembled cylindrical or spherical microdomain layer. The first layer of the cylindrical copolymer can vary from cylindrical to spherical morphology with a few nanometer decrease in film thickness. In the case of spherical PS-PB diblock copolymer films, we observe that the ordering of the microdomains is improved in the films on the PMMA substrates, compared to those on the bare substrates. We also demonstrate a successful transfer of the microdomain patterns to silicon nitride substrates by a reactive ion etching technique.

INTRODUCTION

Diblock copolymers consist of two chemically different polymer chains that are joined by a covalent bond at one end. In bulk and at equilibrium they spontaneously self-assemble into microphase separated morphologies due to the incompatibility between the two blocks and connectivity constraints.[1] Commonly observed microdomain morphologies are ordered arrangements of lamellae, cylinders and spheres. The size of these microstructures and their periods are governed by the chain dimensions and are typically on the order of 10 nanometers. These length scales are difficult to achieve by the standard lithography techniques used in the semiconductor industry. We are motivated by the possibility of employing the ordered microstructures of diblock copolymers to produce a large area with well ordered nanometer surface structures.[2,3] For instance, a monolayer of spherical microdomains on a semiconductor substrate can be used as a lithography template for an array of quantum dots or anti-dots.

The most commonly used means of nanolithography is electron beam lithography, which can provide point-by-point control of patterning with the minimum feature size below 100 nm. However, the obtainable periodicity of features is still not much below 100 nm. Also, large-area

Mat. Res. Soc. Symp. Proc. Vol. 461 ©1997 Materials Research Society

lithography would be time consuming since the processing is serial. For periodic patterning and texturing of simple features (e.g., dots or lines) over a large area, the use of diblock copolymers would be advantageous due not only to the nanometer size microdomain structures and periodicity, but also to the spontaneous formation of ordering. Various applications of two dimensional ordered structures include creation of a periodic electric potential in a two-dimensional electron gas system,[4,5] fabrication of quantum dots or anti-dots,[5,6] synthesizing DNA electrophoresis media,[7] and fabrication of high density magnetic recording devices.[8]

In order to successfully use the self-assembled diblock copolymer microstructures as nanolithography templates, it is important to understand and control their behavior in thin films on various surfaces such as semiconductor substrates, resist materials and metal films. One of our main goals is to reliably produce a large area monolayer of well-ordered microdomain structures on substrates. Thin film behavior may differ from the bulk behavior,[9,10] and recent studies[11,12,13] show that the presence of surfaces affects the morphology of diblock copolymers. In our asymmetric polystyrene-polybutadiene (PS-PB) diblock copolymer thin films, we find that the morphology and ordering of microdomain structures are affected by the underlying substrate in at least two significant ways. First, the strong wetting constraints at the interfaces can cause the effective volume fractions of the blocks in thin films to differ from the bulk values, hence inducing a morphology change. Second, the chain mobility can be hindered, hence affecting the ordering.

EXPERIMENTAL DETAILS

The two PS-PB diblock copolymers used in this work were synthesized by standard high vacuum anionic techniques.[14] The molecular weights of the PS and PB blocks in the first copolymer (SB 36/11) are approximately 36,000 and 11,000 gm/mol, respectively; while in the second copolymer (SB 65/10) they are 65,000 gm/mol and 10,000 gm/mol. The first copolymer microphase separates into a cylindrical morphology in bulk, producing hexagonally ordered PB cylinders embedded in PS matrix; while the second copolymer adopts a spherical morphology, producing PB spheres with body centered cubic order. PS-PB diblock copolymer thin films are produced by spin-coating 1-3 wt % solutions of polymer in toluene directly on different types of substrates: bare silicon, bare silicon nitride, and PMMA coated substrates. The films are then annealed at 125 °C, a temperature above the glass transition temperature, for 24 hours under vacuum. The film thickness is controlled by varying polymer concentration and spin speed, and ranges from 10 nm to 60 nm for the samples discussed in this work. The thickness is measured either by an ellipsometer or interferometer.[15] The PMMA coated substrates are prepared by spin-coating an 120 nm thick PMMA film (from 4 % solution of PMMA 495,000 gm/mol in anisole) on top of silicon or silicon nitride substrates. The PMMA films are annealed at 170 °C for a few hours before spin-coating the PS-PB copolymer film.

The two-dimensional microdomain morphology is examined with either a transmission electron microscope (TEM) or scanning electron microscope (SEM). For TEM work, the films are spin-coated on thin (75 nm) silicon nitride windows[16] and imaged directly. In this way, one can easily obtain two dimensional images of the microdomain structures with the substrate intact. Films on thick silicon substrates are directly imaged with a SEM technique by subjecting the top surface to reactive ion etching (RIE).[17] A RIE step is necessary to image the microdomain morphology since the air surface of the diblock film is always covered by a continuous PB layer.[12] The details of the technique are presented elsewhere.[18] For both TEM and SEM micrographs, the contrast between PS and PB microdomains is provided by selectively staining the PB blocks with OsO_4 vapor. In the TEM micrographs, the stained PB domains appear darker than the PS domains due to increased high angle electron scattering; and in the SEM micrographs, lighter due to an increased secondary electron yield. Secondary ion mass spectroscopy (SIMS) is performed on the cylindrical PS-PB diblock films on both bare and PMMA coated silicon substrates in order to obtain depth profiles.

RESULTS AND DISCUSSION

Due to the discrete domain size produced by block copolymer self-assembly, the microphase separated PS-PB diblock copolymer thin films rearrange into discrete film thicknesses[19] upon annealing. Since most spin-coated films are not at these discrete thicknesses, the excess [deficient] material after satisfying a uniform n^{th} discrete thickness results in formation of islands [holes] with the $(n+1)^{th}$ $[(n-1)^{th}]$ thickness. Figure 1 (a) and (b) show the SEM micrographs of the cylindrical copolymer films, both with same average thickness of 35 nm, on bare silicon (a) and PMMA coated (b) substrates. On the bare silicon, we observe micron size islands of the first PB cylinder layer (50 nm thick) in a thinner background (30 nm thick) with no in-plane microstructure. (The thinner layer with no microstructure is referred to as the 0th layer, while the thicker layer containing PB microdomains is termed as the 1st layer.) With the same average film thickness of 35 nm, however, the film on the PMMA substrate exhibits no island formation. Instead, a uniform film of the PB cylinder layer (1st layer) is observed as shown in Figure 1 (b). When the average film thickness is decreased from 35 nm, the film on the PMMA substrate also exhibits a non-uniform topography (i.e., islands or holes), and at 15 nm it becomes a uniform 0th layer. In comparison, for the films on the bare substrate [PMMA substrate] we find the uniform 0th layer at 30 nm [15 nm] and the 1st layer at 50 nm [35 nm]. The films on a bare silicon nitride behave similarly to those on the bare silicon. Also, we observe a similar substrate dependence of the 1st layer thickness in the spherical diblock copolymer films.

Different 1st layer thicknesses are possible on the bare and PMMA coated substrates if different wetting constraints are imposed at the substrate-copolymer interface. Figure 2 shows the proposed schematics of thin PS-PB diblock films on two types of substrates, where the PB block wets the air and bare substrate interfaces and the PS block the PMMA interface. Hence, a thinner film is needed to obtain a monolayer of PB cylinders or spheres on the PMMA substrate (b) than on the bare substrate (a). The dominant wetting constraints at the interfaces also explain the appearance of the 0th layer where no microstructures are seen. The copolymer satisfies the wetting constraints before assembling into PB cylinders or spheres. Figure 3 shows a series of SIMS analysis performed on the osmium-stained uniform 0th and 1st layer films of the cylindrical copolymer on bare silicon (a) and PMMA coated (b) substrates. Since osmium selectively stains the PB blocks, the osmium SIMS trace maps out the PB profile in film depth. The SIMS analysis confirms the proposed schematics of Figure 2.

The spin-coated cylindrical copolymer thin film generally exhibits PB cylinders lying parallel to the substrate in a fingerprint-like pattern,[20] and the spherical copolymer film exhibits

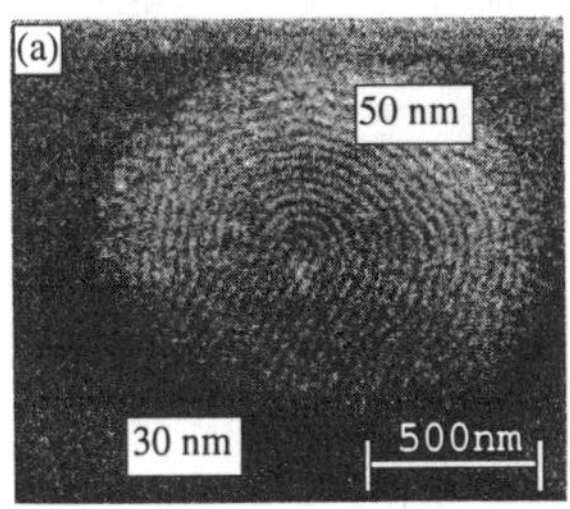

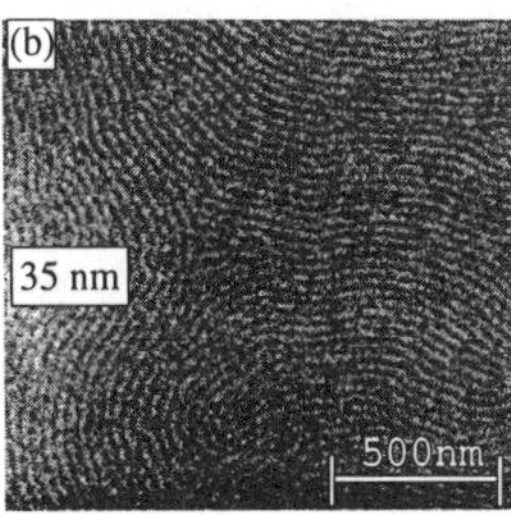

Figure 1. SEM micrographs of the spin-coated ~35 nm thick films of SB 36/11 on (a) the bare silicon substrate and (b) PMMA coated substrate after annealing. The lighter regions correspond to the PB cylinder microdomains. In (a), the islands of the 1st layer (50 nm thick) are formed in the 0th layer background (30 nm thick). In (b), the 1st layer (35nm thick) is uniformly formed.

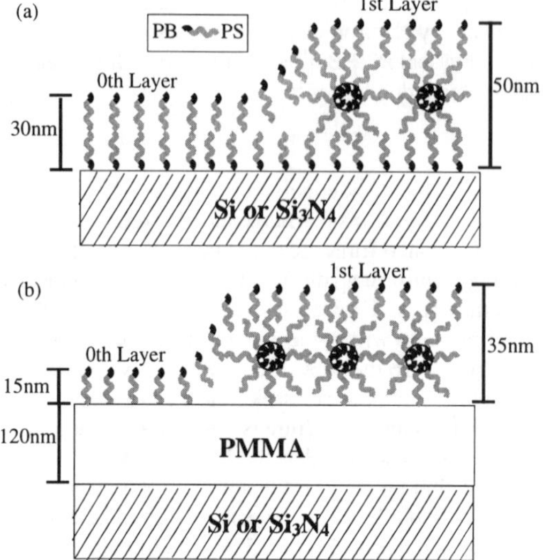

Figure 2. Schematic cross-sectional views of the copolymer films on (a) bare silicon or silicon nitride substrate, and (b) PMMA coated substrate. In (a), PB blocks wet both air and substrate interfaces. In (b), PB blocks wet the air interface and PS blocks the substrate interface.

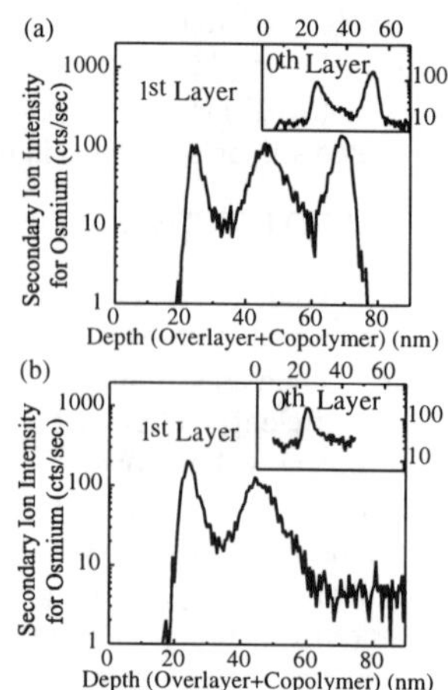

Figure 3. Osmium SIMS profiles of the uniform 1st and 0th (insets) layer films of SB36/11. The PB blocks are selectively stained with osmium. In (a) the films are on bare silicon substrates, and in (b) on PMMA coated substrates. The peaks correspond to the PB interfacial layers and microdomains. Each film has a ~23 nm thick sacrificial homo PS overlayer on top.

PB spheres in a two-dimensional hexagonal order.[2,13] The diameter and period of the PB microdomains are ~15 nm and ~30 nm, respectively, in both types of polymer films. In the case of the cylindrical copolymer film, the morphology of the 1st microdomain monolayer, on both bare and PMMA coated substrates, can vary between cylindrical and spherical. Figure 4 shows the TEM micrographs of the different morphologies observed in the films on bare silicon nitride windows. The equilibrium morphology is cylindrical as shown in Figure 1 (b), but we also observe spherical [Figure 4 (b)] and mixed [Figure 4 (a)] PB microdomain structures. The spherical and mixed state regions of the films are slightly thinner than the regions with the equilibrium morphology, judging from the difference in the transmitted current during TEM micrography. However, the thickness variation must be only a few nanometers since the observed PB microdomain sizes in these films are all ~15 nm. The morphology change is likely due to a decreased PB volume fraction after satisfying the wetting constraints at the interfaces. Hence, qualitatively, the copolymer system will go from a cylindrical phase toward to a spherical phase (and further to a disordered phase) in the phase diagram[1] as the effective PB volume fraction is reduced. We note that similar structures are observed in a free standing film by Radzilowski et al.[9] In the case of the spherical copolymer film, we observe that the grain size -

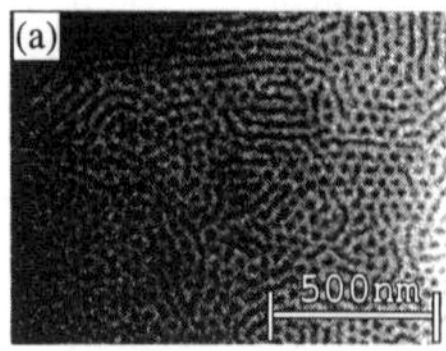

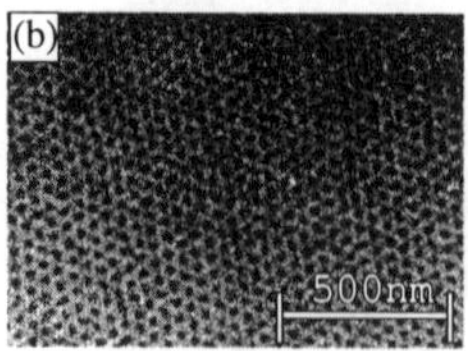

Figure 4. TEM micrographs of thin films of SB 36/11, which forms PB cylinders in bulk. The darker regions correspond to the PB microdomains. The morphology of the first monolayer can vary from cylindrical to spherical. A mixed state morphology as shown in (a) is observed as well as a spherical morphology as shown in (b).

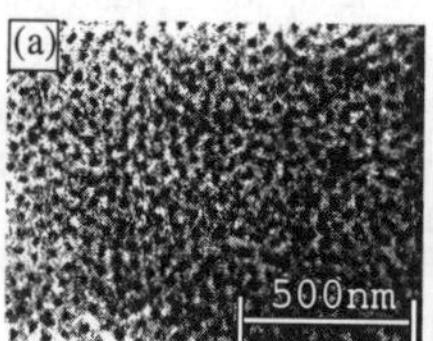
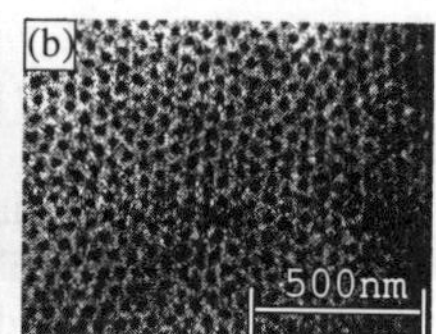

Figure 5. TEM micrographs of thin films of SB 65/10 (a) on a bare silicon nitride substrate and (b) on a PMMA coated substrate. The films are simultaneously prepared with the same processing. Note the improved ordering of the PB microdomains (darker regions) on PMMA.

the size of the region in which the spheres are coherently ordered - in the 1st layer is increased on the PMMA coated substrate [Figure 5 (b)], compared to that on the bare substrate [Figure 5 (a)]. One possible explanation for the differences in grain size lies in differences in chain mobility on the two substrates. Chain mobility can be considerably reduced[21] at hard surfaces due to pinning. By contrast, PMMA provides a thick fluid-like substrate since its glass transition temperature (105 °C) is below the temperature at which the copolymer films were annealed (125 °C). However, it is also possible that the differences in grain size evident in Figure 5 arise from the different wetting constraints for the two cases; both films were prepared simultaneously with the same processing at the same nominal thickness, so after subtracting out the amount of polymer needed to satisfy these wetting constraints, the remaining thickness is unequal in the two cases. Figure 4 has already demonstrated that the thickness alone can influence the degree of order.

Finally, we present preliminary pattern transfer results obtained by using the PS-PB diblock copolymer thin films as nanolithography templates. Figure 6 shows the patterns in silicon nitride, transferred from the cylindrical (a) and spherical (b) PB microdomain monolayers by the use of a CF_4/O_2 RIE technique.[22] The unaltered microphase separated PS-PB film alone does not produce a usable lithography template since the etching rates of the PS and PB microdomains are almost the same under most RIE conditions. To transfer the pattern, etching selectivity between dissimilar microdomains is essential. For Figure 6 (a), selectivity was achieved by staining the PB domains with osmium. This slightly reduces the etching rate of the PB domains, so the PB domain regions are partially masked from the RIE process. For Figure 6 (b), the PB domains were selectively degraded and removed by exposing the copolymer film to ozone prior to RIE. The ozonation step produces a variation in the effective total thickness of the copolymer template (empty spheres in PS matrix), which follows the microdomain pattern; hence the RIE process results in surface corrugations, which also follow the same pattern. The understanding of the monolayer structures as shown in Figure 2 has enabled us to control the etching process more precisely and transfer patterns successfully to the substrate from the copolymer film.

CONCLUSIONS

We have investigated the morphology and ordering of spin-coated asymmetric PS-PB copolymer thin films for use in nanolithography. On different types of substrates, different film thincknesses are required to produce a uniform monolayer (of the microdomains) film with an ordered morphology due to different wetting constraints at the substrate interfaces. In films

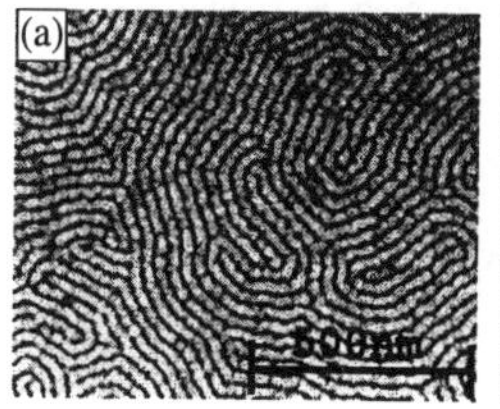
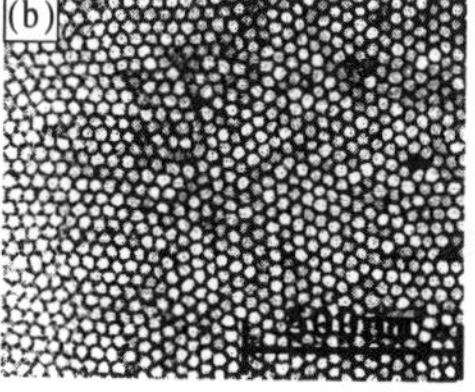

Figure 6. TEM micrographs of the patterns in silicon nitride. In (a), the fingerprint-like pattern is transferred from an osmium stained PS-PB copolymer film. In (b), the hexagonally ordered sphere pattern is transferred from an ozonated copolymer film. The lighter regions are ~10-15 nm deep depressions which have been etched out.

thinner than the monolayer thickness, no in-plane microdomain structures are observed. In the monolayer of the cylindrical copolymer, we observe a morphology change from cylindrical to spherical order. These effects are due to strong wetting constraints imposed at the interfaces. In the spherical copolymer film, the ordering of the PB microdomains is increased on the PMMA substrate. Two possible explanations are discussed – an increased chain mobility on thick fluid-like PMMA, and an effectively thicker film on PMMA due to the different wetting constraints. Finally we have demonstrated that a successful patterning of nanometer scale ordered structures over a large area is possible by using the microphase separated copolymer films as lithography templates.

ACKNOWLEDGMENTS

This project is supported by the NSF under DMR 9400362. Parts of the processing are carried out at the Advanced Technology Center for Photonics and Optoelectronics Materials at Princeton University.

REFERENCES

[1] F. S. Bates, Science, **251**, 898 (1991); F. S. Bates and G. H. Fredrickson, Ann. Rev. Phys. Chem., **41**, 525 (1990).

[2] P. Mansky, P. Chaikin, and E.L. Thomas, J. Mater. Sci., **30**, 1987 (1995).

[3] P. Mansky, C. K. Harrison, P. M. Chaikin, R. A. Register, and N. Yao, Appl. Phys. Lett., **68**, 2586 (1996).

[4] D. Hofstadter, Phys. Rev. B, **14**, 2239 (1976); D. J. Thouless, M. Kohmoto, M. P. Nightingale and M. den Nijs, Phys. Rev. Lett., **49**, 405 (1982).

[5] D. Weiss, M. L. Roukes, A. Menschig, P. Grambow, K. von Klitzing, and G. Weimann, Phys. Rev. Lett., **66**, 2790 (1991).

[6] W. Kang, H. L. Stormer, L. N. Pfeiffer, K. W. Baldwin, and K. W. West, Phys. Rev. Lett., **23**, 3850 (1993).

[7] W. D. Volkmuth and R. H. Austin, Nature 358, 600 (1992); W. D. Volkmuth, T. Duke, M. C. Wu, R. H. Austin, and A. Szabo, Phys. Rev. Lett., **72**, 2117 (1994).

[8] M. H. Kryder, Thin Solid Films, **216**, 174 (1992).

[9] L. H. Radzilowski, B. L. Carvalho, and E. L. Thomas, preprint, submitted to J. Poly. Sci. B (July 1996).

[10] Y. Liu, M. H. Rafailovich, J. Sokolov, S. A. Schwarz, and S. Bahal, Macromolecules, **29**, 899 (1996).

[11] C. S. Henkee, E. L. Thomas, and L. J. Fetters, J. Mater. Sci., **23**, 1685 (1988).

[12] H. Hasegawa and T. Hashimoto, Polymer, **33**, 475 (1992).

[13] E. L. Thomas, D. J. Kinning, D. B. Alward, and C. S. Henkee, Macromolecules, **20**, 2934 (1987).

[14] M. Morton, L. J. Fetters, Rubber Chem. Technol., **48**, 359 (1975).

[15] An ellipsometer is used to measure the average thickness in samples with uniform thicknesses, whereas an interferometer with 1 μm probe size is used to measure the local thickness variations in samples with a non-uniform topography (i.e. islands or holes).

[16] Silicon nitride windows have been previously used in TEM micrography instead of copper grids for specimen support. For example, see T. L. Morkved, M.Lu, A. M. Urbas, E. E. Ehrichs, H. M Jaeger, P. Mansky, and T. P. Russell, Science, **273**, 931 (1996). Having a continuous silicon nitride background reduces the resolution and contrast of the image, however on a 10 nm length scale it is not significant.

[17] S. M. Sze, VLSI Technology, 2nd ed., McGraw-Hill, New York, 1988, pp. 184-232.

[18] C. K. Harrison, M. Park, P. M. Chaikin, R. A. Register, D. H. Adamson, and N. Yao, preprint, submitted to Polymer (July 1996).

[19] G. Coulon, D. Ausserre, and T. P. Russell, J. Phys. France, **51**, 777 (1990).

[20] The microdomain orientation and ordering are sensitively dependent on film preparation. For example, see references 3 and 12.

[21] J. G. Van Alsten, B. B. Sauer, and D. J. Walsh, Macromolecules, **25**, 4046 (1992).

[22] The RIE parameters used for the work presented in Figure 6 are 40 mTorr, 20 sccm, 20W, ~170 V_{dc}.

MODIFICATION OF INTERFACIAL PROPERTIES OF POLYMER BLENDS WITH DIBLOCK COPOLYMER

S. KIM and C. C. HAN
Polymers Division, National Institute of Standards and Technology, Gaithersburg, MD 20899

ABSTRACT

The effect of diblock copolymer on the phase-separation process of polymer blends has been investigated by using light scattering and optical microscopic observations. To quench the system into the two phase region, a shear-jump technique is employed instead of the conventional temperature-jump technique. The samples studied are blends of low-molecular-weight polystyrene and polybutadiene with and without added styrene-butadiene block copolymer as a compatibilizer. It was observed that the addition of diblock copolymers could accelerate the phase separation kinetics depending on the shear history. As the concentration of diblock copolymer increases, the distribution of domain sizes becomes narrower and the growth rate slows down. The extent of slowing-down depends on the molecular weight and concentration of the copolymer. The time dependence of domain growth is clearly observed with optical microscopy.

INTRODUCTION

One of the important applications of block copolymers is their use as a compatibilizer in polymer blends. Block copolymers can be viewed within this application as an analog to a nonionic surfactant [1]. The interfacial activity of a block copolymer added to a homopolymer blend is displayed by the reduction in the interfacial tension between two coexisting phases. Such a reduction in the interfacial tension is believed to be a result of the tendency for diblock copolymer molecules to accumulate at the boundary between the immiscible homopolymer phases. This has been demonstrated by experimental determination of interfacial tension [2] and has also been predicted from theoretical considerations in equilibrium [3-5].

Light scattering studies of the phase-separation kinetics of low-molecular-weight polystyrene-polybutadiene blends reveal that the rate in the late stage of phase-separation was reduced when a sufficient amount of styrene-butadiene diblock copolymer was added [6]. The effect depends on the concentration and molecular weight of the diblock copolymers [7]. The result was interpreted as evidence that the diblock copolymer accumulates at the boundaries and consequently lowers the interfacial tension, which in turn reduces the rate of droplet growth. However, it is not clear what the important parameters are which determine whether the block copolymer molecule has enough time and energy difference to diffuse and adsorb at the interface in a dynamic system with simultaneous phase separation. Therefore, it is not necessary that the addition of block copolymer should always lead to a slowing-down in the phase separation kinetics, especially when the mixing due to a shear flow is a factor. It is thus desirable to optically investigate blends with and without diblock copolymer.

If the diffusion of block copolymers to the interface is relatively slow compared to the phase separation kinetics and the mixing process, then the diblock copolymer effect for shear quenched systems will not be easily recognized. Experimental results under steady shear showed that diblock copolymer effects were not noticeable at high shear rates (e.g., >10 s^{-1}). However, in the case of low shear rates, very different morphology was observed [8].

In this report, we use light-scattering and optical-microscopy techniques to study the effect of diblock copolymer on the phase-separation kinetics of the polymer blends. Unlike

Mat. Res. Soc. Symp. Proc. Vol. 461 © 1997 Materials Research Society

conventional experiments for this type of study which use temperature-quench techniques for introducing phase-separation, we employed "shear-quench" technique. In other words, we first bring the system from the two-phase region to the one-phase region by "shear-induced mixing" [9], then, we remove the shear field and let the system quench back to the two-phase region abruptly. The advantage of this technique is the fast quench process and the possibility of studying the dependence of phase separation processes for a system starting from various morphological structures (shear history). This could provide a new route to controlling the morphological structure of a polymer blend through shear flow.

EXPERIMENT

MATERIALS. The polystyrene was purchased from TOSOH Co.[10] with M_w = 2,630 g/mol [11] and M_w/M_n = 1.05. The polybutadiene was purchased from Goodyear Chemical Co. with M_w = 2,800 g/mol and M_w/M_n = 1.2. The polystyrene-polybutadiene diblock copolymer (PS-PB, the amount of PS by mass is ca. 55 % and the total M_w = 6,300 g/mol, M_w/M_n = 1.04) and polystyrene-d_8-polybutadiene diblock copolymer (PSD-PB, the amount of PSD is ca. 50 % by mass and total M_w = 22,500 g/mol) were synthesized and characterized by Dr. J. W. Mays, University of Alabama, Birmingham. The experimental condition of the tested polymer blends are listed below.

<u>Tested Polymer Blends</u>

system #	nominal M_w of diblock	mass % of diblock	T_c (°C)[#]	ΔT_c (°C)[##]
A		0%	134.2 ± 0.1	
B	3k-3k	2.5%	128.2 ± 0.1	6.0 ± 0.1
C	3k-3k	5%	119.7 ± 0.1	14.5 ± 0.1
D	10k-10k	2.5%	126.4 ± 0.1	7.8 ± 0.1

[#] ± represents the expanded uncertainty of the measurements.
[##] ΔT_c = T_c without diblock copolymer - T_c with diblock copolymer.

INSTRUMENT. The light-scattering/optical microscopy instrument was designed and constructed in our laboratory. The structure and capability of the instrument is described elsewhere [12].

The polymer blend, consisting of polystyrene and polybutadiene in the mass ratio of 60 to 40 and containing additional amounts of copolymers by mass, of 0 %, 2.5 % or 5 % was heated under vacuum to 140 °C which is at least 6 °C higher than the cloud point of the prescribed polymer blends. After thorough mixing, it was transferred to the preheated sample compartment of the instrument.

RESULTS AND DISCUSSIONS

For the four systems studied, the morphologies were similar to each other at high shear rates (e.g., >10 s^{-1}). All systems gave no observable light scattering and blank micrographs indicating a shear induced homogenization at 500 s^{-1}, however, we realize that the concentration fluctuation dynamics are not necessarily identical to a true equilibrium miscible condition. Quench depth was fixed to 1.5 °C for all studies in this report. Fig. 1 shows the effect of adding 3k-3k diblock copolymer on the evolution of the phase-separated domain size after cessation of shear flow. The unit of q_{max} is μm^{-1} and was calculated from the circular average of the spinodal pattern of light scattering after cessation of shear. The dominant size, d, of the droplets can be calculated from the equation,

$$d = 2\pi/q_{max} \qquad\qquad (1).$$

In the case of system B, it appears that the phase-separation kinetics are not affected by the addition of diblock copolymers compared with the blend without any additives in the beginning stage. However, the domain growth is accelerated in the later stage of phase-

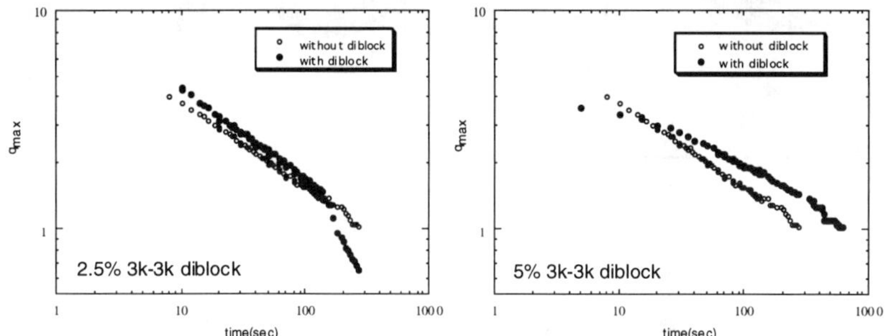

Fig.1. q_{max} vs. time plot of 3k-3k diblock copolymer. The relative standard uncertainty in q_{max} is less than 5 %.

separation as is shown in Fig. 1. If the added diblock copolymer is well partitioned and mixed into the phase separating domains, then this phenomenon of similar rates of phase separation are expected in the beginning stage. When the concentration of the added copolymer is increased to 5 %, the kinetics of the phase-separation seems to be changed significantly. The droplet size of system C grows faster in the presence of diblock copolymer in the beginning and the overall kinetics slow down later. This experimental result can be interpreted as a simultaneous processes of diffusion and phase separation. In the beginning stage, the adsorbed copolymers are partitioned fairly uniformly everywhere, the effective influence on the phase separation kinetics are either due to the modification of free energy function or of the viscosity and effective interdiffusion rate. As the block copolymers are accumulating at the interface, the formation of a interfacial layer could induce a frustrated-fusion mechanism as has been observed in thin films [13].

The above interpretation is also evidenced by the light scattering data which were analyzed according to the theoretical model of Furukawa [14]. The scaled structure function is given by

$$I/I_{max} = \frac{3(q/q_{max})^2}{2 + (q/q_{max})^6} \qquad\qquad (2).$$

For large values of q/q_{max}, Eq. 2 is reduced to Porod's law [15],

$$I/I_{max} \propto (q/q_{max})^{-4} \qquad\qquad (3).$$

For a clear comparison, all data were normalized and displayed in Figs. 2 and 3. For all four sets of experimental conditions, the normalized structure factors are self-similar as a function of time in the beginning. In the case of pure blend and the blend with 2.5 % 3k-3k diblock, it shows the slope of -4 after q_{max}. According to Porod's law, this implies that the boundaries of the phase-separated domains are sharp compared with the wavelength of light. On the other hand, when more diblock was added (i.e., 5 %) or when the molecular

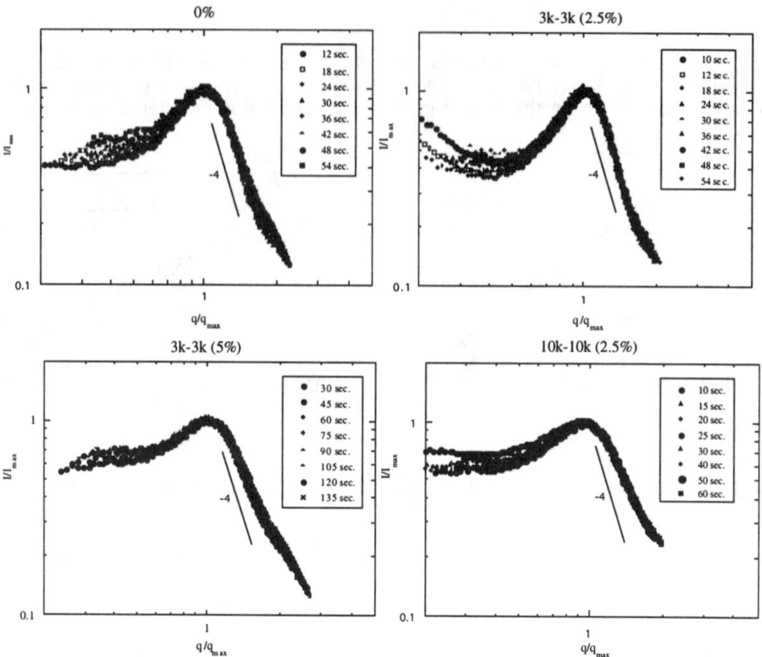

Fig.2. Normalized I/I_{max} vs. q/q_{max} plot in the beginning stage of phase-separation. The relative standard uncertainty in I/I_{max} is less than 5 %.

weight of diblock copolymer is higher (i.e., 10k-10k), the slope became less steep than in the two previous cases. This can be interpreted as an indication of the formation of fuzzy boundaries caused by the accumulation of diblock copolymers at the surface of phase-separated domains. At later times after the shear quench, however, each system revealed its own characteristic behavior in the structure factor (Fig. 3). In the case of the 0 % diblock system, domain distribution is gradually broadened with time, as is reflected from the broadening of the normalized light scattering structure factor. The 2.5 % diblock system showed some unexpected behavior. Although there was no noticeable difference in the q_{max} vs. time plot (Fig. 1), the distribution of the domain sizes is much broader than the 0 % diblock case. Obviously, the system was affected by copolymer and the interfacial tension was lowered to some extent although it did not change the most probable size of the droplets significantly. It seems that lower interfacial tension broadens the dispersity of particle sizes significantly.

In the micrographs of Fig. 4, the morphology of the pure blend is compared with that of 2.5 % 3k-3k diblock system. The faster growth of domains observed by light scattering in the later stage is readily identified. The size of droplets is similar in the beginning (top of Fig. 4), but becomes bigger in the blend with diblock copolymers in the later stage. When more diblock copolymers were added (i.e., 5 %), bigger droplets were observed in the micrograph compared to that of 0% diblock copolymer sample in Fig. 5. This result is not consistent with the result shown in Fig. 1 because this result implies faster growth of droplets for the sample with 5 % of 3k-3k diblock copolymer. However, as was also observed and is shown in the lower left micrograph in Fig. 6, the droplet size becomes smaller for this 5 % diblock copolymer sample in the later stage. The reason why q_{max} vs.

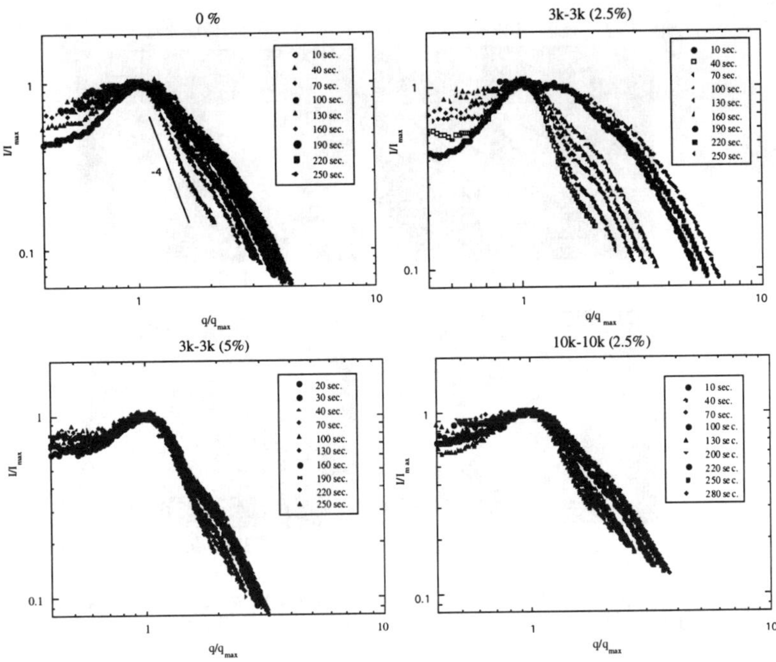

Fig.3. Normalized I/I_{max} vs. q/q_{max} plot in the later stage. The relative standard uncertainty in I/I_{max} is less than 5 %.

t plot obtained from the light scattering experiment shows a different result can be explained as follows. Although we did not observe any structure with our light scattering detector or our phase contrast microscope before shear quenching at 500 s^{-1}, large domain structure in either droplet or string shape [16] with small concentration difference may still exist. The shear quench will form phase separation both inside and outside (in the matrices) of these pre-existing domains with complementary structures and compositions. This is probably

why we did not observe the same power law growth exponents ($q_{max} \propto t^{-1/3}$, t^{-1} and $t^{-1/3}$) as those normally observed in a temperature jump experiment. The light scattering measurements could be measuring the phase separation within the original droplet if this mechanism became important. On the other hand, the phase contrast microscopy is only measuring the outside parameter of the large droplet which was growing from the pre-existing structure at the time of shear quench. Only in the case of higher concentration of block copolymer (5 % of 3k-3k), the large droplet starts to break up after 100 sec. due to the abundance of block copolymers in the large droplet, which helps to break up and stabilize this large droplet into smaller ones.

In the micrograph, it is also observed that the boundaries of the phase-separated domains are irregular and broad (lower right micrograph of Fig. 5). This implies that the concentration of the diblock copolymer at the boundaries is high enough to lower the interfacial tension significantly. As the domains inside a big domain grow, the big domains are disintegrated to smaller ones. It is reasonable to consider the repulsion between diblock copolymer layers formed at the boundaries when the concentration of diblock is high

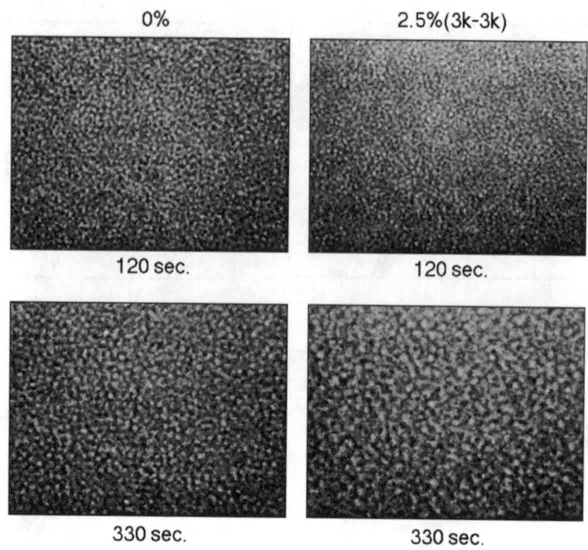

Fig.4. Comparison of micrographs of pure blend and with 2.5 % 3k-3k diblock copolymer. Horizontal dimension of micrographs is 200 μm.

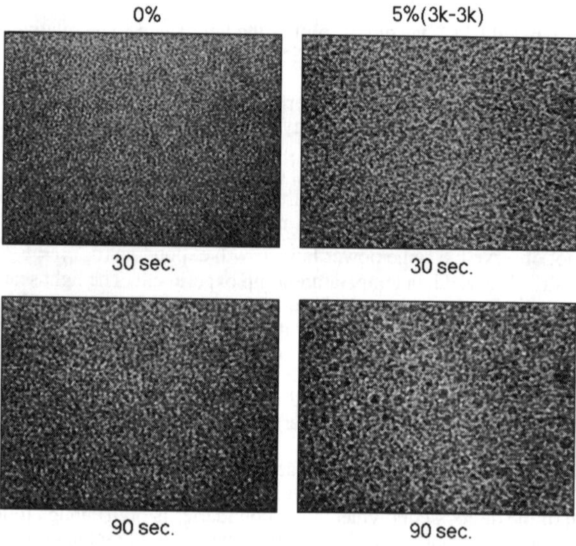

Fig.5. Comparison of micrographs of pure blend and with 5 % 3k-3k diblock copolymer. Horizontal dimension of micrographs is 200 μm.

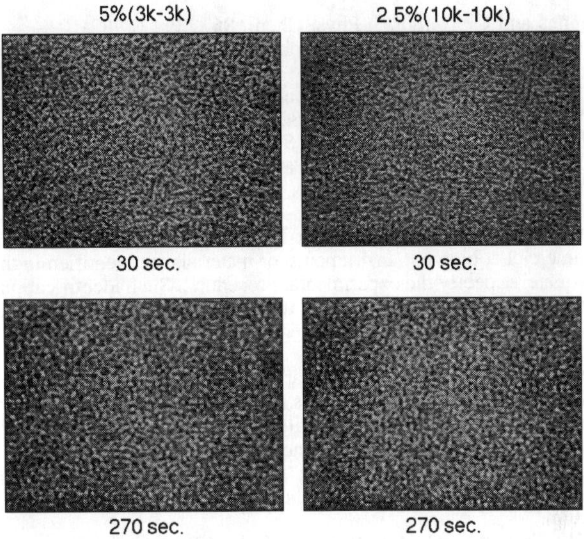

Fig.6. Comparison of micrographs of polymer blend with 5 % 3k-3k and 2.5 % 10k-10k diblock copolymers. Horizontal dimension of micrographs is 200 μm.

enough [13]. As the concentration of the diblock copolymer at the boundaries increases, small and round domains are formed from the large droplets and the dispersity of sizes becomes narrower.

With 10k-10k, 2.5 % diblock copolymer added to the system, the overall trend is the same as that of the 3k-3k, 5 % diblock case (Fig. 6). The same irregular boundaries of phase-separated domains are observed as in the case of 5 % diblock copolymer. This result confirms that in order to observe the same diblock copolymer effect on the phase separation kinetics of a polymer blend, the concentration of the copolymer needed is lower for higher molecular weight copolymers.

CONCLUSIONS

It is widely believed that at a given time after phase-separation has begun the size of the phase-separated domains become smaller with the addition of diblock copolymers. In this experiment, it is found that the domain size depends on the shear history, the amount and molecular weight of the diblock copolymer used. From the microscopic observations, the droplet size often becomes larger right after shear quenching and then stops growing or even decreases. This is probably due to the simultaneous spinodal decomposition inside and outside of these droplet produced from the shear history. It is also observed that diblock copolymer will result in bigger droplets and broader dispersity at the beginning of the phase separation process. It is clear that a very different morphology (compared to simple temperature quench) and growth mechanism can be obtained with the combination of using shear flow and adding diblock copolymer as interfacial modifiers.

REFERENCES

1. P. G. de Gennes and C. Tanpin, J. Phys. Chem. **86**, 2294 (1982).
2. S. H. Anastasiadis, I. Gancarz and J. T. Koberstein, Macromolecules, **22**, 1449 (1989).
3. J. Noolandi and K. M. Hong, Macromolecules, **17**, 1531 (1984).
4. L. Leibler, Macromolecules, **15**, 1283 (1982).
5. L. Leibler, Macromol. Chem., Macromol. Symp, **16**, 1 (1988).
6. L. Sung and C. C. Han, J. Polymer Sci., Part B, **33**, 2405 (1995).
7. R. J. Roe and C. M. Kuo, Macromolecules, **23**, 4635 (1990).
8. unpublished result.
9. T. Hashimoto, T. Takebe and K. Asakawa, Physica A, **194**, 338 (1993).
10. Certain commercial equipment, instruments, or materials are identified in this article in order to adequately specify the experimental procedure. Such identification does not imply recommendation or endorsement by the National Institute of Standards and Technology, nor does it imply that the materials or equipment identified are necessarily the best available for the purpose.
11. Accroding to ISO 31-8, the term "Molecular Weight" has been replaced by "Relative Molecular Mass", symbol M_r. Thus, if this nomenclature and notation were to be followed in this publication, one would write $M_{r,n}$ instead of the historically conventional Mn for the number average molecular weight, with similar changes for M_w, M_z and M_v, and it would be called the "Number Average Relative Molecular Mass. The Conventional notation, rather than the ISO notation, has been employed for this publication.
12. S. Kim, J.-W. Yu and C. C. Han, Rev. Sci. Inst. (1996) in press.
13. A. Karim, J. F. Douglas and C. C. Han, Macromolecules, submitted.
14. H. Furukawa, Physica, **123A**, 497 (1984).
15. G. Porod, Kolloid Z., **124**, 83 (1951).
16. E. Hobbie, S. Kim and C. C. Han, Phys. Rev. E, **54**, R5909 (1996).

LASER INDUCED DECOHESION SPECTROSCOPY: A NEW TECHNIQUE FOR MEASURING POLYMER INTERFACIAL ADHESION

J.S. Meth, D. Sanderson, C. Mutchler, S. J. Bennison
DuPont Co., Central Science and Engineering,
P.O. Box 80328
Wilmington, DE 19880-0328

ABSTRACT

We present a new technique, laser induced decohesion spectroscopy (LIDS), which is capable of measuring the practical work of adhesion G between a transparent polymer film and an opaque substrate. In LIDS, a laser pulse directed onto the sample creates a blister at the film/substrate interface. The blister's internal pressure depends on the laser pulse energy, and at a critical pressure the sample fractures. We have derived a theoretical analysis of this experiment based on elasticity theory and fracture mechanics, and present the results. By measuring physical variables such as the thickness of the transparent polymer, the blister radius, and the blister curvature, it is possible to deduce G between the two coatings. Here we report G for a matrix of automotive finish systems consisting of four opaque basecoats of various colors (black, white, red, green) coated with a clearcoat of various thicknesses. We expect G to be a system parameter for each basecoat independent of clearcoat thickness. The values of G for the different basecoats yield the relative adhesion of the various pigmented paint formulations.

INTRODUCTION

Quantitative adhesion measurements between successive polymer layers in a coating system are difficult [1-3]. Currently, the adhesion in such systems is measured with the crosshatch-peel test, or some variation thereof. This test is not quantitative and has a limited dynamic range. Many samples will "pass" the test, yet there are real differences in the adhesive strength that go undetected. We have developed laser induced decohesion spectroscopy (LIDS) to perform such measurements quantitatively, and present preliminary results here.

LIDS is essentially a new modification of the blister test [4-9]. LIDS employs photothermal ablation as the mechanism for generating internal pressure. Generally in photothermal ablation, a high power, temporally short, laser pulse incident on an opaque material causes it to heat up extremely rapidly, whereupon bonds break in that material. When a transparent layer covers the opaque one, the ablation process produces an internal pressure. In our experimental configuration, Figure 1, the opaque polymer is ablated, and a pressure is generated between the opaque and transparent polymers. The close proximity of a rigid substrate prevents significant deformation of the opaque layer. This internal pressure deforms the transparent coating, producing a blister directly above the ablated area. By measuring the curvature of the blister at its top, center, the internal pressure of the blister can be deduced. By itself, this process does not involve any debonding between the two layers outside of the ablated region. This is analogous to the small initial debond or hole that is necessary in the standard blister test. As the laser pulse energy is increased, more ablation occurs, the internal pressure increases along with the associated strain energy stored in the blister. At some critical laser pulse energy, the Griffith criterion for crack propagation [10] is exceeded (namely that a crack will propagate when the decrease in strain energy with increasing crack length is balanced by the increased surface energy of the newly formed surfaces). The blister expands radially into the region of the sample that was not ablated, producing an annular debond area. At this critical laser pulse energy, there exists a critical internal pressure. This pressure is then related to the work of adhesion.

Mat. Res. Soc. Symp. Proc. Vol. 461 ©1997 Materials Research Society

There are several advantages to the LIDS technique. The geometry of the LIDS experiment allows testing of film systems without special sample preparation – no holes need to be drilled. The initiating laser pulse controls the fracture rate. The rise time of the blister is ~1µs. At these rates, the response of the polymer is determined by the glassy modulus, and viscous effects are not important during the blister creation. The pressures generated by the ablation process can easily be in excess of 100 MPa, allowing large values of adhesion to be measured.

THEORY

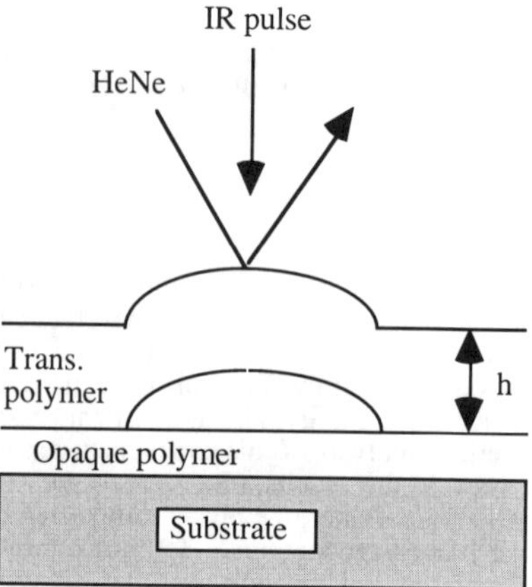

Figure 1. Schematic of LIDS experiment. An IR pulse creates a blister at the interface while a HeNe beam measures the curvature.

The theoretical analysis of the LIDS experiment consists of three links between four parameters. Gaussian beam propagation theory is used to relate the measured spot size of a continuous wave Helium-Neon (HeNe) beam reflected off the blister to its radius of curvature. Next, the biharmonic equation for the mechanical deformation in a model system is solved. From this analysis we are able to theoretically relate the curvature of the blister to the internal pressure, q. We are also able to calculate the strain energy in the system. By applying Griffith's criterion for crack propagation, we are able to relate the internal pressure to the adhesion parameter G, which is the material property of interest.

To measure the curvature of the blister, we measure the intensity of light that passes through a circular aperture placed in the beam path of the HeNe laser. The curvature is extracted by solving the complex algebraic equations for Gaussian beam propagation through the system [11].

To relate the curvature to the internal blister pressure, we create a theoretical model which consists of a thick, circular disk, to which a uniform pressure is applied at the lower surface. We use this as the model for the blister, and solve the problem using stress potential functions [12]. To account for the fact that the blister is part of a larger system, we apply various boundary conditions at the edge of the model disk. The boundary conditions vary from clamped (case 3) to simply supported (case 2), along with an intermediate mixed boundary condition (case 4). The clamped and simply supported boundary conditions are the ends of the spectrum of possible forces, in this way the model provides limiting cases for the pressure-curvature relationship. The mixed boundary condition states that the radial displacement at the lower edge of the blister is zero (to model its attachment to the substrate), while at the upper edge of the blister, the radial stress is zero (preliminary numerical simulations support this hypothesis). Along with these solutions, we also examine the textbook solution for a thin disk (case 1) and a thick disk (case 5) [13]. All data presented here has been analyzed using case 4. The choice of boundary conditions does not have an extreme effect on the results. In all cases, the relationship between pressure and curvature may be summarized by:

$$q = \frac{16\,D}{R_c\,b^2}\left\{k_1 + k_2\,(h/b)^2\right\}^{-1} \qquad (1)$$

$$D = \frac{E\,h^3}{12\,(1 - v^2)} \tag{2}$$

D is the flexural rigidity of the clear coat, E is the modulus, v is Poisson's ratio, h is the thickness, b is the blister radius, R_c is the measured radius of curvature, and k_1 and k_2 are constants depending on the particular boundary conditions, summarized in Table 1. For the work presented here, we take E = 3 GPa (from DMA measurements) and v = 1/3.

Table 1 - Summary of parameters for pressure-curvature relationship

Theory	k_1	k_2
case 1 - clamped thin plate	1	0
case 2 - simply supported disk	$\dfrac{(3+v)}{(1+v)}$	$\dfrac{4(2+v)}{5(1+v)}$
case 3 - clamped disk	1	$\dfrac{2(2-v)}{3(1-v)}$
case 4 - mixed b.c. disk	$\dfrac{(2+v)}{(1+v)}$	$\dfrac{2(2-v)}{3(1-v)}$
case 5 - clamped thick plate	1	$\dfrac{2}{(1-v)}$

We also derive the relationship between the dimensionless parameter q^2b/EG and the aspect ratio h/b. Previous work has shown [6,7] that this relationship can be characterized by a function f(h/b).

$$\frac{q^2b}{E} = G\,f(h/b) \tag{3}$$

When data is analyzed, a plot of q^2b/E versus f(h/b) can be fit to a straight line through the origin, whose slope will be G. To derive f(h/b), we calculate the strain energy in the model disk (the far-field energy), add to that the strain energy at the crack tip (the near-field energy) [14], then differentiate with respect to increasing crack area. We find that f(h/b) can be expressed as a polynomial with constant coefficients:

$$f(h/b) = \frac{(h/b)^3}{a_1 + a_2(h/b)^2 + 16/9\pi(h/b)^3} \tag{4}$$

The values of a_1 and a_2 depend on the boundary conditions. Table 2 collects the various values from different theories, assuming v = 1/3.

Table 2 – Comparison of f(h/b) coefficients for different boundary conditions

Theory	a_1	a_2
case 2	11/24	1097/1225
case 3	1/12	2/5
case 4	11/24	31/90
case 5	1/12	1/2

EXPERIMENT

The samples were prepared by standard techniques for automotive panels. The substrates were 4"x12" phosphated cold-rolled steel panels with electrocoat and primer layers. The opaque base coat was hand sprayed to a dry thickness of ~25 μm (1 mil). Four different base coat colors were tested: black, white, red, and metallic green. The metallic aspect of the green paint comes from aluminum flakes dispersed in the formulation. The panels were sprayed in a vertical orientation with a clearcoat on an automatic spraying machine and cured horizontally. Different film builds were prepared for each basecoat, nominally 1.0, 1.5, 2.0, 2.5, 3.0, 3.5 mils. Not all film builds resulted in unique thicknesses due to some uncertainty in the preparation process, but nevertheless five or six unique thicknesses were obtained for each color basecoat. The thickness of the clearcoat is a very important variable to know accurately. We used optical microscopy of the panels in cross section to measure thickness using a calibrated reticle. A small piece of the panel was cast into an acrylic mold and polished, which made thickness measurements relatively straightforward. By averaging across the sample, a reliable thickness could be extracted (error ±5%).

The ablating laser pulse is produced by a cavity-dumped regenerative oscillator running at 10 Hz, pumped by a CW, mode-locked Nd:YLF laser ($\lambda = 1.053$ μm). The output energy is stable to ±3%, with a maximum energy of 1 mJ. A shutter system isolates a single pulse. The laser is then focused onto the sample to a $1/e^2$ intensity radius of 50 μm. The sample itself is mounted vertically at the focal plane of the IR laser pulse on a motorized translation stage which can be positioned reproducibly to 1 μm via computer control. A blister is formed by exposing the sample to a single IR laser pulse, and recording the reflected HeNe intensity on a digital oscilloscope. The sample is then translated 1 mm before another laser pulse is incident on the sample. No position on the sample is ever exposed to more than a single laser pulse for data acquisition. The radius of curvature of the blister is measured as a function of the laser pulse energy. The laser pulse energy is varied from its threshold energy value, where a blister is just detectable, to the critical energy value, where a debond (crack) starts to propagate radially. The critical energy is determined by exposing the sample, without measuring the curvature, to a range of pulse energies, and then using an optical microscope to determine at what energy the decohesion process begins. Many exposures are performed at each energy to reduce errors. Plots are then constructed of curvature versus pulse energy. These plots can be fitted to a straight line, so we can calculate the curvature at the critical energy. It is also necessary to know the blister radius at the critical energy. This is measured with a microscope using a calibrated reticle examining laser exposures just below the critical energy.

RESULTS AND DISCUSSION

In Figure 2, we present a typical curvature vs. pulse energy plot for a representative panel with red basecoat and 1.5 mil clearcoat, along with a linear fit. The critical pulse energy at which debonding occurs in this sample is 175 μJ. In Figure 3, we present the data for the red basecoat by plotting the parameter q^2b/E vs. $f(h/b)$, using the constants a_1 and a_2 predicted by case 4, along with the best fit line through the origin. The slope of this line is taken as a measure of the adhesion parameter $G = 204 \pm 25$ J/m^2. The data points include error bars of ±25%. In Table 3, we present the variation in G (J/m^2) obtained by using the different theories described above, along with the standard errors.

Regardless of which theory is used to analyze the data, we find that the ordering of the adhesion parameters is unchanged. The black and red paint have roughly the same initial adhesion, with the green intermediary, and white the weakest. This robustness is assured by the fact that the measured critical pressures are larger for the strongly adhering systems than for the weaker ones. This proves that the LIDS technique is sensitive to the strength of the interface. The adhesion values for the various pigment systems are different because the pigments themselves vary, the pigment concentrations vary, and the pigment dispersants vary.

Table 3 – Comparison of G for different models.

Theory	G black	G white	G green	G red
case 2	278 ±67	69 ±14	133 ±19	261 ±41
case 3	221 ±46	58 ±9	124 ±19	236 ±33
case 4	193 ±42	50 ±8	101 ±15	204 ±25
case 5	82 ±17	24 ±3	53 ±9	102 ±12

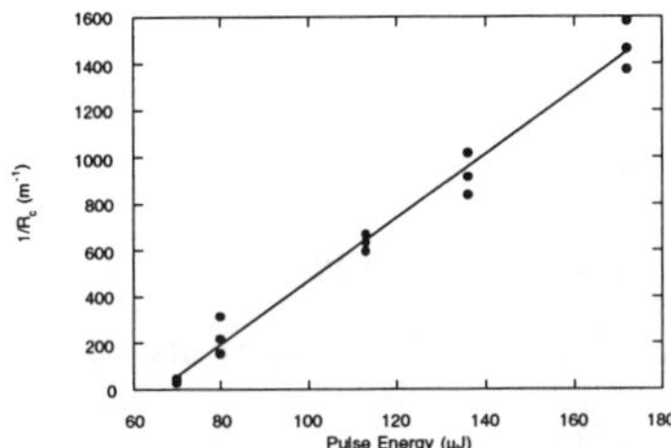

Figure 2. Typical plot of curvature vs. pulse energy, showing linear relation.

Figure 3. Plot of q^2b/E vs. f(h/b), the slope is G.

By allowing the aspect ratio to approach infinity in the theoretical analysis, and using the measured values of G, b and E, one can calculate the critical pressure necessary to induce debonding in a semi-infinite sample:

$$q_\infty = \sqrt{\frac{9\pi\,G\,E}{16\;\;b}}$$

(5)

Since b varied slightly for the different aspect ratios, the average over all aspect ratios was used for a particular basecoat color. Also from these values, we can calculate the critical stress intensity factor for mode I opening from the relationship:

$$K_{Ic} = \sqrt{G\,E}$$

(6)

In Table 4, the values of G, q_∞ , and K_{Ic} are collected for the four basecoat colors. Again, these numbers were obtained using the solution from mixed boundary conditions (case 4).

Table 4 - LIDS adhesion values for basecoat-clearcoat adhesion.

Basecoat color	G (J/m^2)	q_∞ (MPa)	K_{Ic} (MPa m$^{1/2}$)
Black	193 ±42	175±19	0.76±0.09
White	50 ±8	81±10	0.39±0.05
Green	101 ±15	102±12	0.55±0.06
Red	204 ±25	150±17	0.78±0.09

We are currently pursuing a numerical approach to the mechanical deformation and fracture in this system. Shear yielding would be manifested by a nonlinear curvature-pulse energy relationship, which is not seen in this data. While the ablated region does heat up considerably, the time scale of the experiment is too fast for thermal diffusion to occur.

CONCLUSIONS

We have demonstrated how the LIDS technique may be used to derive the practical work of adhesion for systems consisting of a transparent polymer coated onto an opaque polymer. The initial adhesion between an automotive clear coat and four different colored base coats has been measured, and we find that the red and black colors have the strongest initial adhesion, while the white has the weakest, and the green intermediate. Although approximate analytic models were used, the derived values of the adhesion parameter could be bounded by these models, and the results are qualitatively unchanged.

REFERENCES

1. K. L. Mittal, Ed., <u>Adhesion Measurement of Films and Coatings</u> (VSP, Utrecht, 1995).
2. G. P. Anderson, S. J. Bennett, K. L. DeVries, <u>Analysis and Testing of Adhesive Bonds</u> (Academic, New York, 1977).
3. R. Lambourne, Ed., <u>Paint and Surface Coatings: Theory and Practice</u> (Ellis Horwood, Ltd., West Sussex, England, 1987).
4. H. Dannenberg, J. Applied Polymer Science, **5**, 124 (1961).
5. M. L. Williams, J. Applied Polymer Science, **13**, 29 (1969).
6. S.J. Bennett, K. L. DeVries, M. L. Williams, Int. Journ. of Fracture, **10**, 33 (1974).
7. E. H. Andrews, A. Stevenson, Journ. Mat. Sci., **13** , 1680 (1978).
8. M. G. Allen, S. D. Senturia, J. Adhesion, **25**, 303 (1988).
9. Y. Chang, Y. Lai, D. A. Dillard, J. Adhesion, **27**, 197 (1989).
10. A. A. Griffith, Phil. Trans. Royal Soc., **221** (1921) 163-198.
11. A. E. Siegman, <u>Lasers</u> (University Science Books, Mill Valley, CA 1986).
12. S. Timoshenko, J. N. Goodier, <u>Theory of Elasticity</u> (McGraw-Hill, New York, 1970).
13. S. Timoshenko, S. Woinowsky-Krieger, <u>Theory of Plates and Shells</u> (McGraw-Hill, New York, 1959), p. 74.
14. I. N. Sneddon, M. Lowengrub, <u>Crack Problems in the Classical Theory of Elasticity</u> (Wiley, New York, 1969) p. 134.

MEASURING POLYMER MICROSTRUCTURE USING SPATIALLY-RESOLVED EELS IN THE STEM

K. Siangchaew, D. Arayasantiparb, and M. Libera
Department of Materials Science and Engineering
Stevens Institute of Technology, Hoboken, N.J. 07030

ABSTRACT

Image contrast for the examination of multiphase polymers in the transmission electron microscope (TEM) usually requires differential staining by a heavy element (Os, Ru, U). Staining methods have provided a wealth of microstructural information in polymers, but there are situations, particularly where high resolution is needed, where staining is undesirable or impossible. This research has collected microstructural information from multiphase polymers without heavy-element stains using spatially-resolved electron energy-loss spectroscopy (EELS) in a scanning transmission electron microscope (STEM). The technique is known as spectrum imaging. The principal problems facing spectrum imaging in polymer applications are: (1) the identification of spectral fingerprints distinguishing different polymer phases; and (2) the extraction of meaningful microstructural data from large data sets where the signal is weak due to instrumentation and materials constraints. This paper describes applications of spectrum imaging to PE/PS and HDPE/Nylon 6 blends. The aim is to identify adequate spectral features and establish data acquisition and extraction procedures for applications to general, unstained, multiphase polymers.

INTRODUCTION

Traditionally, multiphase polymers must be differentially stained when they are to be observed in the transmission electron microscope (TEM) [1]. Differential staining is necessary in order to provide a contrast mechanism by which various polymer phases can be distinguished. A differentially-stained polymer carries such information as the size, shape, and distribution of the various phases within the microstructure. However, differential stains are not available for all multiphase polymer systems [2], and stains can often lead to artifacts in the microstructure [3], particularly in the area of interphase boundaries. These factors can undermine the accurate and precise measurement of microstructures in multiphase polymers.

Ideally, microstructural information should be obtained from a specimen in its natural unstained state. This would avoid artefacts due to the staining process and would enable a wide range of polymer systems to be studied including ones for which differential stains are not available. In addition, one would prefer to collect both structural and chemical information with high spatial resolution. A combination of STEM and EELS in the technique of spectrum imaging [4,5] satisfies both of these criteria. This paper presents a description of the spectrum imaging technique along with its application to two homopolymer blends of high density polyethylene (HDPE) / Nylon 6 and polythylene (PE) / polystyrene (PS).

EXPERIMENTAL PROCEDURE

Inelastic scattering involves energy transfer from an incident high-energy electron to atomic electrons in a specimen. The energy transfer associated with valence-shell and core-shell electron

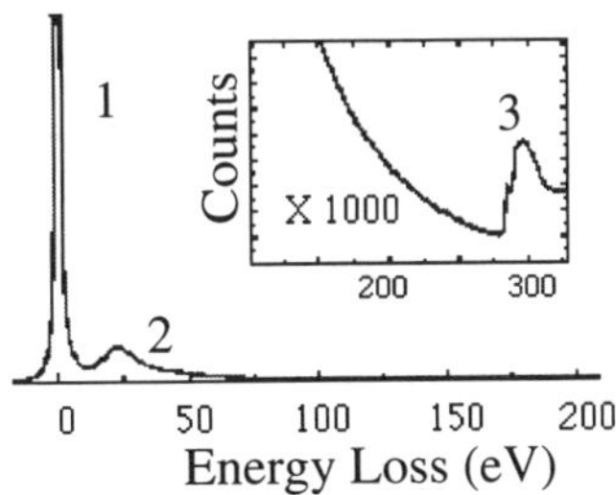

Fig 1. Three main features in a typical electron energy-loss spectrum: (1) The zero-loss peak accounts for those incident electrons which did not suffer inelastic scattering; (2) energy losses in the range of 0-50 eV arise from the excitation of the valence shell electrons; (3) by amplifying the signal in the range of >100 eV core-shell edges are seen which measure the binding energy of the core-shell electrons and carry both compositional and chemical information

Mat. Res. Soc. Symp. Proc. Vol. 461 ©1997 Materials Research Society

excitations produces distinctive spectroscopic features depending on the characteristic properties of the material. In general, these excitations provide information concerning both the atom species and interatomic bonding. The energy of electrons inelastically scattered by a specimen can be measured with EELS. The physics and application of EELS are described by Egerton [6]. Figure 1 illustrates the general features and information contained in a typical EELS spectrum.

Spatial resolution at nanometer length scales can be achieved by focusing an incident electron beam into a fine probe with a Gaussian intensity distribution. Spectroscopic image data can be collected by digitally rastering this focused probe across a specimen such that an entire energy-loss spectrum is collected at each x-y pixel to produce a 3-D data set (Fig. 2). The digital raster can be performed with controllable pixel resolution and dwell time to optimize the spatial resolution and the spectroscopic signal, respectively.

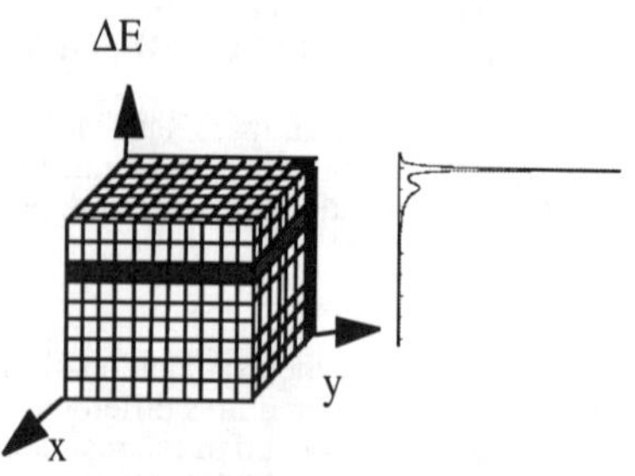

Fig. 2 A 3-D data set is formed by collecting a complete energy-loss spectrum at each position of the probe.

This research collected spectrum images using a Philips CM20 FEG TEM/STEM with a Schottky field-emission electron source. The energy spread of the electron beam is 1 eV (FWHM). The digital raster and coupling to a Gatan 666 PEELS spectrometer used an EMiSCAN data acquisition system.

The following steps were taken to obtain spectrum images. Specimens of melt-mixed homopolymer blends were cryotomed to approximately 70-90 nm thickness. Two blended polymer systems were studied: HDPE(matrix)/nylon 6 and PE(matrix)/PS. Specimens were cooled to -130 ^{0}C during data acquisition in order to minimize the effects of specimen-born contamination and radiation damage. Individual low-loss and core-loss test spectra were collected from specific point locations within each polymer phase. Background fitting parameters and energy windows were selected based on these test spectra to best resolve the spectroscopic fingerprints following Leapman [7] and Hunt [8]. Each spectrum within the measured 3-D data set was then subjected to the optimized data-processing parameters to form spectrum images.

RESULTS AND DISCUSSION

<u>Melted-mixed HDPE / Nylon 6 homopolymer blend</u>

Figure 3 shows typical core-loss spectra from Nylon 6 (fig. 3a) and HDPE (fig. 3b). The Nylon 6 spectrum is characterized by a large edge at 284 eV energy loss which corresponds to the carbon K-shell electron excitation. There are two additional edges at 401 and 530 eV. These correspond to nitrogen and oxygen K-edge

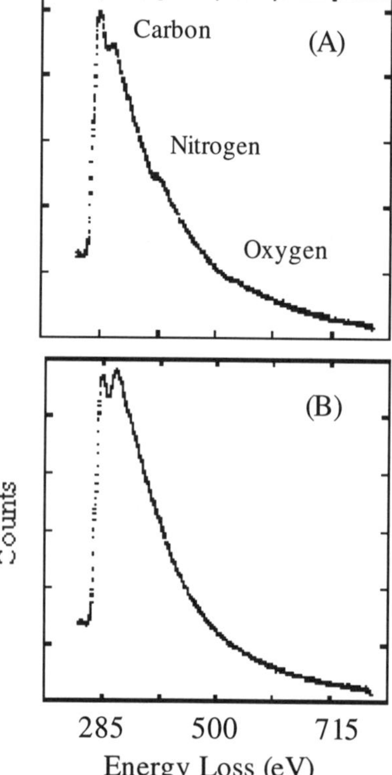

Fig. 3 (A) Core-loss spectrum from pure nylon 6. (B) Core-loss spectrum from pure HDPE. Nitrogen and oxygen edges are evident in nylon 6 but not HDPE

excitations, respectively. Nitrogen and oxygen come from the amide group of the nylon. The relatively large carbon edge indicates an abundance of carbon versus nitrogen and oxygen in this material. In a similar energy-loss range, a spectrum from HDPE shows only the C-K edge (fig. 3b). The carbon post-edge structure (at ~293 eV) corresponds to the 1s -> σ^* transition. The fact that it is more prominent in the HDPE than in the nylon suggests that a higher concentration of σ bonding is present in HDPE. Thus the presence of nitrogen and oxygen edges as well as differences between the carbon edges can be used as spectral fingerprints to differentiate the nylon and the HDPE. Figure 4 shows a typical spectroscopic image using the nitrogen signal to differentiate the nylon and the HDPE. The pre-edge nitrogen signal was modeled by using a power-law fit to the background [4]. Careful background modeling and subtraction is essential in order to confidently recover the signal. This image shows that the Nylon/HDPE interface is rough in this particular interfacial area. The amount of nitrogen present at the interphase can be derived from the contrast level of this elemental map.

   The highest spatial resolution can be achieved in 1-D line scans (spectrum profile) rather than in 2-D area scans (spectrum image). A line scan reduces the total acquistion time as well as the degrading effects of specimen drift and electronic fluctuations. Figure 5 shows a typical result across an HDPE/nylon interface. The annular-dark-field (ADF) STEM profile (fig. 5a) plots the intensity of electrons scattered to high angles as a function of position. It, as well as the carbon profile (fig. 5b), show a transition between the two phases, but the nylon side of the interphase boundary displays a region of lower signal extending for several nanometers prior to assuming a relatively constant value characteristic of the bulk nylon. The transition is much more abrupt in both the nitrogen (fig. 5c) and oxygen (fig. 5d) profiles. The nitrogen and oxygen peaks contain many fewer counts than the carbon peak, hence the corresponding spectrum profiles are noisier. Neither show a reduced intensity on

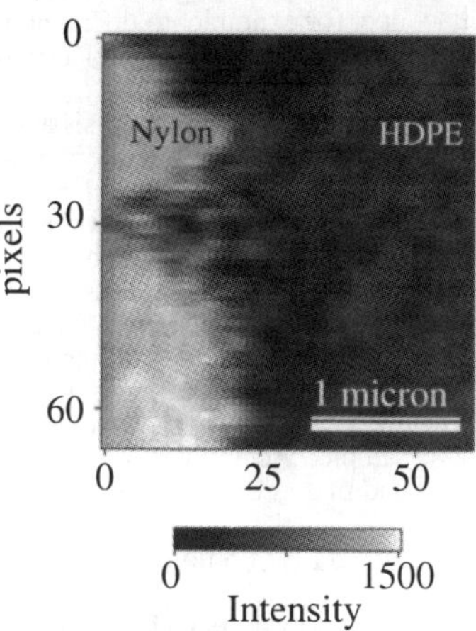

Fig. 4 Nitrogen map in HDPE/Nylon blend. Light contrast represents area with higher concentration of nitrogen.

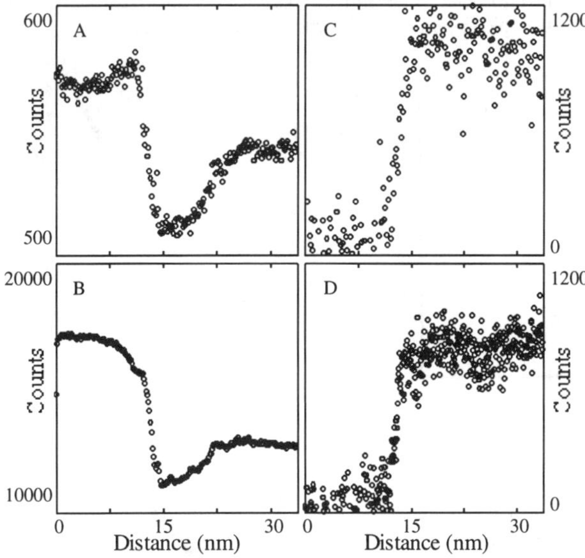

Fig. 5 Spectroscopic signal from a line scan across the HDPE/Nylon 6 interphase. (A) ADF, (B) Carbon, (C) Nitrogen and (D) Oxygen map

the nylon side of the interface like the ADF or C signals, however. Further work is needed to determine whether the signal-depleted region is characteristic of HDPE/nylon interfaces or is an artifact due, for example, to different elastic response of the two phases to microtomy. This feature is unlikely to be due to electron-beam damage, since the dose used for each pixel in the profile was constant.

Quatitative EELS analysis and the effects of electron beam damage on the nitrogen K-edge in nylon and chlorinated polymers have been previously investigated by Briber and Khoury [9]. One of the primary concerns with radiation damage from the experimental standpoint is the ability to obtain good signal statistics before the signal of interest is changed or disappear. Minimization of irradiation usually means a poor signal-to-noise ratio. During this investigation it has been observed that a good signal statistics of weak edges can be obtained when the specimen was cooled. Although it is uncertain whether cooling the specimen minimizes the effect of radiation damage, it is observed that cooling the increases the signal-to-noise ratio of the N-K and O-K edges. Such improvement can in part be attributed to the minimization of the mobility of the hydrocarbon contaminants which are often seen to aggregate at the region where the electron beam hits the sample. Accumulation of contaminants locally increases the thickness of the region of interest and makes extraction of a weak signal difficult [6].

<u>Melted-mixed PE / PS homopolymer blend</u>

Since polystyrene (PS) and polyethylene are pure hydrocarbon polymers, spectrum imaging based on core-loss excitations is less attractive than in a case like HDPE/nylon. There are, however, distinguishing spectral features due to valence-electron excitations. Figure 6a shows typical low-loss spectra from PS and PE. Both materials display a broad peak circa 20 eV corresponding to the σ–σ* transition [10]. σ bonds are present in both PS and PE, and this spectral feature does not provide a good distinguishing fingerprint. The conjugated π bonding associated with the aromatic ring in PS produces a peak at 7 eV energy loss corresponding to the π–π* transition (fig. 6b) [8,10-11]. This peak is absent from the PE spectrum (fig. 6c). This 7 eV peak can thus be used as spectroscopic fingerprint to characterize the microstructure of this blend.

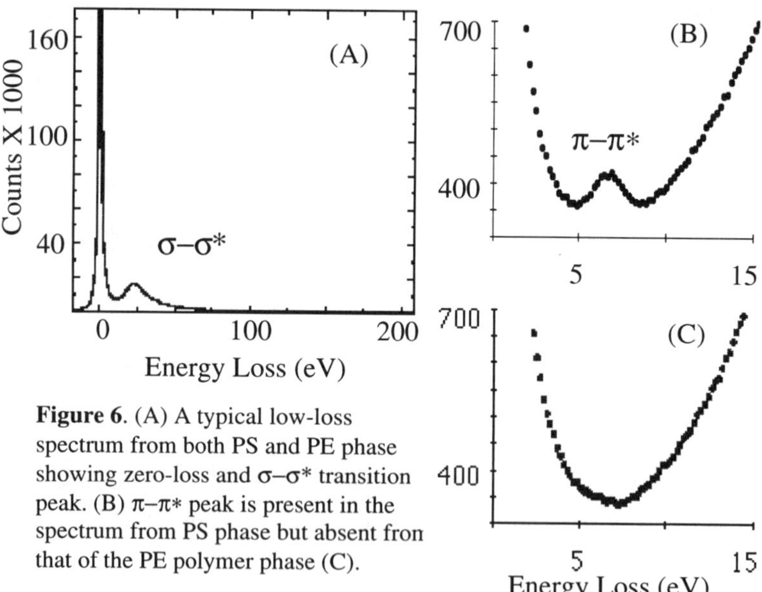

Figure 6. (A) A typical low-loss spectrum from both PS and PE phase showing zero-loss and σ–σ* transition peak. (B) π–π* peak is present in the spectrum from PS phase but absent fron that of the PE polymer phase (C).

Figure 7 shows the results of a PE/PS spectrum image which maps the $\pi-\pi^*$ peak intensity as a function of position. The $\pi-\pi^*$ peak was extracted by modeling the background with a 2nd order polynomial fit to both pre-edge and post-edge data adjacent to the 7 eV peak. The 3-D data set was collected at a 1 sec dwell time with a 10 nm pixel spacing. The probe current was 0.007 nA and the resulting dosage per pixel was approximately 10 Coulomb/cm^2. This amount of dose is relatively high [12] but meaningful signal is still present in these spectra.

Since the inelastic scattering cross-section of the low-loss signals is governed by the material's dielectric function, further data processing of low-loss data such as that in figure 6 may provide yet another avenue with which to quantify the nature of polymer-polymer interfaces. The fact that the complex dielectric properties of a material can be recovered from low-loss EELS data by a Kramers-Kronig analysis [6] is well established. Collecting such data using a focused-probe STEM instrument enables these dielectric properties to be mapped as a function of position. Recent developments in the application of the London dispersion transform to low-loss EELS data [13] enables the calculation of the dispersion force across an interphase boundary. Such analysis in a heteropolymer system may enable one to measure the spatially-varying secondary forces which lie at the heart of polymer-polymer adhesion.

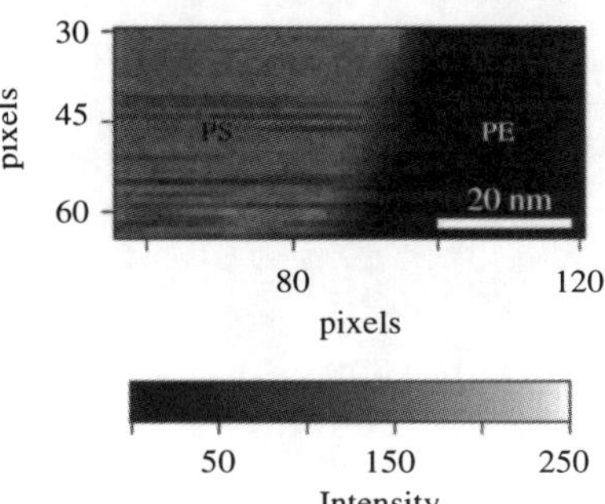

Fig. 7 Map of $\pi-\pi^*$ bond distribution

CONCLUSION

Spectrum imaging provides a new way to study the microstructure of polymer blends without heavy-element stains. Meaningful spectroscopic signals can be identified in both the low-loss and core-loss regions. Continued development of this method will make it particularly attractive in studies of polymer-polymer interfaces where stains do not necessarily reflect the intrinsic interfacial structure with good integrity as well as in lower-resolution applications involving multiphase polymers where differential stains are not available.

REFERENCES

1. L. Sawyer and D. Grubb, *Polymer Microscopy*, Chapman and Hall, N.Y. 1987.
2. J.S. Trent, J. Scheinbeim and P.R. Couchman, Macromolecules **16**, (1983) pp. 589.
3. R.W. Horobin, J. Microscopy **131** pt.2 (1982) pp. 173.
4. J.A. Hunt and D.B. Williams, Ultramicroscopy **38** (1991) pp. 47.
5. C. Jeanguillaume and C. Colliex, Ultramicroscopy **28** (1989) 252.
6. R.F. Egerton, *Electron energy-loss spectroscopy in the electron microscope*, Plenum Press, N.Y., 1989.
7. R. Leapman, *Transmission electron energy-loss spectroscopy in material science*, ed. M.M. Disko, C.C. Ahn, and B. Fultz, TMS, (1992) pp. 47-83.
8. J.A. Hunt, M.M. Disko, S.K. Behal, and R.D. Leapman, Ultramicrocopy **58** (1995) pp. 55.
9. R.M. Briber and F. Khoury, J. Polymer Science B: Physics **26** (1988) pp. 621
10. M.J. Issacson, Chem. Phys. **56** (1972) pp. 1803.
11. A.P. More, A.J. McGibbon, and D.W. McComb, Inst. Phys. Conf. **119** Section 8 (1991) pp. 353.
12. L. Reimer, *Transmission electron microscopy 3rd ed.*, Springer-Verlag, N.Y., 1993
13. R. French, R.M. Cannon, L.K. DeNoyer, and Y.-M. Chiang, Solid State Ionics **75** (1996) pp. 13

ULTRA-HIGH RESOLUTION SMALL-ANGLE NEUTRON SCATTERING INVESTIGATIONS OF LIQUID-LIQUID PHASE SEPARATION IN LINEAR LOW-DENSITY POLYETHYLENE

M. M. Agamalian (a), R. G. Alamo (b), J. D. Londono (a), L. Mandelkern (c), S. Spooner (a), F. C. Stehling (d) and G. D. Wignall (a).

(a) Oak Ridge National Laboratory, Oak Ridge, TN 37831-6393; (b) Florida Agricultural and Mechanical University and Florida State University College of Engineering, Department of Chemical Engineering, Tallahassee, FL 32310-6046; (c) Institute of Molecular Biophysics, Florida State University, Tallahassee, FL 32306-3015; (d) Plastics Technology Division, Exxon Chemical Company, Baytown, TX 77520 (current address: 214 Post Oak, Baytown, TX 77520).

ABSTRACT

Previous small-angle neutron scattering (SANS) studies [1] of heterogeneous ethylene-hexene linear low-density polyethylene (LLDPE) copolymers have confirmed the existence of a dispersed minority phase (volume fraction $\phi \sim 10^{-2}$) consisting of highly branched, amorphous material. However, these experiments were conducted via a pinhole SANS spectrometer with an upper resolution limit $\sim 10^3$ Å, whereas microscopy indicates that the dimensions of the disperse phase extend to the μm-range. We have therefore complemented these investigations via a Bonse-Hart ultra-small angle neutron scattering (USANS) instrument which increases the instrumental resolution in reciprocal space by a factor ~ 100, and thus particle size up to 30 μm can be resolved. The sensitivity of the USANS camera has recently been increased by two orders of magnitude by using the modified channel cut crystals [2], and the performance is therefore comparable to the best x-ray Bonse-Hart cameras. Xylene extraction removes the highly branched molecules and hence the volume fraction of the disperse phase is higher ($\phi \sim 0.3$) in the extracted material.

INTRODUCTION

It is well established [1, 3-9] that small-angle neutron scattering (SANS) can be used to determine the melt compatibility of mixtures of linear and branched polyolefins, including high density (HD), low density (LD) and linear low density (LLD) polyethylenes. HDPE is the most crystalline form of polyethylene (PE) because the chains contain very little branching. LDPE contains some short chain branches (typically 1-3 per 100 backbone carbon atoms), as well as long chain branches (typically 0.1-0.3 per 100 backbone carbon atoms). Linear low density polyethylene (LLDPE) is produced by catalytically co-polymerizing ethylene with an alpha-olefin such as hexene. LLDPEs can have a wide range of branch contents, depending primarily on the type of catalyst and concentration of added comonomer. SANS data [3] indicate that for HDPE/LDPE blends with molecular weights $\sim 10^5$, the melt is homogenous, after proper accounting for H/D isotope effects [4]. Similarly, mixtures of HDPE and LLDPE are homogenous in the melt [5-7] when the branch content is low (i.e. < 3 br./100 backbone carbons). However, when the branch content is higher (> 10 br./100 C), the blends phase separate.

In previous SANS experiments [5-7], the LLDPE's were simulated by hydrogenated (or deuterated) polybutadienes, because these materials may be prepared as nearly monodisperse ($M_w/M_n < 1.1$) molecules with a homogenous branch distribution within each chain. Thus, these studies were not affected by polydispersity effects, either in the branch content or molecular weight. However, for typical LLDPEs prepared with heterogeneous-type Ziegler-Natta catalysts, it is well known that the multi-site nature of catalysts typically used in the preparation of such materials leads to a wide distribution of chain compositions [10], with the low molecular weight chains exhibiting the most branching [11,12]. Heterogeneous LLDPE may therefore be thought of as a "blend" of different species and when the composition distribution is broad enough the multicomponent system can, in principle, phase separate. Thus, even when the average branch content is low (e.g. 1-3 br./100 C), there may be a small fraction of highly branched chains

Mat. Res. Soc. Symp. Proc. Vol. 461 © 1997 Materials Research Society

(e.g. > 10 br./100 C), which are incompatible with the lightly branched molecules and will thus phase separate. To our knowledge, this was first suggested by Mirabella and co-workers [12,13], based on scanning electron microscopy (SEM) investigations of the solid state. Subsequently, Nesarikar and Crist [14] performed a thermodynamic calculation of the equilibrium melt state using a distribution of chain branching estimated from temperature rising elution fractionation [13,14]. The calculations predicted a second phase consisting of highly branched amorphous material, and the volume fraction ($\phi \sim 10^{-2}$) was in reasonable agreement with the SEM findings [13,14].

Previous SANS studies have examined the structure of the liquid directly, though because the linear and branched molecules have virtually the same scattering power, there is no contrast between the phases and the melt structure would not be manifested via x-ray or light scattering. A fraction of deuterated linear polymer (20%) was therefore added prior to examining the blends by neutron scattering. Based on previous studies, the linear material should be incompatible with the minority phase, but should mix homogeneously with the predominantly low branched matrix. Thus, the addition of HDPE-D should provide SANS contrast between the phases, without perturbing the predicted two-phase morphology.

SAMPLE PREPARATION AND CHARACTERIZATION

A commercial ethylene-hexene-1 (EH) copolymer, prepared with a Ziegler-Natta catalyst, was dissolved in p-xylene (1 wt/v%) at 170°C during approximately 2 hours under N_2 purge. The solution was allowed to cool to room temperature, and the high molecular weight, lightly branched material crystallized out of the solution. These crystals were filtered, washed with methanol and dried under vacuum. The lower molecular weight, more highly branched component (~ 8%) was recovered from the filtrate by evaporating the solvent and drying at 50°C in vacuum.

The molecular weight and butyl branching composition were analyzed following conventional GPC and ^{13}C NMR techniques. The melting temperatures and degrees of crystallinity were obtained by differential scanning calorimetry (DSC) on rapidly quenched samples. These data are listed in Table I, along with those of the deuterated linear HDPE-D.

Table I. Characterization of Original Polydisperse Ethylene-Hexene-1 Copolymer and its Fractions

Sample	$10^{-3}M_w$	$10^{-3}M_n$	Br./100 C	T_m (°C)	Crystallinity
Original EH	120	30	1.7	124	0.29
High MW fraction	124	37	1.4	123.5	0.27
Low MW fraction	53	8.5	3.1	55-65	0.05
Deuterated HDPE-D	139	23	-	123.1	0.42

Mixtures of the original EH copolymer and the two fractions (each with 20% deuterated linear polyethylene) were prepared by dissolving the components in o-dichlorobenzene. The solution was rapidly quenched into chilled methanol, and after washing with methanol, the samples were dried overnight in vacuum oven. Disks ~ 1 mm thick were obtained via compression molding in a Carver Press at 190°C and quenching into ice-water. Single melting peaks and degrees of crystallinity similar to he protonated species found for the 80/20 blends of the original and high molecular weight fraction were consistent with a high amount of co-crystallization of both species. However, double peaks with melting temperatures corresponding to the initial components were found in the blend of the deuterated HDPE with the low molecular weight fraction.

SANS AND USANS INSTRUMENTS

The original data [1] were collected on the W. C. Koehler 30m SANS facility [15] at the Oak Ridge National Laboratory with a minimum value of the scattering vector of $Q = 4\pi\lambda^{-1}\sin\theta \approx 4\times10^{-3}$ Å^{-1}, where 2θ is the angle of scatter. After correction for instrumental backgrounds, the net intensities were converted to an absolute ($\pm 3\%$) differential scattering cross section per unit sample volume [$d\Sigma/d\Omega(Q)$ in units of cm^{-1}] by comparison with pre-calibrated secondary standards [16]. Details of transmission measurements, data collection and correction procedures for instrumental and incoherent backgrounds have been given previously [17,18].

The USANS measurements were carried out on the Bonse-Hart Double-Crystal Diffractometer equipped by triple-bounce Si(111) channel-cut monochromator/analyzer crystals. These have been modified by adding a cadmium absorber in the middle of the long wall [2] to prevent propagation of radiation inside the transparent crystals (fig.1). This reduces the wings of the rocking curve, improving the ultimate sensitivity by two orders of magnitude (fig.2). The sensitivity of the ORNL USANS camera is now similar to the best ultra-small angle x-ray instruments[19, 20] and the Q-range ($2\times10^{-5} < Q < 5\times10^{-3}$ Å^{-1}) allows the determination of inhomogenieties with the dimensions up $\sim$ 30 μm. The wavelength of the primary beam is $<\lambda> = 2.59$ Å, and the cross-section of the beam is 2cm x 4cm. Absolute calibration was accomplished via a dilute D_2O solution of monodisperse polystyrene latexes of the diameter 5 μm as a standard sample.

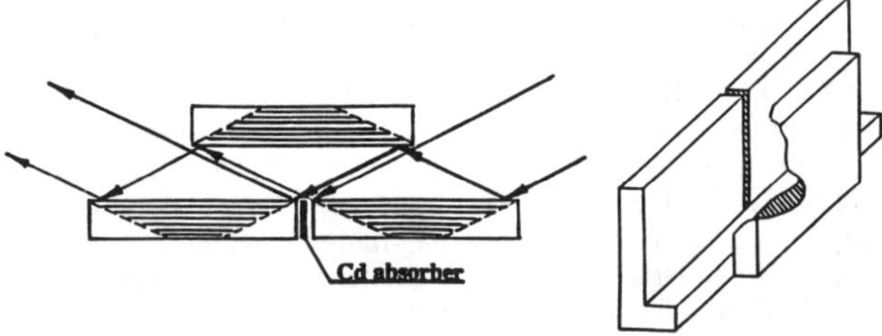

Fig. 1. Modified channel-cut crystal with additional groove for Cd absorber and optical diagram of triple-bounce reflection.

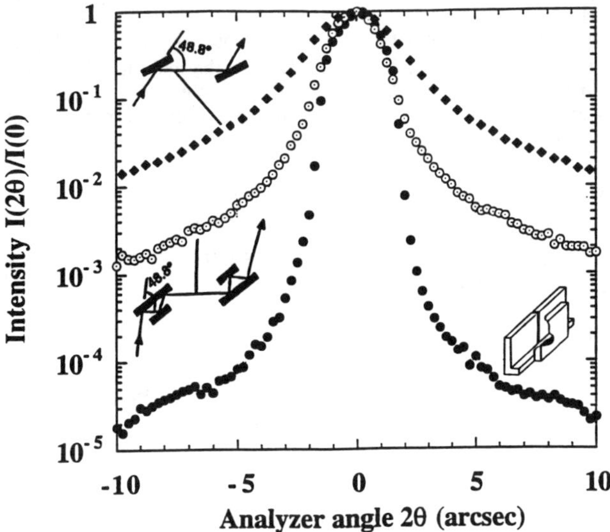

Fig. 2. Comparison of rocking curves for single-/single-bounce scheme, triple-/triple-bounce scheme and that with modified channel-cut crystals.

RESULTS AND DISCUSSION

Figure (3) shows a log-log plot of the cross section in the melt (T = 160°C) for the original ethylene-hexene LLDPE, along with the same sample after xylene extraction. Both samples blended with 20 wt.% linear D-HDPE (to provide SANS contrast). The existence of multiple phases should be reflected in the SANS data which clearly exhibit two regions where the scattering curve varies as $Q^{-3.8}$ and $Q^{-1.7}$ for the lowest and highest Q-values respectively. These values are close to the Porod limit (Q^{-4}), expected for separate phases with sharp boundaries [21], and to the Gaussian (Debye) coil limit (Q^{-2}), expected for individual random coil molecules [22, 23]. It may be seen that the extraction procedure, which is expected to remove the highly branched and low molecular weight components, only changes the scattering at the lowest Q-values, and that the data superimpose over most of the range ($Q > 0.01$ Å^{-1}). Xylene extraction effectively removes the component, which varies as $\sim Q^{-4}$.

The original LLDPE has an average branch content [Table (1)] of 1.4 mole % butyl branches (or 1.4 br./100 C), though a small fraction of the distribution will be highly branched, and chains with > 10br./100 C should phase separate from the lightly branched matrix. As the disperse phase (volume fraction, $\phi \sim 10^{-2}$) is incompatible with lightly branched chains, it is to be expected that it is also incompatible with the linear (deuterated) material which is added to provide SANS contrast. If these domains consisted of particles with relatively sharp boundaries, this would naturally give rise to the Q^{-4} variation observed in the low-Q (Porod) limit. Conversely, it is well known [1,3-5] that lightly branched material is compatible with linear HDPE-D, so the matrix should consist of an homogenous mixture of HDPE-D and LLDPE-H chains, and such a morphology would also give rise to the observed Q^{-2} variation over most of the Q-range. Thus the 2-phase hypothesis [13,14], accounts qualitatively for the general features of the scattering. The volume fraction of the disperse phase can be estimated [24] from the SANS invariant:

$$Q_0 = \int Q^2 d\Sigma/d\Omega(Q)dQ = 2\pi^2\phi_1\phi_2(\rho_{1n} - \rho_{2n})^2 \qquad (1)$$

where ϕ_1, ϕ_2 and ρ_{1n}, ρ_{2n} are the volume fractions and neutron scattering length densities of the two phases respectively.

The scattering from the minority (disperse) phase is manifested [fig. (3)] below $Q \sim 0.01$Å^{-1}, where it is superimposed on the cross section of the majority of the sample consisting of a homogenous mixture of HDPE-D and LLDPE-H. We have removed this coherent "background" by subtracting the cross section of the xylene-extracted material, which superimposes on the scattering of the original blend for $Q > 0.1$ Å^{-1}.

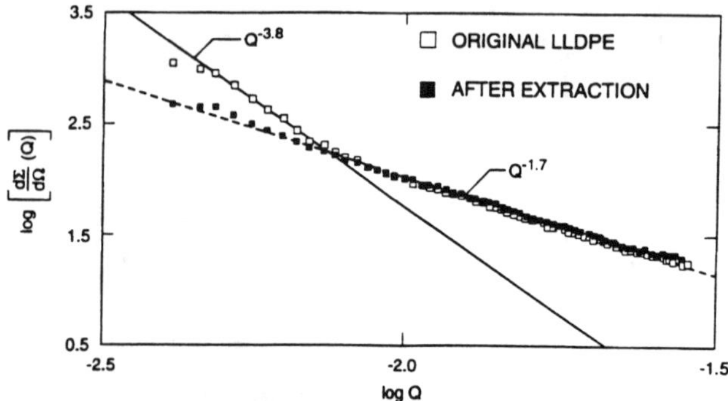

Fig. 3. Log-log plot for compositionally heterogeneous etylene-hexene copolymer before and after xylene extraction (both samples blended with 20 wt.% HDPE-D to provide SANS contrast).

In order to estimate the portion of the area below the minimum observable $Q_{min} \approx 0.004$ Å^{-1}, we assumed initially [1] that the missing low-Q data follow the Guinier approximation, to give an integrated area of $Q_o \approx 3.3 \times 10^{-5}$ cm^{-1}Å^{-3}, which lead to an estimate of $\phi \approx 0.021$ for the original sample via the equation (1).

The xylene-extract formed 8% of the original polymer 2% of which was phase separated, thus we might expect that the volume fraction of the disperse phase in the xylene extract would be $\phi \approx 0.20$, when blended with 20% HDPE-D. Invariant analysis leads to $\phi \approx 0.20$, after correction for the coherent "background" from scattering from the homogeneously mixed material. The close agreement between these estimates is probably fortuitous in view of the fact that the scattering from particles with dimensions > 2000 Å appears at Q-values < 0.004 Å^{-1}, which is beyond the resolution limit of the experiment. However, the USANS measurement allows us to fill in the missing portion of the data and improve the estimate of ϕ. Fig. 4 and 5 show the overlap of the pinhole SANS and USANS data and also the Kratky plot of the combined data. The area under the curve in fig. 5 is the invariant Q_o [equation (1)], which leads to a volume fraction of the disperse phase of $\phi = 0.30$. This may be compared with the previous determination ($\phi \approx 0.20$) derived from the pinhole SANS data with $Q_{min} \approx 4 \times 10^{-3}$ Å^{-1}. Thus USANS substantially improves the estimate of ϕ by filling in the previously inaccessible portion of the SANS invariant.

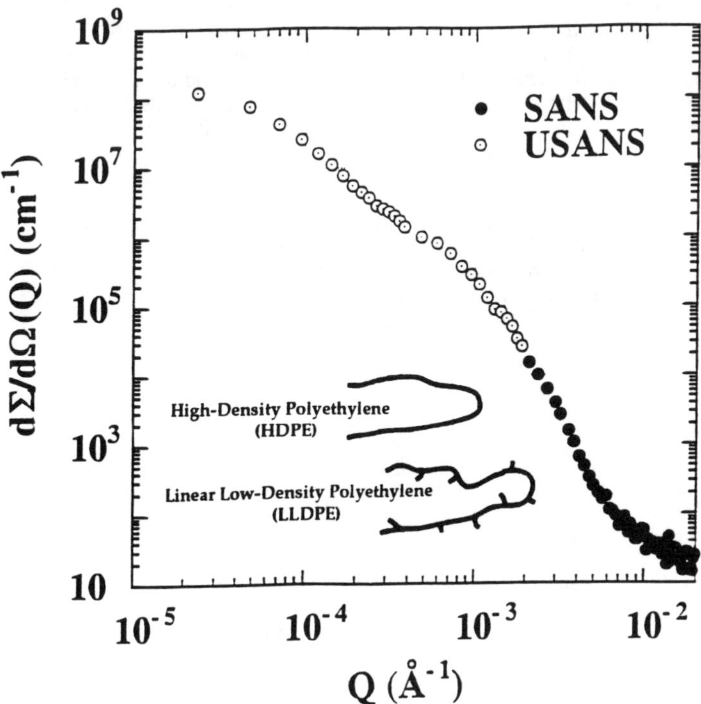

Fig. 4. Overlap of SANS and USANS data obtained for 20% linear HDPE-D blended to branched LLDPE-H at T = 160°C.

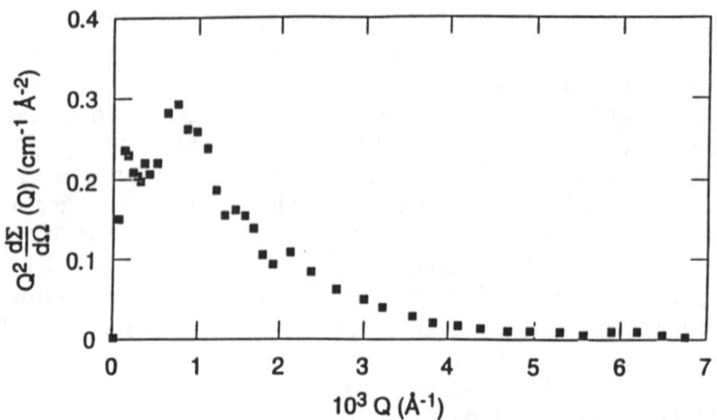

Fig. 5. Kratky plot of SANS/USANS data (T=160ºC) for linear HDPE-D in branched LLDPE-H.

ACKNOWLEDGEMENTS

The research at Oak Ridge was supported by the Division of Materials Sciences, U. S. Department of Energy under Contract No. DE-AC05-96OR22464 with Lockheed Martin Energy Research Corporation and in part by appointment to the Oak Ridge National Laboratory postdoctoral research associates program administered jointly by the Oak Ridge National Laboratory, the National Institute of Standards and Technology and the Oak Ridge Institute for Science and Education. The work at Florida State was supported by the National Science Foundation Polymer Program (DMR 94-19508), whose aid is gratefully acknowledged.

REFERENCES

1.   Wignall, G.D., et al., Macromolecules, 1996,29, 5332.
2.   Agamalian, M., Wignall, G.D., & Triolo, R., J.Appl.Cryst, in the press.
3.   Alamo, R.G., et al., Macromolecules, 1994, 27, 411.
4.   Londono, J.D., et al., Macromolecules, 1994, 27, 2864.
5.   Krishnamoorti, R, et al., Macromolecules, 1994, 27, 3073.
6.   Graessley, W.W., et al., Macromolecules 1994, 26, 1137; 1994, 27, 2574, 3073, 3896.
7.   Nicholson, J.C., et al., Macromolecules, 1991, 24, 5665.
8.   Bates, F.S., et al., Macromolecules, 1992, 25, 5547.
9.   Tashiro, K., et al., Macromolecules, 1995, 28, 8484.
10. Karbashewski, E., et al., Appl. Polym. Sci., 1992, 44, 425.
11. P. Schouteren, et al., Polymer, 1987, 28, 2099.
12. Mirabella, F.M., & Ford, E.A., J. Polym. Sci., Polym., Phys., Ed., 1987, 25, 777.
13. Mirabella, F.M., et al., J. Polym. Sci., 1988, B26, 1995.
14. Nesarikar, A., & Crist, B., J. Polym. Sci., 1994, B32, 641.
15. Koehler, W.C., Physica (Utrecht) 1986, 137B, 320.
16. Wignall, G.D., & Bates, F.S., J. Appl. Cryst. 1986, 20, 28.
17. Londono, J.D., et al., Mater. Res. Symp. Proc., in the press.
18. Wignall, G.D., et al., Macromolecules, 1995, 28, 3156.
19. Bonse, U., & Hart, M., Appl.Phys.Lett., 1965, 7, 238.
20. Matsuoka, H., & Ise, N., Chemtracs-Macromolecular Chemistry, 1993, 4, 59.
21. Porod, D., Small-Angle X-Ray Scattering, Academic Press, 1982, 30.
22. Debye, P., J.Appl.Phys., 1944, 15, 338.
23. Flory, P.J., Principles of Polymer Chemistry, Cornell University Press, 1969, 295.
24. Wignall, G.D., "Neutron and X-Ray Scattering", Polymer Properties Handbook,
     ed. J.E. Mark, American Institute of Physics, in the press.

Morphology Characterization in Multicomponent Polymer Systems using Scanning Probe Microscopy

Ravi Viswanathan, Jing Tian
Raychem Corporation, Menlo Park, CA 94025

David W.M. Marr
Chemical Engineering and Petroleum Refining Department
Colorado School of Mines, Golden, CO 80401

Abstract
A newly available scanning probe microscopy, phase imaging, has been used to image polyethylene (PE) blends and composites. By providing not only a topographical map of the surface but also information on material properties, this technique has allowed ready identification of individual components within the mixtures investigated.

Introduction
Since its development, the field of atomic force microscopy (AFM) has undergone significant change. Initially touted as a tool to resolve atomic and molecular species across a surface [1,2], the AFM/STM has been incorporated into the larger field of scanning probe microscopy. The advantages of high-resolution topographic imaging through AFM/STM can now be supplemented by examining other interactions. Material properties of systems such as frictional forces [3], electrostatic [4,5], magnetic forces [6], and moduli of elasticity [7,8] can now be mapped.

AFM techniques that rely solely on topographic mapping suffer from difficulties in interpretation of data and artifacts that can result from imaging. From a topographic map, it can be difficult to definitively state the location and composition of features. As such, analysis of multicomponent systems can be problematic. To overcome this difficulty, we have used a facile and nondestructive scanning probe microscopy, phase imaging, to distinguish components in polymer mixtures. The technique is an extension of the tapping mode AFM where the cantilever is excited into resonance oscillation with a piezoelectric driver and the oscillation amplitude used as a feedback signal to measure topographic variations of the sample. In phase imaging, the phase lag of the cantilever oscillation relative to the signal sent to the cantilever's piezo driver is simultaneously monitored and recorded. This lag is very sensitive to variations in material properties such as adhesion and viscoelasticity. Therefore, unlike very specific SPM techniques, differences in a number of material properties can give rise to phase image contrast. In addition, because phase imaging highlights edges and is unaffected by large-scale height differences, it provides for clearer observation of fine features [9]. Its ease of operation and the fact that a tapping mode image is simultaneously produced makes phase imaging an ideal tool for examining the multicomponent systems presented in this study.

We examine two systems here. The first is a blend of polyethylene and polydimethylsiloxane (PDMS). To tune both the domain size and interfacial properties in the system, a block copolymer of PE and PDMS was synthesized and utilized as a compatibilizer. Varying the system structure in such a fashion provides a test of the phase imaging capabilities. In addition, a mixture of polyethylene and carbon black was also prepared and used to test the phase imaging capability of tuning the contrast, either making polyethylene crystalline and amorphous regions visible or appear as of a single component.

Experimental
Blends of polyethylene (M_w = 66000, M_w/M_n = 2.0) and polydimethylsiloxane (M_w = 186000, M_w/M_n = 1.4) were prepared in a Brabender Plasti-Corder blender at a temperature 30°C above the polyethylene melting point. The polymer mixture was mixed at 50 rpm for 5 mins in a 30cc bowl. Blends including compatibilizer were prepared in a similar fashion: the

Mat. Res. Soc. Symp. Proc. Vol. 461 © 1997 Materials Research Society

desired amount of PE was mixed first with the compatibilizer for 2 mins followed by the addition of PDMS, and then the mixture blended for an additional 5 minutes. The blends unloaded from the mixer were compression molded into 1mm thick sheets at 160°C in a hydraulic press, and cooled in a cold press to room temperature. The composites of polyethylene and carbon black were prepared in a similar manner.

A commercial Digital Instruments Nanoscope Multimode AFM [10] with a Phase Extender Box was used for imaging the blends in Tapping and Phase modes. The films were imaged in air at ambient conditions. "Height" mode (tip-to-sample distance is adjusted by feedback to keep the force constant) was used for imaging with microfabricated single-crystal silicon tips (5nm radius of curvature). A 128μm X 128μm scanner was used for imaging.

Results & Discussion

Figure 1a shows a 20μm TMAFM image of a blend of polyethylene/PDMS (90/10 weight ratio). From this image it would appear that the lighter circular-shaped areas are the dispersed PDMS phase. This interpretation suffers from a number of ambiguities however; it can be difficult to clearly distinguish the two phases. By examining the corresponding phase image shown in Figure 1b, however, distinguishing different phases is readily accomplished. The image quality is far superior to that obtained through TMAFM, as SPM is extremely sensitive to phase lags caused by differences in material properties. The darker regions in the phase image correspond to the dispersed PDMS domains, with the surrounding lighter regions representing the PE matrix.

Though the phase image confirms many of the features observed in the TMAFM image, the utility of the technique can be verified by several subtle features. First, the phase image is able to resolve smaller-sized PDMS domains than appear in the topographic image. Also, some of the raised (lighter) features, initially thought to be PDMS regions, actually are shown to be clumps of PE. Hence, phase imaging can be a powerful complement to normal tapping mode AFM, and the two techniques can simultaneously map out the morphology and component behavior of blends and/or composite systems.

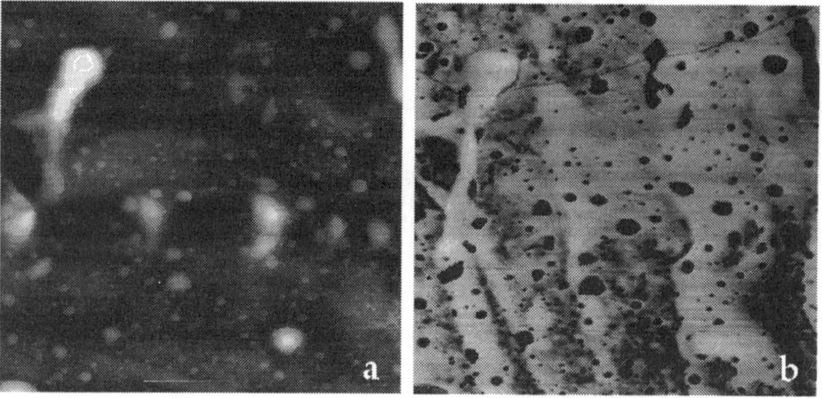

Figure 1: 15μm images of PE/PDMS blends using a) tapping mode b) phase mode.

To further demonstrate the utility of TMAFM/phase imaging in studying polymer blend morphology, we introduce a third component, a compatibilizer, and observe its effects on polymer blend microstructure. This compatibilizer, a copolymer of PE & PDMS, migrates to the interface during processing, mitigates interfacial energetics, and results in lower domain size and different interfacial structure. The effect of the compatibilizer can be quantified by comparing one-dimensional line scans taken across phase images of the with and without compatibilizer systems and the sharpness of the transitions in traversing from one component to another can be compared. Figures 2 and 3 show phase images and line scans of the systems with- and without-compatibilizer. In these figures, the phase units shown on the y-axis cannot be interpreted absolutely as they are not tied to any specific material property or effect; the line scans can only be compared relatively. In Figure 2 it can be seen that the line scan of the without-compatibilizer case shows abrupt transitions between PE and PDMS and deep wells in going from PE to PDMS. Figure 3, which represents the with-compatibilizer scenario, shows distinct differences, in particular, a more gradual transition between phases. From these line scans, one can estimate the transition (interfacial) width via SPM doubles in size when this particular compatibilizer is added.

The line scans from the two systems also show that the transition heights measured between phases are not as great with the addition of the compatibilizer. Though the exact cause of this is unclear, it does indicate that the physical properties of the two phases have become more similar with the addition of compatibilizer. One expects a certain miscibility of the copolymer in each phase which could certainly lead to this. Though it is clear from this that the properties are different in each case, it does emphasize the current non-quantitative nature of this technique.

Finally, in Figure 4 we show tapping and phase mode images of a polyethylene/carbon black composite. We illustrate here the usefulness of varying the drive amplitude in discerning various features within the composite. Under moderate conditions we obtain contrast between the carbon black and the polyethylene as if it were a single amorphous phase. With increase in the drive amplitude however, structure within the polyethylene matrix phase becomes readily apparent. This ability to tune the contrast in the system will prove useful in analysis of such complicated systems.

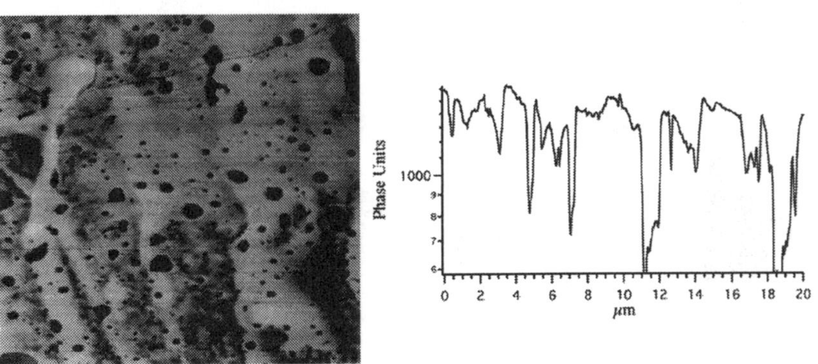

Figure 2: Line scan of image showing sharp transitions between phases.

Summary
This study establishes the utility of SPM techniques in examining complicated polymer mixtures. Using phase imaging together with tapping mode, one can now readily establish the surface morphology of multicomponent systems such as these. We have examined blends of PE/PDMS and quantified the effects of adding an interfacial modifier. In addition, we have studied mixtures of PE and carbon black and have shown the technique allows one to tune the contrast in complicated systems.

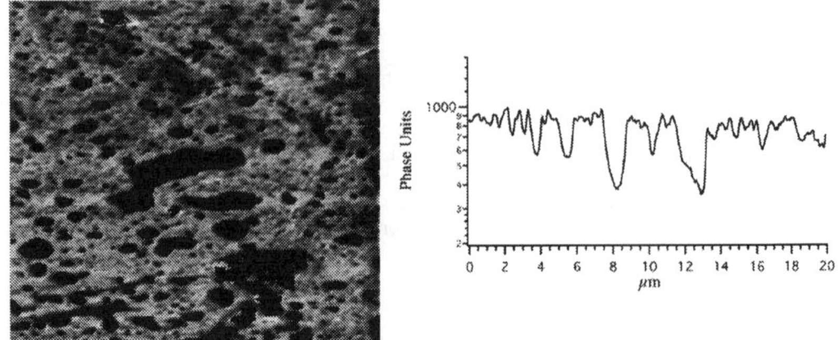

Figure 3: Line scan of compatibilized blend showing smoother transitions between phases.

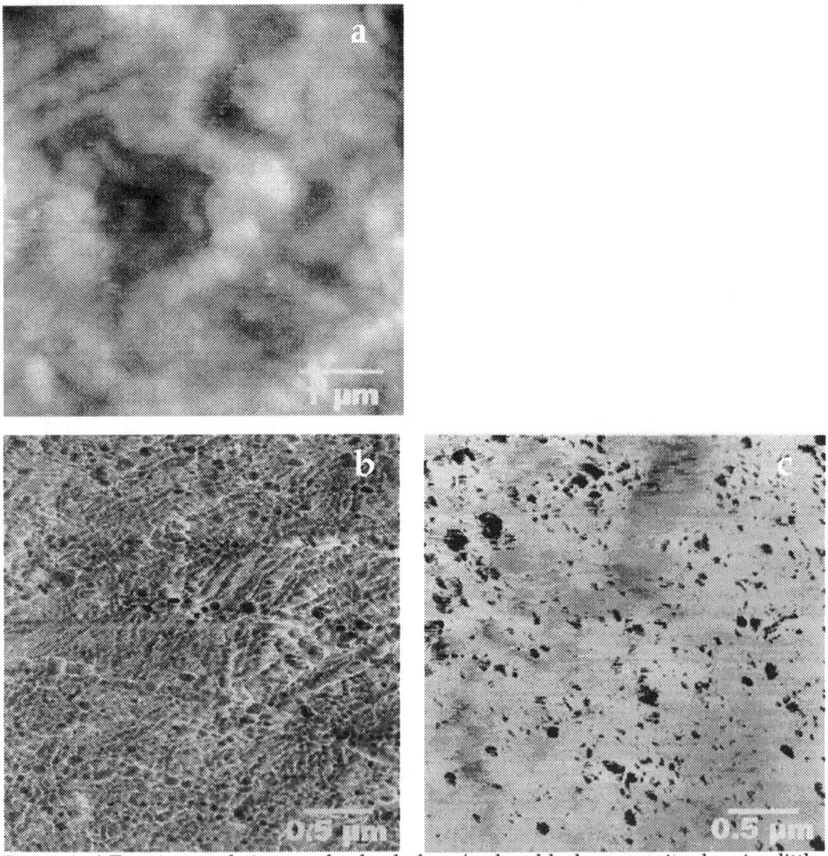

Figure 4: a) Tapping mode image of polyethylene/carbon black composite showing little structure. Phase imaging highlights the presence of b) polyethylene structure at high drive amplitudes and c) carbon black structure at lower amplitudes.

References

[1] Binnig, G.; Rohrer, H.; Gerber, Ch.; Weibel, E. *Phys. Rev. Lett.* **1982**, *49*, 57.
[2] Binnig, G.; Quate, C.F.; Gerber, Ch. *Phys. Rev. Lett.* **1986**, *56*, 930.
[3] Ruan, J. A.; Bhushan, B. *Trans. ASME, J. Tribol. (USA)* **1994**, *116*, 378.
[4] Huber, C.A.; Huber, T.E.; Sadoqi, M.; Lubin, J.A.; Manalis, S.; Prater, C.B. *Science* **1994**, *263*, 800.
[5] Viswanathan, R.; Heaney, M.B. *Phys. Rev. Lett.* **1996**, *75*, 4433.
[6] Grutter, P. *MSA Bull.* **1994**, 24, 416.
[7] Maivald, P.; Butt, H.J.; Gould, S.A.C.; Prater, C.B.; Drake, B.; Gurley, J.A.; Elings, V.B.; Hansma, P.K. *Nanotechnology* **1991**, 2, 103.
[8] Radmacher, M.; Tillmann, R.W.; Gaub, H.W. *Biophys. J.* **1993**, *64*, 735.
[9] Babcock, K.L.; Prater, C.B. Application Notes -- "Phase Imaging - Beyond Topography" Digital Instruments, Santa Barbara, CA.
[10] Digital Instruments, Santa Barbara, CA.

3D MICRO-TOMOGRAPHIC IMAGING AND QUANTITATIVE MORPHOMETRY FOR THE NONDESTRUCTIVE EVALUATION OF POROUS BIOMATERIALS

R. MÜLLER[1,3], S. MATTER[2], P. NEUENSCHWANDER[2], U.W. SUTER[2,] P. RÜEGSEGGER[3]

[1]Orthopedic Biomechanics Laboratory, Beth Israel Deaconess Medical Center and Harvard Medical School, 330 Brookline Avenue, Boston, MA 02215, USA. [2]Institute of Polymers, Swiss Federal Institute of Technology (ETH), ETH Zentrum, 8092 Zürich, Switzerland. [3]Institute for Biomedical Engineering, University of Zürich and Swiss Federal Institute of Technology (ETH), Moussonstrasse 18, 8044 Zürich, Switzerland

ABSTRACT

Micro-computed tomography (μCT) is a new and emerging technique for the nondestructive assessment and analysis of the three-dimensional trabecular bone architecture. The applications of μCT with respect to the analysis of bone are manyfold. Nevertheless, it also holds high promise for the microstructural measurement and analysis of porous biomaterials. For the purpose of the study, a desk-top μCT providing a nominal isotropic resolution of 14 μm was used. Since the polymeric material has a very low X-ray absorption coefficient, the scaffolds were stained prior to measurement using a commercial X-ray contrast agent. This allowed not only to acquire important microstructural features of the three-dimensional scaffold but also to compute standard structural indices such as BV/TV, BS/BV, Tb.N, Tb.Th, Tb.Sp and the degree of anisotropy (DA) using mean intercept length measurements. The preliminary results show that different types of scaffolds can be distinguished both qualitatively (visualization) and quantitatively (morphometry) provided an adequate X-ray staining technique is used. It can be concluded that, in the future, μCT may be of considerable help in basic as well as in applied research and development.

INTRODUCTION

Engineering new tissue by transplanting cells on degradable three-dimensional scaffolds is one approach to alleviate the vast shortage of donor tissue [1-3]. With this respect degradable polymeric materials hold high promise for medical applications. The new class of DegraPol®-polymers, based on a small number of chemical base entities, in which by minor changes in chemical composition the elastic modulus is widely variable (5 MPa - 1 GPa) and degradation rate can be controlled independently (degradation times in the range 1 week - 1 year), has been developed [4-5]. Such materials are designed to degrade in vivo in a controlled manner over a predetermined implantation period. Porous DegraPol®-scaffolds have also been tested for cell-biocompatibility [6-8]. The encouraging results from the in vitro culture experiments with chondrocytes and osteoblasts indicate clearly an application of these highly porous DegraPol®-scaffolds in the tissue engineering of bone and cartilage.

To further the development of the foaming process, a fast and nondestructive characterization method is needed providing a strong feedback to tailor the scaffold morphology of this very elastic material in a cell-preferred shape. Quantitative morphometry is a method to assess structural properties of porous materials. Morphometry has traditionally been assessed in two dimensions, where the structural parameters are either inspected visually or measured from sections, and the third dimension is reconstructed on the basis of stereology [9,10]. The conventional approach to morphologic measurements typically entails substantial preparation of the specimen, including embedding in polymethyl methacrylate, followed by sectioning into slices. Although the method offers high spatial resolution and high image contrast, it is tedious and time-consuming. Particularly limiting is the destructive nature of the procedure, preventing the probe from being used for other measurements such as the analysis in different planes. To overcome some of the limitations of the analysis of two-dimensional sections, several three-dimensional measurement and analysis techniques are under investigation.

Mat. Res. Soc. Symp. Proc. Vol. 461 © 1997 Materials Research Society

Micro-computed tomography (μCT) is a new and alternate approach to image and quantify complex structures in three dimensions. Where the early implementations of 3D micro-tomography focused more on the technical and methodological aspects of the systems and required equipment not normally available to a large public [11-13], a more recent development [14] emphasized the practical aspects of micro-tomographic imaging. The project aimed to enlarge the availability of the technology in basic research as well as in clinical laboratories.

Historically, the application of micro-tomographic imaging was more focused on the nondestructive assessment and analysis of the three-dimensional bone architecture. The purpose of this study was to investigate whether micro-tomographic imaging could also be employed for the microstructural measurement and analysis of porous biomaterials, nowadays successfully used for bone substitution or to accelerate bone in-growth around implants. In this respect, μCT measurements would be especially useful to monitor the processing of polymer scaffolds.

EXPERIMENT

Preparation of DegraPol®-Scaffolds

The new polymer DegraPol® represents a class of phase-segregated multi-block copolymers prepared by co-condensation of two macrodiols coupled with a chain-extender, which form, after phase separation, a crystalline and an amorphous phase. The soft segments can be used to monitor the degradation rate of the polymers because the mechanical properties of our polymers are dominated by the amount and the properties of the crystalline part and the chemical nature of the soft segments is of minor importance. Through incorporation of flexible soft segments based on diglycolide/ϵ-caprolactone oligomers differing in chemical composition, microstructure, and molecular weight, we have obtained polymers that disintegrate initially by a hydrolytically dominated glycolid ester bond cleavage process, followed finally by biological removal of the left-over polymer fragments [9].

The polymer was processed into a highly porous foam by a thermoinduced phase-separation-process (TIPS) [10-12], which is a much more suitable method for hydrolytically sensitive

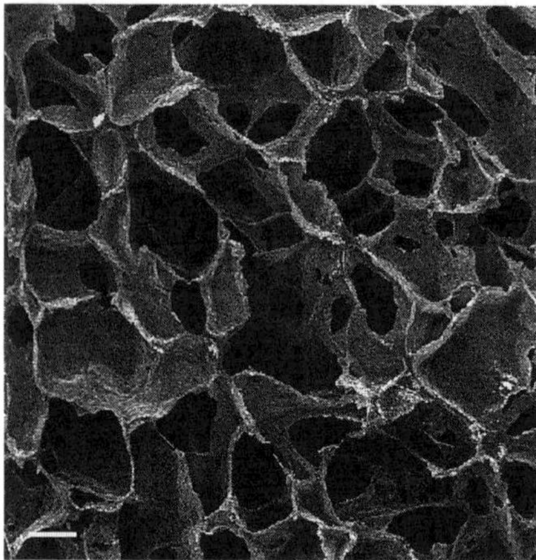

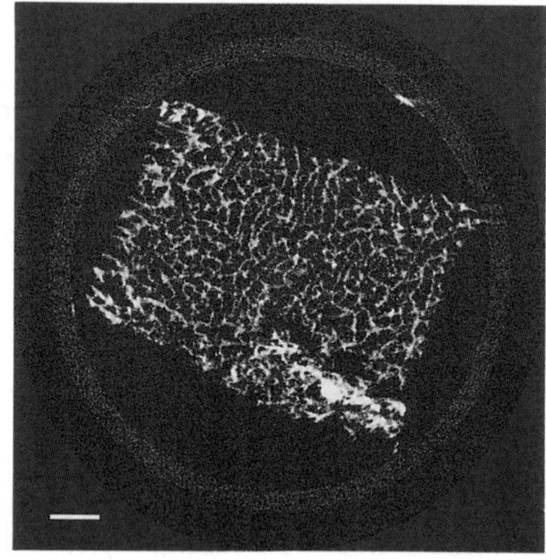

Figure 1. Scanning electron micrograph (SEM) of the top surface of the polymeric scaffold (bar represents 100 μm). The pore size of the foam is on the order of 50 - 100 μm with wall thicknesses of about 10 μm.

Figure 2. Micro-tomographic image of a cross-section of the polymeric scaffold (bar represents 1 mm). Prior to measurement the polymer was X-ray-stained with a commercial contrast agent in order to make it radiopaque.

polymers than the often used salt-leaching technique [13]. This process produces highly porous scaffolds and allows drug loading of the polymeric structure. The pore size can be tailored for special applications. In essence, the polymer (DegraPol®/btc13(44)) was dissolved in a solvent and then the solution temperature was lowered until liquid-liquid phase separation was induced. The solvent in the demixed polymer system was afterwards removed by freeze-drying, leaving behind the highly porous polymeric scaffold. Using this phase-separation-technique, it was possible to produce very thick samples of up to one centimeter thickness. Figure 1 shows a scanning electron micrograph (SEM) of the top surface of such a polymeric disc.

Micro-Tomographic Imaging

The micro-tomographic system described here [14] is based on a compact fan-beam type tomograph, also referred to as desk-top μCT. A microfocus X-ray tube with a focal spot of 10 μm is used as a source. The imaging chain includes a detector with integrated acquisition electronics to measure the intensity of the transmitted X-rays. To perform a measurement, the object is mounted on a turntable that can be shifted automatically in the axial direction. A total of 600 projections are taken over 216° (180° plus half the fan angle on either side). A standard convolution-backprojection procedure with a Shepp and Logan filter is used to reconstruct the CT images in 1024 x 1024 pixel matrices. An AlphaWorkstation (Digital Equipment Corporation, Maynard, MA, USA) is used to steer the stepping motors, to synchronize rotation, axial shift, and data recording, as well as for the image reconstruction. The whole system is housed in a standard 19-inch rack allowing for an extremely compact design that fits on every laboratory table. In order to be able to select different fields of view ranging from 8 mm to a maximum of 18 mm, the turntable is mounted on a linear slider, which permits to position the turntable freely between source and detector. The spatial resolution of the system is defined by the 10% contrast-level of the modulation transfer function (MTF) resulting in a spatial resolution of 28 μm in plane [14].

The desk-top μCT was especially designed for the nondestructive measurement and analysis of unprocessed surgical bone biopsies. It allows to automatically assess two- and three-dimensional morphometric indices without having to deal with the costly and time-consuming

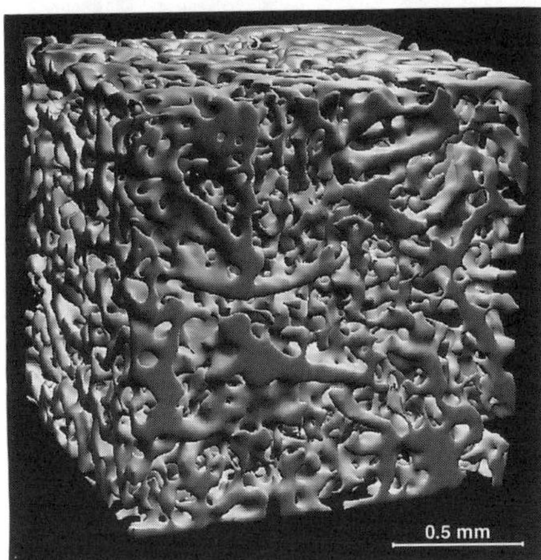

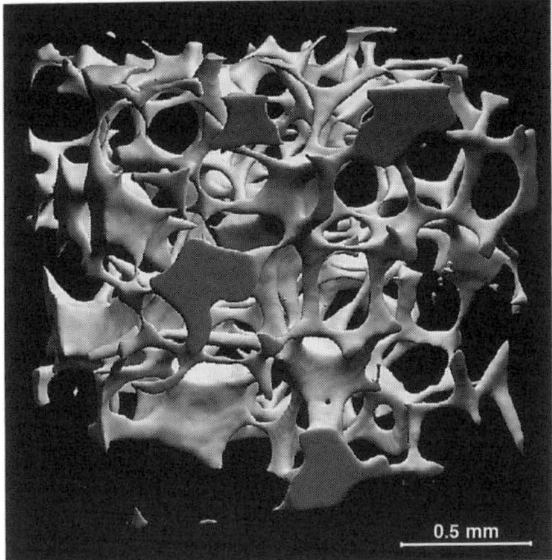

Figure 3. Three-dimensional reconstruction of porous polymer foam (DegraPol®) with small pores (61 μm) and high volume density (39%). The structure was measured with a desk-top μCT providing a nominal resolution of 14 μm.

Figure 4. Same as figure 3 but for an open structure with large pores (514 μm) and rather low volume density (11%). The two figures illustrate the capabilities of the μCT system to measure and analyse porous foams.

processing necessary for the histologic preparation of the specimens. Since the polymeric material typically has a very low X-ray absorption coefficient, it was necessary to contrast the polymeric structure prior to measurement. In order to obtain a radiopaque structure, the DegraPol®-scaffolds were soaked in a contrasting solution consisting of 15% Lipiodol® and 85 vol.-% isopropanol. Once completely filled with the solution, the scaffolds were then cooled down to -196° C in liquid nitrogen before freeze-drying for 2 days.

A typical measuring procedure consists of a scout view, selection of the examination volume, automatic positioning, measurement, off-line reconstruction and evaluation. For the measurement, the scaffolds were located in a cylindrical sample holder of perspex. The sample holders were machined with different tube diameters to guarantee a snug fit of the specimen with varying sample diameters. Additionally, they were marked with an axial alignment line on the outside of the tube to allow for consistent positioning of the specimens within the tube, which is important for consecutive measurements of the foam or for a morphometric comparison with standard sectioning and microscopic analysis. Together with the alignment notch to connect the sample holder with the turntable, this permits a precise positioning of the specimen within seconds.

RESULTS

For each sample a total of 286 micro-tomographic slices were acquired. Measurements were stored in three-dimensional image arrays. The total examination time per specimen was approximately three hours for a voxel size of 14 μm³. Figure 2 gives an example of a micro-tomographic image representing a two-dimensional cross-section of the contrasted foam. For the subsequent analysis, a subvolume of the originally measured data was selected. This selected volume of interest (VOI) was located in the center of the specimen to eliminate preparation artifacts on the surface. The size of the VOI was 2.8 x 2.8 x 2.8 mm³ (200³ voxels). A constrained gaussian filter was used to partly suppress the noise in the volume. The foam was separated from the background with the help of a thresholding procedure. The threshold was chosen in such a way to maintain the foam topology. For the three-dimensional visualization, we used an extended marching cubes algorithm [20]. The algorithm, which is based on a divide-and-conquer approach, triangulates the surface of any given voxel array in a fast and secure way. It does not only allow perfect triangulation but also smooths the surface effectively. In addition to the visual assessment, morphometrical indices can be

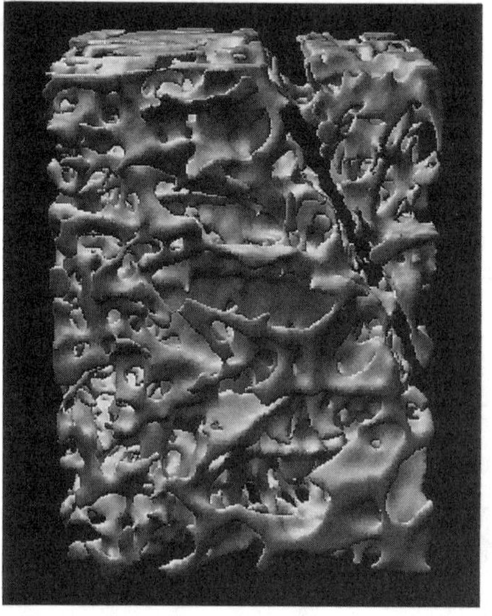

Figure 5. Three-dimensional reconstruction of a DegraPol®-scaffold. Inhomogeneities can be detected easily for quality control and process monitoring without destroying the structure.

determined quantitatively from the micro-tomographic examinations (nomenclature according to the conventions for bone morphometry [21]). Volume density (BV/TV) and the trabecular plate number (Tb.N) were determined directly from the binarized VOI. From these two indices, other indices were derived [9] such as the trabecular plate thickness (Tb.Th), the trabecular plate separation (Tb.Sp) and the surface density (BS/BV), which defines the ratio between surface and volume of the solid foam. In addition, the anisotropy of the specimens was determined using mean intercept length (MIL) measurements [22]. MIL denotes the average distance between object/background interfaces and is measured by tracing test lines in different directions in the examined VOI. From this measurement, a MIL tensor can be calculated by fitting the MIL values to an ellipsoid. The eigenvalues of this tensor can then be used to define the degree of anisotropy (DA) which denotes the maximum to minimum MIL ratio [23].

Table I. Basic three-dimensional morphometric indices assessed from µCT.

Variable	Sample A	Sample B	Sample C
BV/TV (%)	39	11	30
BS/BV (mm²/mm³)	52	33	40
Tb.N (1/mm)	10	2	6
Tb.Th (µm)	38	61	50
Tb.Sp (µm)	61	514	115
DA (mm/mm)	1.16	1.12	1.09

Basic descriptive results from the quantitative morphometry of samples A to C (figures 3 to 5) are given in table I. Figures 3 to 5 show different examples of foams with different pore sizes or with different degrees of homogeneity respectivly. Where the structure in figure 3 depicts a typical DegraPol®-scaffold with small pore sizes (61 µm) and a high volume density (39%), the foam depicted in figure 4 represents a more open structure with more separated structural elements (514 µm) and a much lower volume density (11%). The foam depicted in figure 5 represents the same type of foam as the one in figure 3 but displays a more inhomogeneous structure due to the central artifacts introduced with the foaming process. This is also expressed in the structural indices where the average pore size is almost twice as high (114 µm) and the volume density is considerably reduced (30%). The values for the assessment of the degree of anisotropy vary only little between the three measurements with a range from 1.09 to 1.16.

CONCLUSIONS

To monitor the processing of customized porous scaffolds, a nondestructive measuring system is mandatory in order to control the processing as well as to characterize the foams visually and, even more important, quantitatively. Desk-top µCT is a new and emerging technique that facilitates such a nondestructive approach and offers easy access to the three-dimensional nature of porous polymeric foams. It allows to automatically generate internal views of the porous structure as well as to assess basic three-dimensional morphometric indices which permit to discriminate different types of foam. In that sense and as illustrated in figure 5, micro-tomographic imaging enables the tissue engineer to control the outcome of the design and to detect irregularities and inhomogeneities in the structure immediately after synthesis without having to deal with the tedious and time-consuming preparation necessary for standard histomorphometry. Although the preliminary results from the micro-tomographic analysis look very promising, it has to be mentioned that standard sectioning offers the advantage of even higher resolutions - in the submicron region - and therefore a superior accuracy can be anticipated for the assessment of structural indices in two dimensions. This is especially true for the computation of the average wall thickness (below 10 µm) of the structural elements, which cannot be assessed accurately using desk-top µCT (14 µm nominal resolution). The lack of resolution is also expressed in an underestimation of surface density, which has been found to peak around 200 mm²/mm³ for a series of foams using histological preparation (unpublished data), whereas the corresponding values predicted from the µCT measurements turned out to be much lower, around 60 mm²/mm³.

In conclusion, micro-tomographic imaging in combination with an adequate X-ray staining technique is a nondestructive, fast and precise procedure to characterize porous biomaterials both qualitatively (visualization) and quantitatively (morphometry). In the future, µCT may be of considerable help in basic as well as in applied research and development.

ACKNOWLEDGMENTS

This work was supported by the Swiss National Sciences Foundation (Project No 3100-029971). The authors would also like to thank Drs. Bruno Koller and Tor Hildebrand for their contributions with respect to the development of the desk-top µCT.

REFERENCES

[1] R. Langer and J.P. Vacanti, *Science* **260**, 920 (1993).

[2] G.J. Beumer, C.A. van Blitterswijk, D. Bakker and M. Ponec, *Biomaterials* **14**, 598 (1993).

[3] C.H. Rivard, C.J. Chaput, E.A. DesRosiers, L.H. Yahia and A. Selmani, *J. Appl. Biomat.* **6**, 65 (1995).

[4] D. Hirt, P. Neuenschwander and U.W. Suter, *Makromol. Chem. Phys.* **197**, 1609 (1996).

[5] P. Neuenschwander, G.K. Uhlschmid, U.W. Suter, G. Ciardelli, T. Hirt, O. Kaiser, K. Kojima, A. Lendlein and S. Matter, Europ. Pat. No. 0696605.

[6] B. Saad, G. Ciardelli, S. Matter, M. Welti, G.K. Uhlschmid, P. Neuenschwander and U.W. Suter, *J. Biomed. Mat. Res.* **30**, 429 (1996).

[7] B. Saad S. Matter, G. Ciardelli, G.K. Uhlschmid, M. Welti, P. Neuenschwander and U.W. Suter, *J. Biomed. Mat. Res.*, submitted (1996).

[8] B. Saad, G. Ciardelli, S. Matter, M. Welti, G.K. Uhlschmid, P. Neuenschwander and U.W. Suter, *J. Mat. Sci: Mater. in Med.* **7**, 56 (1996).

[9] A.M. Parfitt, C.H.E. Mathews, A.R. Villanueva, M. Kleerekoper, B. Frame and D. S. Rao, *Calcif. Tiss. Int.* **72**, 1396 (1983).

[10] R.K. Schenk and A.J. Olah, in *Handbuch der inneren Medizin VI/1A, Knochen Gelenke Muskeln*, edited by F. Kuhlencordt and H. Bartelheimer (Springer, Berlin, 1980) pp. 437-494.

[11] L.A. Feldkamp, S.A. Goldstein, A.M. Parfitt, G. Jesion and M. Kleerekoper, *J. Bone Min. Res.* **4**, 3 (1989).

[12] J.H. Kinney, T.M. Breuning, T.L. Starr, D. Haupt, M.C. Nichols, S.R. Stock, M.D. Butts and R.A. Saroyan, *Science* **260**, 789 (1993).

[13] U. Bonse, F. Busch, O. Günnewig, F. Beckmann, R. Pahl, G. Delling, M. Hahn and W. Graeff, *Bone Min.* **25**, 25 (1994).

[14] P. Rüegsegger, B. Koller and R. Müller, *Calcif. Tiss. Int.* **58**, 24 (1996).

[15] B. Saad, S. Matter, G.K. Uhlschmid, T. Hirt, O.A. Trentz, P. Neuenschwander and U.W. Sutter, *Langenbecks Archiv für Chirurgie*, Forumband, 65 (1995).

[16] S.Q. Liu and M. Kodama, *J. Biomed. Mat. Res.*, **26**, 1489 (1992).

[17] H. Lo, S. Kadiyala, S.E. Guggino and K.W. Leong, *Proc. Mat. Res. Soc.* **331**, 41 (1994).

[18] K. Whang, C.H. Thomas, K.E. Healy and G. Nuber, *Polymer* **36**, 837 (1995).

[19] A.G. Mikos, G. Sarakinos, S.M. Leite, J.P. Vacanti, R. Langer, *Biomaterials* **14**, 323 (1993).

[20] R. Müller, T. Hildebrand and P. Rüegsegger, *Phys. Med. Biol.* **39**, 145 (1994).

[21] A.M. Parfitt, M.K. Drezner, F.H. Glorieux, J.A. Kanis, H. Malluche, P.J. Meunier, S.M. Ott and R.R. Recker, *J. Bone Min. Res.* **21**, 595 (1987).

[22] T.P. Harrigan and R.W. Mann, *J. Mater. Sci.* **19**, 767 (1984).

[23] R.W. Goulet, S.A. Goldstein, M.J. Ciarelli, J.L. Kuhn, M.B. Brown and L.A. Feldkamp, *J. Biomech.* **27**, 375 (1994).

MICROSTRUCTURE OF THERMALLY CROSSLINKABLE
POLY(ETHYLENE TEREPHTHALATE) (PET-co-XTA)
BENZOCYCLOBUTENE FUNCTIONALIZED COPOLYMERS

Brendan J. Foran, Elizabeth Pingel, Gary E. Spilman, Larry J. Markoski, Tao Jiang,
and David C. Martin, Department of Materials Science and Engineering,
University of Michigan, Ann Arbor, MI 48109

ABSTRACT

The microstructure and thermal properties of copolymers of polyethylene terephthalate
(PET) containing a crosslinkable terephthalic acid, 1,2-dihydrocyclobutabenzene 3,6 dicarboxylic
acid (XTA) are reported. Wide angle x-ray scattering (WAXS) show that the addition of XTA
does not alter the PET crystal structure in copolymers at low XTA contents. However, the degree
of crystallinity drops for higher XTA levels. WAXS profiles show that PET-co-XTA 50% is
amorphous, and that PEXTA homopolymer has a different crystal structure. Thermal data from
DSC and TGA show that crosslinking of the benzocyclobutene groups (~350°C) occurs at
temperatures between melting (~250°C) and degradation (~400°C), making it possible to melt
process the copolymers into fibers before the onset of crosslinking. Limiting oxygen index (LOI)
measurements show that increased oxygen concentrations are required to sustain a stable flame in
PET-co-XTA copolymers; whereas unmodified PET had an LOI value of ~18%, the copolymers
had LOI values near 32%. Further, while unmodified PET melts and drips as it burns, XTA
copolymers formed a stable char that inhibiting flame propagation. An increased char was
observed in optical micrographs for XTA containing polymers, and crystalline domains were
observed near the burn surface in transmission electron micrographs.

INTRODUCTION

Benzocyclobutene (BCB) moieties are well known as crosslinking and grafting agents in
polymer chemistry.[1] Our group, in collaboration with J.S. Moore has recently studied several
polymeric systems into which BCB groups were introduced by means of a new monomer,
1,2-dihydrocyclobutabenzene 3,6 dicarboxylic acid or "crosslinkable terephthalic acid" (XTA).[2]
XTA copolymers have been used to increase compressive strength, solvent resistance, and creep
resistance in high performance extended chain polymer fibers.[3] XTA has also recently proven
effective in producing thermally crosslinkable thermotropic copolyesters.[4] In this proceedings we
present recent work in the characterization of PET copolymers with varying amounts of XTA
content. This copolymer is represented in Figure 1.

Thermal properties were
measured by differential scanning
calorimetry (DSC), and thermal
gravimetric analysis (TGA), while
morphological and microstructural
characterization was performed by
hot-stage optical microscopy, UV-
Vis. spectroscopy, wide-angle x-ray
scattering (WAXS) and
transmission electron microscopy
(TEM). By incorporation of XTA
into PET copolymers, we have
produced thermoplastic polymers
that can be thermally crosslinked at
temperatures after melt processing.

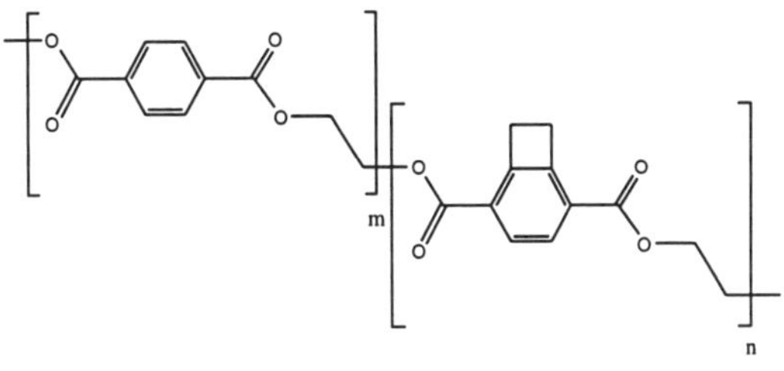

Figure 1. PET-co-XTA

Mat. Res. Soc. Symp. Proc. Vol. 461 © 1997 Materials Research Society

METHODS

PET and PET-co-XTA copolymers were synthesized by melt transesterification using zinc acetate as a catalyst at 180°C. Ethylene glycol was removed by heating to 280°C with antimony oxide (Sb_2O_3). Triphenyl phosphate was used to prevent discoloration at high temperatures.

Thermal properties were studied by differential scanning calorimetry (DSC) and thermal gravimetric analysis (TGA) using Perkin-Elmer "Series 7" instruments. Degradation temperature was measured as the temperature at 5% weight loss in TGA runs, while "% Residue" was determined as the residual weight percent at 800°C. These TGA runs were performed both in air and N_2 environments. Flammability was studied using a Limiting Oxygen Index test, following ASTM D-2836, using an LOI instrument manufactured by Atlas Electric Devices. Wide-angle x-ray scattering was recorded on a Rigaku theta-2theta diffractometer with a 12 kW rotating anode source, monochromated Cu Kα, λ=1.542Å in reflection mode. Hot-stage optical microscopy was performed on a Nikon Optiphot-POL equipped with a Linkham LH 1600 hot stage, video overlay system, and Macintosh Quadra computer with RasterOps Video board. The images were analyzed using Adobe Photoshop and NIH Image software. UV-Visible spectroscopy were performed on a Perkin Elmer UV/Vis spectrometer. Transmission electron microscopy (TEM) (200 kV JEOL 2000 FX and 400 kV JEOL 4000 EX) was done on ~60 nm thin sections (Riechert-Jung ultramicrotome with DDK diamond knife) of burned copolymer embedded in epoxy (Pelco Eponate-12 from Ted Pella Inc.).

RESULTS AND DISCUSSION

<u>Thermal Properties</u>

The composition, intrinsic viscosity (IV), and residual weight percent (from TGA in an N_2 environment) of the polymers studied are tabulated in Table 1.

Polymer	IV (dl/gm)	LOI (% Oxygen)	% Residue (at 800°C)
PET	0.52	18	13.0
PET-co-XTA 1%	0.48	22	14.4
PET-co-XTA 4.8%	0.47	23	14.8
PET-co-XTA 10%	0.16	25	15.5
PET-co-XTA 20%	0.23	32	15.7
PET-co-XTA 50%	0.1		20.8
PEXTA 100%			31.5

Table 1. Composition of polymers studied showing intrinsic viscosity (IV), LOI and % Residue.

LOI tests reveal a systematic increase in the oxygen content of the atmosphere required to sustain a stable flame. Furthermore, there is a dramatic change in the appearance of the burned samples. Unmodified PET melts and drips while it burns, and the melting continues to feed the propagating flame. However, XTA copolymers develop a stable char which effectively prevents further burning. Figure 2 shows polymer pieces that were burned in air. In Figure 2, two samples of PET are shown above the marking (a), the one on the left is a commercial PET (Celanese 39114), while the other is PET made in our laboratory. In both cases, there is little char at the burn surface, and a solidified drip can be seen. For these neat PET samples the burning polymer would drip, and the dripping plastic

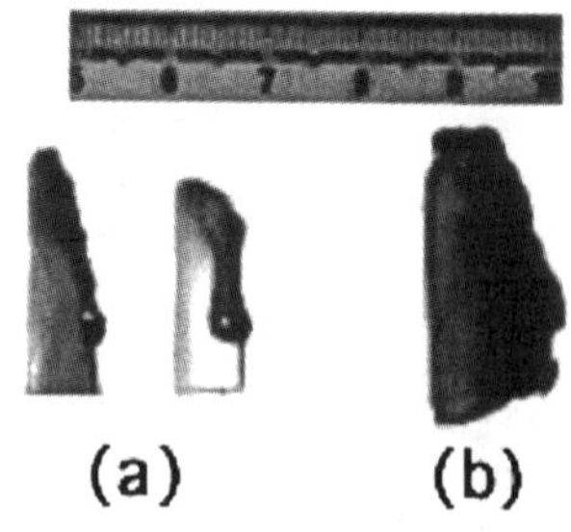

Figure 2. Burned samples of (a) neat PET and (b) PET-co-XTA 20%.

would continue to burn carrying the flame to where it fell. Figure 2(b) shows a burned piece of PET-co-XTA 20%. Here a black char can be seen which formed before the plastic extinguished itself; in this piece there was no dripping of burning-melted polymer. Despite the appearance of char on samples burned in air, residual char contents from TGA experiments were below weight percents that could be accurately measured. TGA experiments run in N_2 environment show a systematic increase in residue weight with increasing XTA contents (as shown in Table 1).

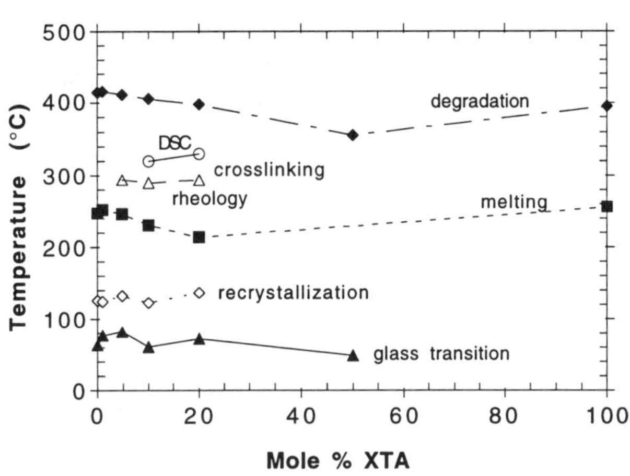

Figure 3. Thermal Data Summary for PET-co-XTA copolymers.

The characteristic transition temperatures (glass transition, recrystallization, melting, BCB reaction or crosslinking, and 5% weight loss degradation are shown in the Figure 3. Td decreases from 420°C to 360°C with increasing XTA content except that PEXTA homopolymer is raised back to 395°C. The Tm follows a similar trend, after an initial small rise from 0 to 1% XTA, melting temperatures decrease with increased XTA content, except again that PEXTA homopolymer is higher at 355°C. These trends are common for partially miscible mixtures. Glass transition temperatures increase from 65 to 82°C (from 0 to 4.8% XTA), followed by decreasing Tg's as further XTA is added.

<u>Hot-Stage Optical Microscopy and UV-Vis Spectroscopy</u>

Samples melted repeatedly when heated to only 10°C above the melting point; the melt was clear and relatively colorless. On cooling, materials recrystallized with a slight brownish color. Heating near crosslinking temperatures caused an increased greenish tint, while heating further resulted in orange-brown coloring. As evidence of crosslinking, the polymer would no longer recrystallize on cooling or flow when subsequently reheated.

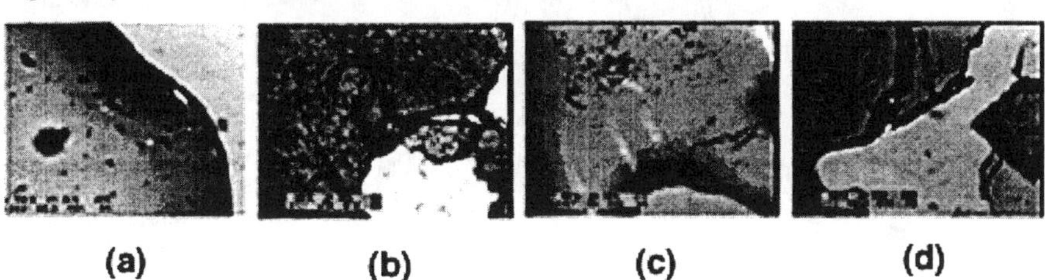

(a)   **(b)**   **(c)**   **(d)**

Figure 4. Selected images from hot-stage optical microscopy of PET-co-XTA 4.8%. (a) First melt, heated to 260°C. (b) Cooled and fracturedwhen solidified at room temperature. (c) Second melt showing flow. (d) Heated further to 350°C. (e) Crosslinking has prevented flow in further heating cycles. Image size is 0.5 x 0.8 mm.

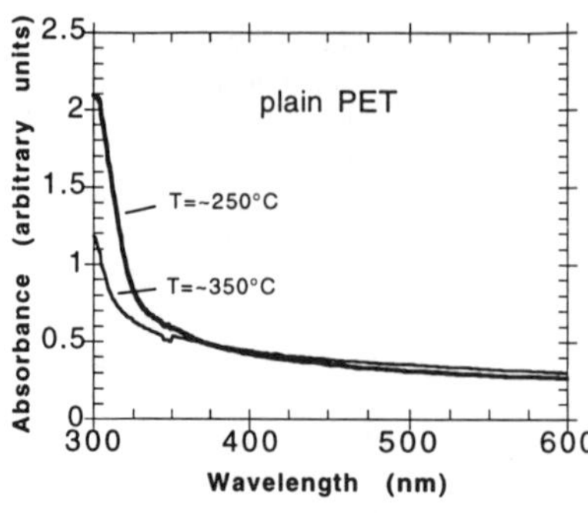

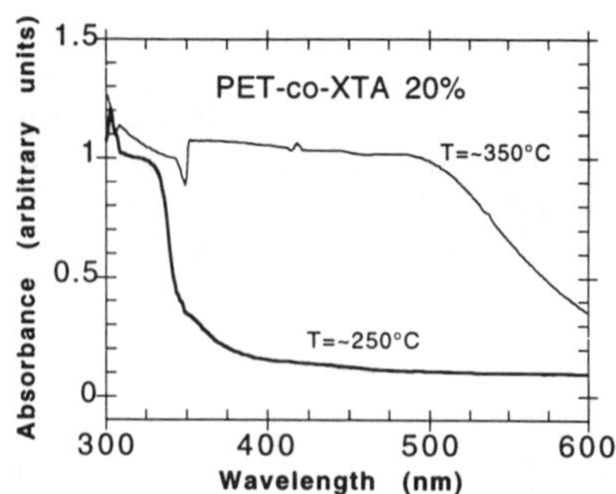

Figure 5. UV-Vis spectra for neat PET and PET-co-XTA 20% heated to melt and BCB-crosslinking temperatures.

After crosslinking, increased absorbance is found from 350-550 nm for copolymers containing XTA. This change in absorbance spectra may be associated with increased lengths of aromatic conjugation . Such increases are in agreement with one model for BCB crosslinking which results in the formation of a stilbene bridge (shown in Figure 5).

Figure 6. Stilbene-bridge model for BCB crosslinking.

<u>Wide Angle X-ray Scattering</u>

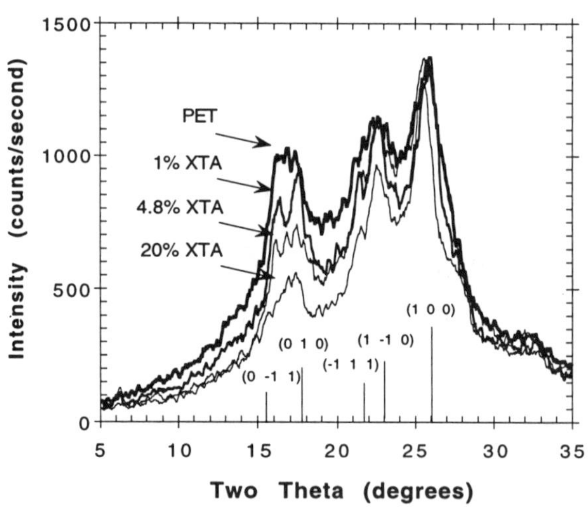

PET-co-XTA copolymers with low XTA content show x-ray diffraction patterns similar to that of the neat PET. In Figure 7, the scattering patterns show three main peaks which are well described by the crystal structure of neat PET.[5] The d-spacings for copolymers with XTA contents below 20% do not vary significantly. These results suggest that XTA is excluded from PET crystalline domains.

The WAXS data shows that the crystallinity of the PET-co-XTA copolymers is decreased with additional BCB content. However,

Figure 7. WAXS scans for PET and low XTA content copolymers; indices are shown for the known PET structure.

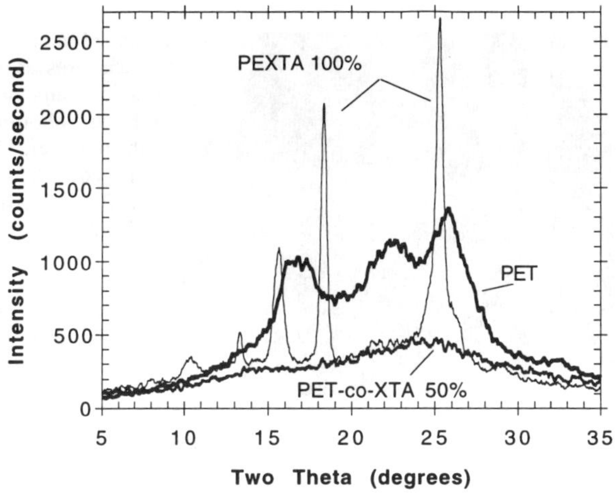

Figure 8. WAXS scans for higher % XTA polymers, the pattern of PET is included as a reference.

considerable crystallinity can still be achieved in even the 20%-XTA copolymers.

In Figure 8, scattering paterns are shown for 50% XTA copolymer as well as the PET-co-XTA 100% (PEXTA) homopolymer. The sharpness of the peaks for the PEXTA homopolymer most likely reflects the low molecular weight obtained for this material. However, it is clear that the crystal structure for PEXTA is different from that of neat PET.

Transmission Electron Microscopy

The burned surface of the 10% XTA copolymer shows a highly crystalline zone that is not evident in the PET homopolymer. As can be seen in figure 9(a), for PET-co-XTA 10%, this burned region contains dense particles with an average diameter of 300Å. Within these particles are smaller crystallites as shown in the dark field image of Figure 9(c). The depth of the "char" region was observed to vary in depth from 2-12 μm. Voids were evident at the char-bulk interface. Below the interface, some voids were observed at interspherulitic boundaries.

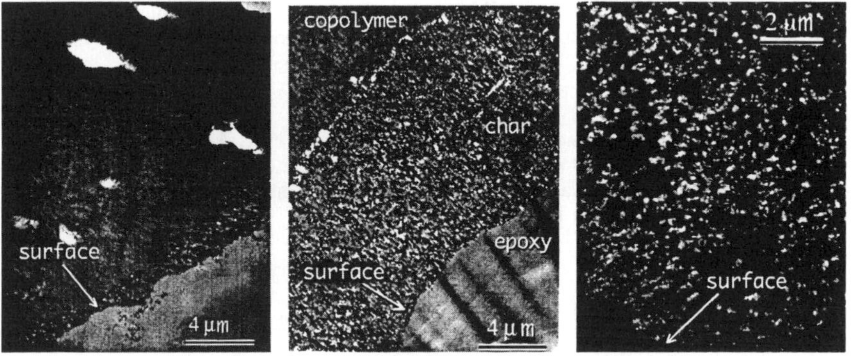

Figure 9. Two bright-field TEM micrographs at the surface of burned PET-co-XTA 10%. The third micrograph is a dark field image and shows highly crystalline microcrystalline particles.

In neat PET, the burn surface shows less obvious structural reorganization than in the PET-co-XTA 10% shown. However, degradation is suggested by regions near the burn surface

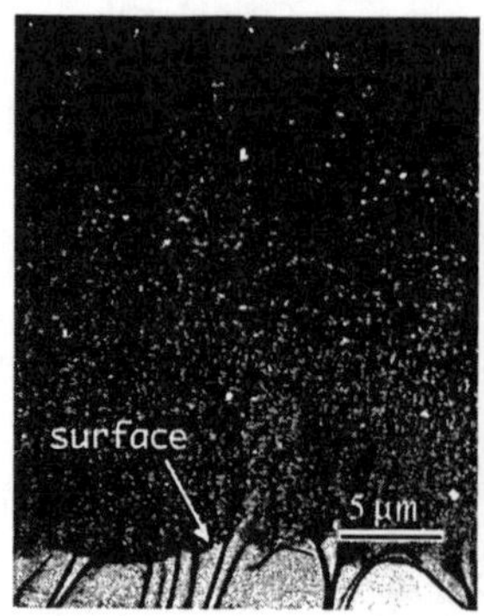

Figure 10. Two micrographs of the burned surface of neat polyethylene terephthalate (PET).

where the polymer fails under the physical stresses imparted by ultramicrotomy (seen in Figure 10, as ribbon-like bands across the polymer. These regions penetrate to depths of 0.1 mm. Dark field images of neat PET did not exhibit increased crystallinity near the burn surface.

Rheological studies by P.T. Mather and A. Romo-Uribe have shown that the complex viscosity of the PET-co-XTA copolymers increases as the crosslinking reaction proceeds. Above the melting point, the viscosity decreases with increasing temperature for PET homopolymer and 1%-XTA copolymer. The viscosity remains relatively constant for the 4.8%-XTA copolymer and increases significantly with temperature for 10% and 20% copolymers.

CONCLUSIONS

PET-co-XTA copolymers can be synthesized by melt esterification to reasonable molecular weights. The synthetic technique requires more care than for neat PET to avoid crosslinking before obtaining high molecular weight polymer. The PET-co-XTA copolymers show systematic variations for the glass transition, recrystallization, melting and degradation temperatures that reveal the influence of the BCB functionalities. These thermal properties provide a processing window between the melt and degradation temperatures within which the XTA copolymers can be melt processed. Systematic trends were also found for LOI and percent residue values. X-ray scattering data suggest that XTA is excluded from PET crystalline domains which are able to crystallize well at XTA contents below 20%. Crystallinity observed for PEXTA homopolymer indicates a new unit cell. TEM has proved to be a powerful technique for studying the microstructure of polymers near burn surfaces, and has revealed the presence of a crystalline char region at the burn surfaces of the PET-co-10%XTA copolymers.

ACKNOWLEDGMENTS

Rheological measurements were conducted by P.T. Mather and A. Romo-Uribe at the Phillips Laboratory, Edwards Air Force Base, CA. Funding for this project was provided by National Institute of Standards and Technology (NIST).

REFERENCES

[1] (a) M.F. Farona, Prog. Polym. Sci., **21**, 505 (1996). (b) R.A. Kirchoff, and K.J. Bruza, Prog. Polym. Sci. **18**, 85 (1993).
[2] K.A. Walker, L.J. Markoski, J.S. Moore, Synthesis, 1265 (1992).
[3] M.-C. Jones, T. Jiang, and D.C. Martin, Macromolecules, **27**, 6507 (1994). (b) T. Jiang, J. Rigney, M.-C. Jones, L.J. Markoski, G.E. Spilman, D.F. Mielewski, and D.C. Martin, Macromolecules, **28**, 3301 (1995).
[4] P.T. Mather, K.P. Chaffee, A. Romo-Uribe, G.E. Spilman, T. Jiang, and D.C. Martin, Polymer, in press.
[5] Y. Fu, B. Annis, A. Boller, Y. Jin, B. Wunderlich, J. Polym. Sci. **32**, 2289 (1994).

COMPATIBILIZERS MADE FROM COPOLYMERS THAT EXHIBIT STRONG INTERACTIONS

Barry J. Bauer and Da-Wei Liu, Materials Science and Engineering Laboratory, National Institute of Standards and Technology, Gaithersburg, MD 20899.

ABSTRACT

Random copolymers of poly(methyl methacrylate-d8) (PMMA) and poly(methacrylic acid) (PMAA) were synthesized and blended with polyethylene oxide (PEO), and SANS was used to estimate the strength of interaction between the various polymer pairs. The addition of PMAA greatly reduced the scattering of the blend, giving a large negative Flory-Huggins interaction parameter. Copolymers of the type PMAA-g-PS, PMAA-r-PS, PMMA-g-PS, and PMMA-r-PS (g = graft, r = random) were used as compatibilizers for the melt blending of mixtures containing 80% PS and 20% PEO. Random copolymers containing half PMMA and half PS or half PMAA and half PS were ineffective compatibilizers while graft copolymers of these polymers produced a much finer dispersion of PEO in PS when blended at 190°C. Graft copolymers made with 90% PS were different, however, with PMMA grafts having only a small effect, while PMAA grafts having a large effect. Random and graft copolymers of PS and PMMA were also used to compatibilize blends of PS and PMMA. The results were similar to the same compatibilizers with blends of PS and PEO, with the graft copolymer causing a finer dispersion.

INTRODUCTION

Blends of immiscible polymers often require the addition of a polymer that improves the dispersion and adhesion of the phases [1]. Such "compatibilizers" are commonly block copolymers of the two immiscible polymer types that are to be blended. The function of the compatibilizer is not to make the immiscible polymer pair miscible, however. Rather, its function is to control physical properties by modifying the two phase morphology. If the block copolymer goes to the interface between the two blended polymers, it may act as a surfactant, lowering the interfacial energy and allowing the shearing action of a melt blending process to break the polymer droplets into a finer size scale. Block copolymers at the interface may also inhibit coalescence of the mechanically dispersed phases, causing the dispersed phases to remain stable after shearing has stopped.

Since the blocks are identical to the blended polymers, they will be miscible, and copolymers at the interface should have the blocks anchored in the phase of the same polymer type. This interpenetration places covalent bonds between the phases, strengthening the interface. Therefore, there are many useful functions of block copolymers used as compatibilizers.

The two polymers in the block copolymer need not be identical to the ones being blended, however. If one of the blocks is made from a polymer that is miscible with a blended polymer, it may have the same effect as a block made from a polymer that is the same as the blended polymer. This "cross compatibilization" would give more possible combinations of polymer types that could be used as effective compatibilizers. The major advantage of cross compatibilization, however, is that there is an additional thermodynamic parameter that can be used to control the compatibilization effect. A block copolymer of an A-B type that is used in an A/B blend, will have athermal interactions between polymers or blocks of the same type with a Flory-Huggins interaction parameter $\chi = 0$. In a blend of a A-C copolymer there is an interaction parameter, χ_{BC}, that can have a wide range of values. If $\chi_{BC} = 0$ then the copolymer might act the same as an A-

Mat. Res. Soc. Symp. Proc. Vol. 461 ©1997 Materials Research Society

B copolymer. If $\chi_{BC} < 0$, there is a favorable interaction between copolymer and the blended polymer. This may result in a very powerful compatibilization effect and represent a new class of compatibilizers.

A number of recent small angle neutron scattering (SANS) studies have examined the effect of polymers exhibiting specific interactions, such as hydrogen bonding [2-3]. Small amounts of such interactions can make polymers that are normally immiscible, form miscible blends. If a block copolymer is made of such polymers, its use as a compatibilizer may be improved.

To measure this effect, graft copolymers have been synthesized from poly(methyl methacrylate) (PMMA) and polystyrene (PS), along with poly(methacrylic acid) (PMAA) and PS. There copolymers were then used as compatibilizers for blends of poly(ethylene oxide) (PEO) and PS. Random copolymers of PMMA-d8 and PMAA were blended with PEO, and SANS was used to estimate the strength of interaction between the various polymer pairs. SANS studies of PMMA-d8/PEO blends have measured the interaction parameter, χ, and it is reported to be between -0.005 and -0.001 depending on the composition. This represents a favorable, but very weak interaction [4].

Once the relative strength of the interactions is determined by SANS, random and graft copolymers of the various polymer combinations can be synthesized and melt blended under a range of compositions and blending conditions. The relative effect of compatibilizer addition is studied by the changes in size of the dispersed phases, as measured by scanning electron microscopy (SEM).

EXPERIMENTAL

Random copolymers of methyl methacrylate -d8 and methacrylic acid were polymerized in bulk at 60°C with azobisisobutyronitrile as initiator. Graft copolymers were formed by copolymerizing 13,000 g/mol methacrylate terminated polystyrene (Scientific Polymer products) [5] with methacrylic acid and/or methylmethacrylate in N-methyl pyrillidone. The PEOs used were obtained from Aldrich Chemical Co. with reported relative molecular mass of 100,000 g/mol (for SANS) and 8,000,000 g/mol (for melt mixing). The PS was obtained from Dow and had a reported Mr,w of 182,000 g/mol.

SANS samples were made by solvent casting the PMMA-PMAA copolymers and equal weight PEO from toluene-isopropanol. The clear films were dried in vacuum for a day at 70°C and pressed into 1 mm thick samples. The SANS measurements were carried out at the NIST 8M facility.

Blends were made by melt mixing in a Haake Mixer/ Extruder with a small volume mixing head for 15 minutes at 160 °C or 190 °C. Samples were pressed at 150°C and fracture surfaces obtained by tensile impact were examined by Scanning Electron Microscopy. Temperature control was not an important factor but was kept within ±5 as is common for these types of operations

RESULTS AND DISCUSSION

Determination of χ by SANS

The Flory-Huggins interaction parameter is calculated trough the use of the random phase approximation. The scattering is fit to equation 1.

$$\frac{K_n}{S(q)} = \frac{1}{v_A N_A \phi_A D_A(q)} + \frac{1}{v_B N_B \phi_B D_B(q)} - 2\frac{\chi}{v_0}$$

In this equation S(q) is the scattered intensity in absolute intensity units as a function of the scattering vector $q = 4\pi/\lambda \sin(\theta/2)$ where λ is the wavelength and θ is the scattered angle. Kn is a neutron contrast factor based on the polymer repeat unit, N is degree of polymerization, v is the volume of a repeat unit, χ is the interaction parameter, and D is a Debye function. Fitting of blend scattering is described elsewhere [8].

When this equation is applied to the scattering from a blend in which one of the polymers is a random copolymer, the value of χ is a composite value of three different interaction parameters. If the blend is between polymers of random copolymer of type A-C and homopolymer of type B then the composite χ is calculated by equation 2.

$$\chi = f\,\chi_{AB} + (1-f)\,\chi_{AC} - f(1-f)\,\chi_{BC}$$

In this equation, f is the fraction of B in random copolymer B-C. Therefore, the composite χ value calculated for blends of different random copolymers plotted against f can be fit with a parabola to calculate the various χ values.

SANS of blends of 1:1 weight ratios of PEO and PMMA-d8-PMAA are shown in figure 1. The uncertainties of the data points are calculated statistically from the number of counts, and are propagated through standard methods [9]. In the plots displayed in this paper, the uncertainties are all less then the size of a data point. The squares show the scattering from the blend of PMMA-d8/PEO. These results are consistent with the results of Ito et al [4] who found a slightly negative, but small value of χ. When only 2% PMAA is present in the copolymer, the scattering considerably is reduced, and the scattering intensity becomes less with increased PMAA content. The reduced scattering is a result of lowered concentration fluctuations in the blend which are a result of a favorable interaction between the PMMA-r-PMAA random copolymer and PEO. The composite interaction parameter being negative and increasing in magnitude as the PMAA content increases. Equation 1 can be used to fit the scattering data to give the composite χ through a nonlinear least squares routing what weights the data points according to their calculated uncertainties.

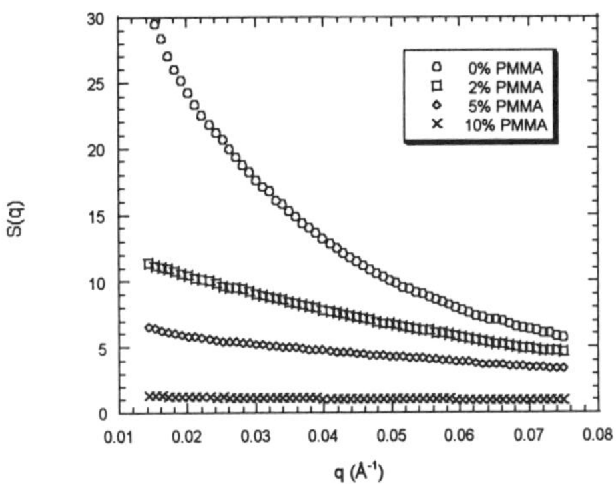

Figure 1. SANS of 50/50 blends of PEO and (PMMA-d8)-r-PMAA.

Figure 2 is a plot of the composite χ values plotted against the volume fraction PMAA, f. It is clear that the composite χ values become extremely negative as the amount of PMAA increases. A fit of equation 2 to the data is shown in the figure also. Values of the three interaction parameters can calculated from the intercepts and curvature of the fit. The uncertainties are standard deviations calculated by standard methods [9]. While accurate values of χ PMMA-PMAA and χ PEO-PMAA cannot be determined from this data, it is evident that the composite interaction parameter is quite favorable for PEO and random copolymers of PMMA and PMAA.

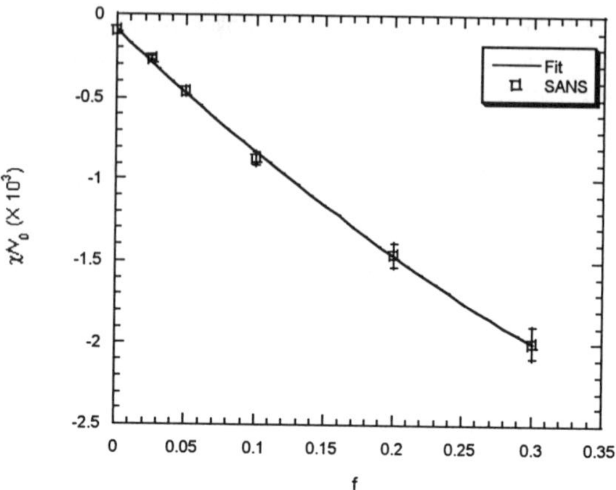

Figure 2. C/v0 versus f for blends of PEO and (PMMA-d8)-r-PMAA.

The upward curvature of the fit in figure 2 is characteristic of a χ PMMA-PMAA > 0. This repulsive interaction between units of the copolymer enhances the miscibility of its blends with other polymers, but causes immiscibility when blended with a random copolymer of a different composition. The value of χ_{AC} is effectively diluted by the copolymerization.

Figure 3 is a plot of the scattering from a blend of PMMA and PMMA(d8) and one of PMMA and PMMA(d8)-r-PMAA with a value of f = 0.02. The blends of h and d versions of PMMA produces a curve typical of a miscible blend, with a limiting power law of -2. The blends of the polymer and copolymer combination give a limiting power law with a larger negative exponent, which is typical of a phase separated, immiscible blend. Qualitatively, this is an indication that $\chi \gg 0$, which is consistent with the upward curvature of figure 2.

The SANS results show that there are four possible combinations of ingredients that have different interactions. For blends of PS/PEO homopolymers, a block copolymer of PS-PMMA would have an interaction parameter χ PMMA-PEO that is slightly negative but very small in magnitude, a block copolymer of PS-PMAA would have an interaction parameter that is negative and larger in magnitude and a block copolymer of PS-(PMMA-r-PMAA) would have an effective interaction parameter that is negative and very large in magnitude. A fourth combination has a blend of PS/PMMA homopolymers with a PS-PMMA block, and since the polymer types in the block and the blend are the same, the interaction parameter is zero.

Blends of many components are very complicated due to the various combinations of contact energy between the components. Some of the interaction parameters have not been

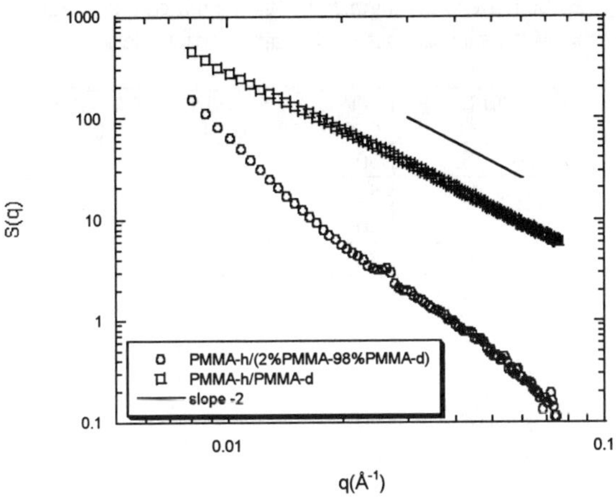

Figure 3. SANS of PMMA-h/PMMA-d and PMMA-h/((PMMA-d)-r-PMAA)

measured in this work such as χ PS-PMMA and χ PS-PMAA , and while these may affect the overall compatibilization results, it is likely that their effect is negligible in comparison to the other interactions.

<u>Melt Mixing Results</u>

All of the melt mixed blends studies had a matrix phase of 80 weight % polystyrene. The minor phase in most cases was poly(ethylene oxide), with polymethylmethacrylate in the last two examples. Compatibilizers were both random and graft copolymers of styrene and either methylmethacrylate and/or polymethacrylic acid with various amounts of copolymer.

The primary method of estimation of the effect of the compatibilizer was SEM of fracture surfaces which gives a relative size of the dispersed phase. The micrographs of the blends that are shown are from typical areas of the surface. No estimate was made of average sizes of the dispersed phase, rather, qualitative descriptions are made based on the micrographs, since in most cases, the effect of the compatibilizer is quite dramatic.

Table I lists the characteristics of the blends studied with the major variable being the compatibilizer type. The weight % PS in the compatibilizer was value calculated from the monomer charge. Conversions were all greater than 90%, and previous experience shows that this gives composition values of ± 5%. The weight % compatibilizer was calculated from the measured weights, giving accuracy typically ± 0.1%. The mixing temperature was ± 5 °C as is typical for melt mixers.

Figure 4 shows the morphology of the fracture surface of a PS/PEO blend blended at 190 °C . Figure 4a is a blend without any added compatibilizer. Figure 4b is a blend with a PMMA-r-PS copolymer the contains 50% styrene by weight and figure 4c is an equivalent blend with a PMMA-g-PS graft copolymer. The magnification in the SEM micrographs of all of the PS/PEO is the same with a bar of 10 μm displayed. The blank shown in figure 4a shows many dispersed particles in the range of 1-10 μm. The size scale is smaller in figure 4b, and in figure 4c the size scale is small enough to be hard to distinguish at this magnification. Higher magnification of the sample in 4c clearly shows the second phase with many particles less than 1 μm in size.

BLEND POLYMERS	COMPATIBILIZER	WT% PS IN COMPATIBILIZER	WT% COMPAT-IBILIZER	MIXING TEMP, °C
PS/PEO	PMMA-r-PS	50	1-10	160
PS/PEO	PMMA-g-PS	50	10	160
PS/PEO	PMAA-r-PS	50	1-10	160
PS/PEO	PMAA-g-PS	50	10	160
PS/PEO	PMMA-r-PS	50	10	190
PS/PEO	PMMA-g-PS	50	10	190
PS/PEO	PMAA-r-PS	50	10	190
PS/PEO	PMAA-g-PS	50	10	190
PS/PEO	PMMA-g-PS	90	10	160
PS/PEO	PMAA-g-PS	90	10	160
PS/PEO	(PMMA-r-PMAA)-g PS	50	10	160
PS/PEO	(PMMA-r-PMAA)-g PS	90	1-10	160
PS/PMMA	PMMA-r-PS	50	10	160
PS/PMMA	PMMA-g-PS	50	10	160

Table I. Compatibilizer and blend types studied

Figure 5 shows the morphology of a compatibilized blend that is similar to figure 4 except that equivalent blends with PMAA were used, PMAA-r-PS in figure 5a and PMAA-g-PS in figure 5b. The random copolymer in figure 5a has not significantly reduced the size of the dispersed phase, while the graft copolymer of figure 5b reduces the size of the dispersed phase in a way similar to the graft in figure 4c. In all of the blends studied here that were made with PMMA or PMAA copolymers, the random copolymers were less effective at reducing the size of the dispersed phase than graft copolymers.

Figure 6 shows blends compatibilized with graft copolymers containing 90 weight % styrene. Figure 6a shows a blend compatibilized with PMMA-g-PS, and the phase size is slightly reduced from the blank. In figure 6b, the phase size is considerably reduced from that of the blank, so that for the 90 % grafts, PMAA is far more effective than PMMA. In this case the polymer combination that has the strongest interaction (a large negative χ) is by far the most effective.

The SANS measurements of the composite χ suggest that the interaction between PEO and PMAA is more favorable than between PEO and PMMA, but the most favorable interaction may be for grafts in which the methacrylate block is a random copolymer of PMMA and PMAA. To test this, two graft copolymers were synthesized from a 50 weight % mixture of PMMA and PMAA along with the PS macromonomer. Figure 7a shows the morphology of a blend made with a 50 weight % PS graft and figure 7b with a 90 weight % PS graft. Both are very effective at reducing the size of the dispersed phase.

a

b

c

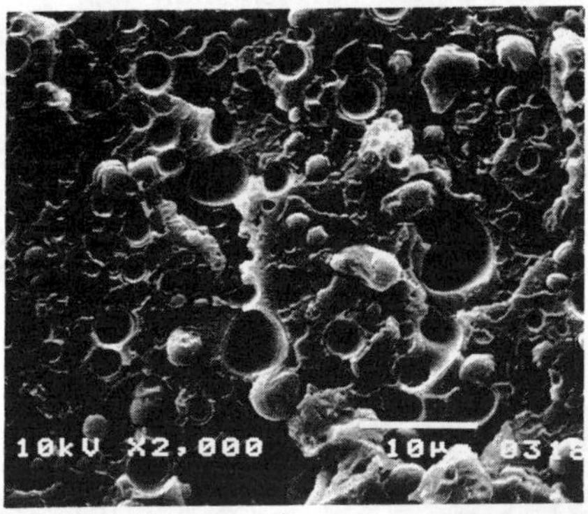

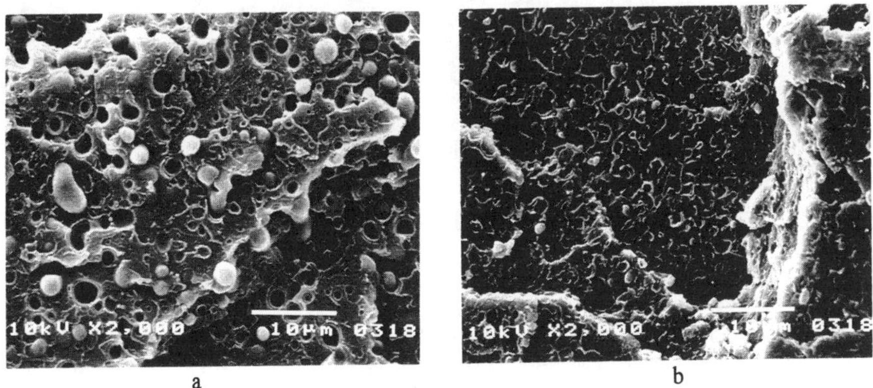

Figure 4. SEM of PS/PEO blends with a. no compatibilizer, b. PMMA-r-PS, c. PMMA-g-PS

a

b

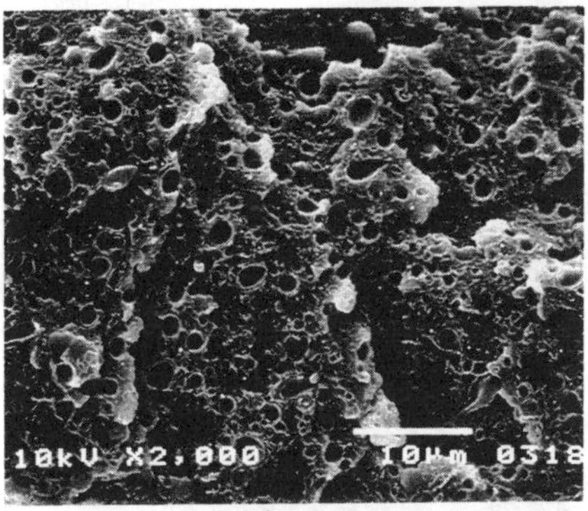

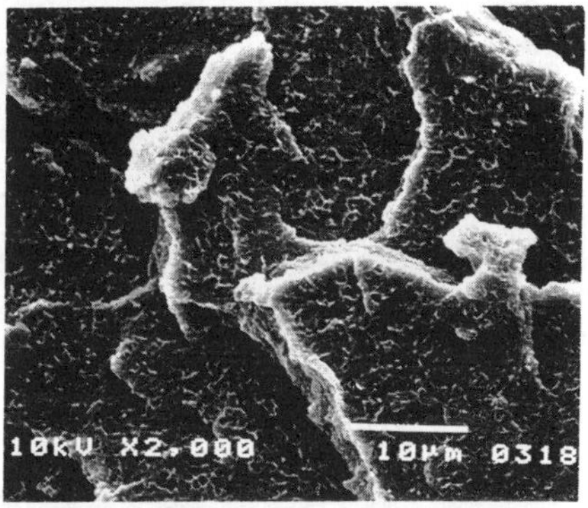

Figure 5. SEM of PS/PEO blends with a. PMAA-r-PS, PMAA-g-PS.

As a control, blends were made of PS/PMMA with and without copolymers of PS-PMMA. This represents a truly athermal interaction between the compatibilizer components and the blend components. It is known [10] that the interaction parameter $\chi_{\text{PS-PMMA}}$ is positive but small, so that blends have borderline miscibility. This makes the blends hard to characterize, since the blank mixture containing no compatibilizer already has a fine dispersion. SEM studies of fracture surfaces has little definition, and it is impossible to judge the phase size. To bring out the surface morphological character, the samples were extracted with acetic acid which is a solvent for PMMA and a non solvent for PS. The SEM micrographs of the PS/PMMA blends are at a higher magnification to show the slight differences.

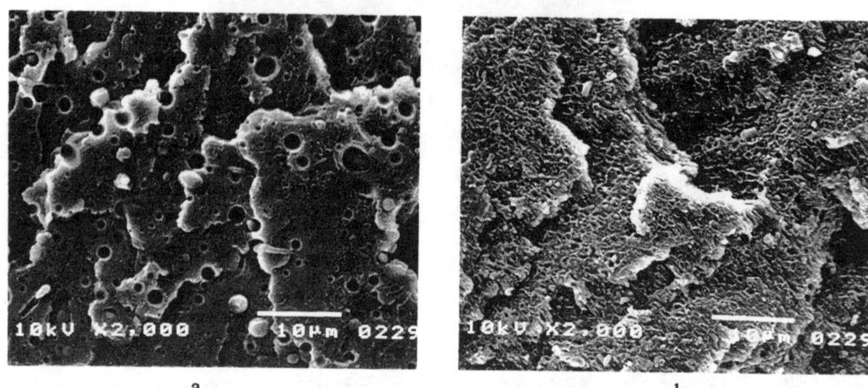

a
b

Figure 6. SEM of PS/PEO blends with 90% PS in graft, a. PMMA-g-PS, b. PMAA-g-PS.

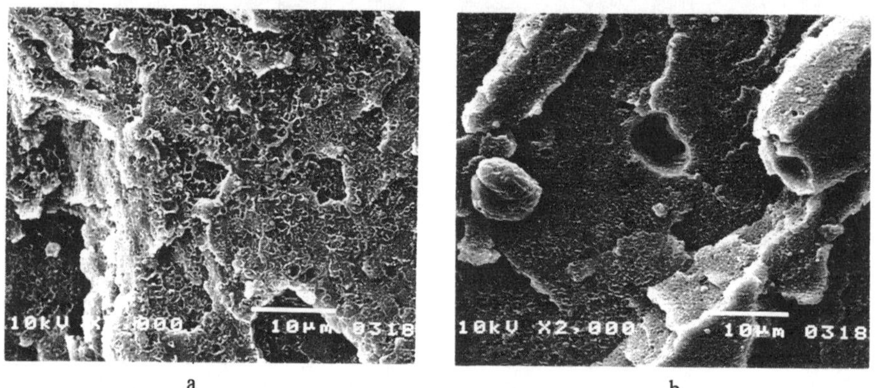

a
b

Figure 7. SEM of PS/PEO blends with (PMMA-r-PMAA)-g-PS, a. 50% PS, b. 90% PS.

Figure 8 shows the extracted surfaces of PS/PMMA blends, with 8a being a blank with no compatibilizer, 8b containing PMMA-r-PS, and 8c containing PMMA-g-PS. As with the samples shown in figure 4, the random copolymer has a slight effect while the graft polymer has a much stronger effect.

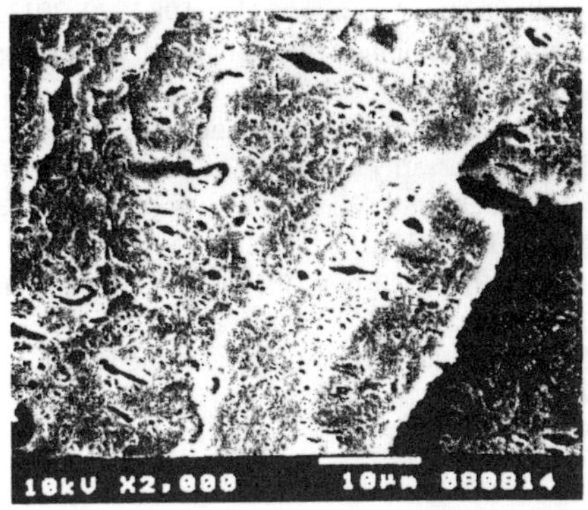

a

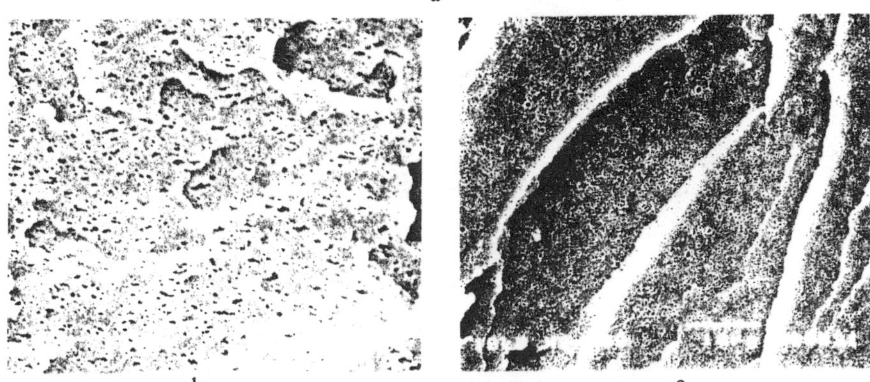

b   c

Figure 8. SEM of PS/PMMA blends with a. no compatibilizer, b. PMMA-r-PS, c. PMMA-g-PS.

CONCLUSIONS

SANS of blends of PEO and random copolymers of PMMA-d8 and PMAA show that there is a very favorable overall χ. Scattering drops significantly $\chi_{\text{PEO-PMMA}}$ is slightly negative and small in magnitude, causing a blend that is nearly athermal. With increased PMAA content, the overall χ becomes negative and large in magnitude. Random and graft copolymers were synthesized and melt mixed with PS/PEO blends to test the effect of cross compatibilization.

Blends using both athermal and strongly interacting compatibilizers show that graft copolymers are more effective than random polymers of the same type. When graft copolymers have 50 weight % of each polymer type, both athermal and strongly interacting compatibilizers are effective, but when

REFERENCES

1. Utraki, L. A., *Polymer Alloys and Blends*, Oxford University Press, **1990**, New York.
2. Hobbie, E. K.; Bauer, B. J.; Han, C. C.; *Physical Review Letters*, **1994**, *72*, 1830.
3. Hobbie, E. K.; Merkle, G.; Bauer, B. J.; Han, C. C.; *Physical Reveiw E*, **1995**, *52*, 3256.

4. Ito, H.; Russell, T. P.; Wignall, G. D.; *Macromolecules,* **1987**, *20*, 2213.

5. Certain commercial materials and equipment are identified in this paper in order to specify adequately the experimental procedure. In no case does such identification imply recommendation by the National Institute of Standards and Technology nor does it imply that the material or equipment identified is necessarily the best available for this purpose.

6. ten Brinke, G.; Karasz, F. E.; MacKnight, W. J.; *Macromolecules,* **1983**, *16*, 1827.

7. Bauer, B. J.; *Polym. Eng. Sci.,* **1985**, *25*, 1081.

8. Han, C. C.; Bauer, B. J.; Clark, J.C.; Muroga, Y.; Matushita, Y.; Okada, M.; Tran-Cong, Q.; Chang, T.; and Sanchez, I; *Polymer,* **1988**, *29*, 2002.

9. Bevington, P. R., Data Reduction and Error Analysis for the Physical Sciences, McGraw-Hill Book Company, **1969**, New York.

10. Benoit, H.; Wu, W.-L.; Mozer, B; Bauer, B. J.; Lapp, A., *Macromolecules,* **1985**, *18*, 986.

AUTHOR INDEX

SUBJECT INDEX